HIGHER EDUCATION AND SUSTAINABLE DEVELOPMENT

Sustainability issues will shape what students do in their future careers, especially within built environment professions. Yet many institutions struggle to adapt programs and coursework to reflect this. Responding to a need for guidance, this book provides a practical resource to quickly – and effectively – update curriculum to meet the emergent sustainability context. The authors use their experiences in engineering education and insights from colleagues and institutions around the world to provide tools to address common institutional challenges and to make the most of emerging opportunities.

The book begins by exploring the rationale for action, discussing why curriculum renewal has been challenging to-date and identifying urgent drivers for change. It then presents a new model for curriculum renewal to deal with these challenges, from course and program design through to stakeholder engagement and organisational considerations. The model shows what educational leaders are beginning to practise: a whole-of-system approach to timely program design and review.

The book uses the model to detail practical ways to move forward, including creating a strategy, identifying graduate attributes, mapping learning outcomes, auditing the program, updating coursework and implementing the renewed curriculum. With extensive case study material from around the world, this resource will assist institutions – from department heads to program convenors – to cost-effectively align offerings with present and future educational demands.

Cheryl Desha has a degree in Engineering (Environmental, first class) and a PhD in rapid curriculum renewal. Her research focuses on building capacity for sustainable development within tertiary education. She is a Senior Lecturer in the Science and Engineering Faculty, QUT, and a Principal Researcher with The Natural Edge Project (TNEP).

Karlson 'Charlie' Hargroves has a degree in Engineering (Civil/Structural) and is undertaking a PhD in carbon structural adjustment. His research focuses on opportunities for transformational change towards sustainable development. He is a Senior Research Fellow in the Curtin University Sustainability Policy Institute and a Principal Researcher with The Natural Edge Project (TNEP).

Cheryl and Charlie are members of The Natural Edge Project, a sustainability think-tank which operates as a collaborative partnership for research, education and policy development on innovation for sustainable development. TNEP's mission is to contribute to and succinctly communicate leading research, case studies, tools, policy and strategies for achieving sustainable development across government, business and civil society. Working with the TNEP team, Cheryl and Charlie have co-authored four books on sustainable development, published in four languages.

"The team from The Natural Edge Project have provided a well argued appraisal of the rationale for rapid curriculum renewal to education for sustainable development. Higher education institutions around the world clearly have significant incentives and a variety of tools to embrace this challenge over the next decade."
Wynn Calder, Co-Director, Association of University Leaders for a Sustainable Future, Washington, DC

"With my experiences as Head of the Department of Mechanical Engineering at the University of Zaragoza, I agree with the idea that the universities who can innovate a process to integrate sustainability content within their existing programs will, in the medium to longer term, attract larger numbers of students and achieve notoriety as leading education institutions. In this respect, and as a first step prior to a deeper integration of sustainable development into the programs, we are promoting EESD at the undergraduate and postgraduate level with specific courses. For this purpose we have used parts of the curriculum developed by the TNEP team, which is excellent and without any doubt we have found it useful in reducing time spent in generating and checking new content."
Professor Luis M. Serra, Head of Department of Mechanical Engineering, University of Zaragoza, Spain

"Engineering education for sustainable development is important to business as the national and global economy gears up for the challenges presented by climate change, resource constraint and greater public engagement with the sustainability agenda. This book highlights that employers are increasingly seeing sustainability, and particularly the emerging carbon economy, as an opportunity rather than just a risk. Business needs graduates with the knowledge and skills to operate in a rapidly emerging market in sustainable engineering services."
Dr Fabian Sack, Principal, Fabian Sack and Associates; Former Group Sustainability Manager, Downer EDI

"This is a wonderful compilation of local and international initiatives that highlight ways of embedding sustainability and sustainable development issues, from the outlying teacher scenario of 'I'll include it if I must' (assuming the staff has heard of the topic and sees any need) to the very core of any teaching, and the needs of the student experience – the raison d'être."
Dr Euan Nichol, Consultant, Victoria University

HIGHER EDUCATION AND SUSTAINABLE DEVELOPMENT

A model for curriculum renewal

Cheryl Desha and Karlson 'Charlie' Hargroves

LONDON AND NEW YORK

First edition published 2014
by Routledge
2 Park Square, Milton Park, Abingdon, Oxon, OX14 4RN

and by Routledge
711 Third Avenue, New York, NY 10017

Routledge is an imprint of the Taylor & Francis Group, an informa business

British Library Cataloguing in Publication Data
A catalogue record for this book is available from the British Library

Library of Congress Cataloging-in-Publication Data
Desha, Cheryl.
Higher education and sustainable development : a model for curriculum renewal / Cheryl Desha and Karlson Charlie Hargroves. – First edition.
pages cm
Includes bibliographical references and index.
1. Education, Higher. 2. Sustainable development. 3. Curriculum planning. 4. Curriculum change.
I. Hargroves, Karlson, 1974– II. Title.
LB2325.D435 2014
378–dc23
2013015806

ISBN13: 978–1–84407–859–2 (hbk)
ISBN13: 978–1–84407–860–8 (pbk)
ISBN13: 978–1–315–88395–3 (ebk)

Typeset in Bembo by
Keystroke, Station Road, Codsall, Wolverhampton

Printed and bound in Great Britain by
TJ International Ltd, Padstow, Cornwall

Many colleagues around the world have shared with us their frustrations and fatigue in persuading peers and senior management to embed sustainable development within their curriculum. We dedicate this book to assisting these colleagues (you know who you are!) and in memory of a mentor and international champion in education for sustainable development, Professor Leo Jansen (1934–2012). May this book, for which he provided encouraging review comments, contribute to his vision of pursuing sustainable development and, in particular, empowering students for their professional future.

With gratitude to Alex and Stacey, and family and friends who have enabled this adventure over the years. With thanks also to our children Aidan, Mia and Kiran, and Grace and Tyson, for giving us inspiration – and ultimate deadlines – to work on the manuscript! To our furry friend Harry who has been with us since the start, thanks for all the cuddles.

CONTENTS

FOREWORD

Barry Grear
AO, HonFIEAust, FIPENZ, FACE, FAICD, MAIES, JP
Former President, World Federation of Engineering Organisations (WFEO) 2007–2009*

> In times of change learners inherit the earth; while the learned find themselves beautifully equipped to deal with a world that no longer exists.
>
> *Eric Hoffer, Philosopher*

I have had the pleasure of mentoring the team from The Natural Edge Project since their beginnings as a special interest project with the Institution of Engineers Australia in 2002. Watching their emergence as an internationally regarded sustainability think-tank that produces rigorous content for capacity building within the engineering profession gives me great hope for the future. In this their latest work I congratulate the team for succinctly presenting the rationale for issues related to sustainable development to be given the highest priority by engineering educators over the coming decade, as we seek to rapidly improve the sustainability attributes of our engineering graduates. Engineering departments must expect that program accreditation requirements related to sustainability will increase and the set of graduate attributes to meet such challenges will quickly become increasingly demanding over the next decade as the profession clarifies its responsibilities.

I have had a long interest in improving the quality of engineering education and training in all member nations, as well as improving the procedures for accrediting engineering courses that provide assurance of the quality of engineering education to meet international standards. I have been involved in many exciting conversations over recent years about how the engineering profession can address sustainability challenges facing humanity this century. In the same period I have also witnessed the reluctance of many heads of engineering departments from around the world to integrate sustainability knowledge and skills within all engineering programs. Despite this inertia to stay the same, I believe 2008–2009 will be historically referred to as a tipping point for action in our profession, where increasing pressures from government, industry and the profession itself, resulted in the commencement of a transition to engineering education for sustainable development (EESD) in many countries. I am particularly pleased with the authors' summary of the key drivers for engineering education to embrace sustainable development, which will assist any departments where such debate has previously stalled action.

The authors have also provided an enlightening discussion of the serious time lag dilemma facing engineering departments to equip the profession with knowledge and skills to meet society's needs this century. Simply put, engineering departments must have a clear understanding of this current context and the risk implications of their decisions today. Globally, the public is becoming increasingly aware that development need not come at the price of a compromised environment. Among the ever-increasing global population that continues to shift to urban areas, engineers are being challenged to meet demands for energy, drinking water, clean air, safe waste disposal, transportation, and infrastructure that does not further diminish our natural systems. We are also being challenged to address built environment complications arising from sea level rise; the increasing regularity and severity of threats such as drought, flooding, heat-waves and hurricanes resulting from climate change phenomena; and perhaps other indirect political threats arising from issues such as oil, food and water scarcity. These issues require truly intra-disciplinary, cross-disciplinary and multi-disciplinary collaborations, where engineering contributions must be well grounded in sustainability principles and practices.

I commend this guide to the engineering education community as a milestone in moving towards sustainable engineering. As the authors make clear, the critical issue for our profession is now *how fast* we can undertake this process of curriculum renewal, in such urgent and challenging times. In the absence of a 'how to' manual, the authors have consulted with an extraordinary international network of accomplished practitioners and academics, to present a strategic yet ultimately pragmatic approach for undertaking rapid curriculum renewal for EESD. The elements of rapid curriculum renewal highlight many opportunities for departments to meet multiple objectives in addressing EESD, including accreditation requirements and recruitment, while also reducing their risk exposure to rapidly shifting market, regulatory and accreditation requirements.

I also highlight this book for the attention of all accrediting institutions internationally. The authors make it clear that professional engineering institutions that are responsible for accrediting university programs are a crucial component in driving accelerated curriculum renewal within the higher education sector. Engineering educators are calling for benchmarks for action with clear time horizons. By incorporating and prioritising explicit sustainability-related attributes for engineering programs in accreditation requirements, accreditation can play a key role in driving rapid and systemic curriculum renewal to EESD.

* The World Federation of Engineering Organisations (WFEO) has a core objective to use the skills and knowledge of the engineering profession for the wider benefit of humanity. With membership comprising national organisations, it represents in the order of 15 million engineers worldwide.

FOREWORD

Professor Goolam Mohamedbhai
Former Secretary-General, Association of African Universities (2008–2010)
Past President, International Association of Universities (2004–2008)

In the mid-1960s, when I was studying civil engineering in the UK, the emphasis of our curriculum was basically on how engineers can use nature's resources and how they can develop technologies for using them for the benefit of mankind. That was before the 1970s oil crisis, before we were aware of the scale of the ozone or greenhouse issues, and at a time when materials and other resources seemed infinite in supply. The term 'sustainable development' had not yet been coined – that was to come two decades later with the Brundtland Commission's report *Our Common Future*.

There is, today, no question that the engineer's work directly affects the environment, whether locally, regionally or globally. Almost every civil engineering activity – from building a dam, to managing traffic in a city, to offshore oil drilling – can be shown to have a direct impact on sustainable development. The civil engineers of today must therefore be made aware of the major challenges facing the world: water scarcity, pollution, depleting resources, increasing population, climate change etc. In designing and implementing their projects, they must be conscious of and be able to assess the environmental, social and economic impact of their work. They must think long-term and consider the local, regional and global effects of their activities. It is therefore imperative that all these aspects be reflected in the curricula of today.

But promoting sustainable development needs more than just changing the curricula. It requires adopting a multi-disciplinary and holistic approach, encouraging teamwork, promoting creativity and innovation, and developing critical and systems thinking in students. This is perhaps where the real challenges lie. The departmental structure of most institutions is hardly conducive to encouraging an inter- or intra-disciplinary approach, and the teaching staff themselves have not been exposed to such an approach. In many institutions students are still 'taught' by transmitting facts rather than encouraged to 'learn' though inquisitiveness, understanding and analysis.

Higher education institutions are, however, gradually responding to change. Some countries and regions have made greater progress than others. In some cases the adoption of a regional approach is bringing about positive results. In Africa, for example, the United Nations Environmental Programme (UNEP) has launched Mainstreaming Environment and Sustainability in African Universities Partnership (MESA), a project that aims at embedding sustainability in all curricula, research and community engagement activities of universities, and in which the Association of African Universities is a partner. Sharing experiences and learning from successful practices are crucial for promoting sustainable development in higher education across the world.

In that context, this publication makes an important contribution. The synthesis of the elements that might help departments to strategically engage in curriculum renewal – for their own viability and in capacity-building the professions to address our 21st-century challenges – is most relevant. In addition to further exploring the problems and challenges that we face, the authors have also provided a sense of optimism by highlighting the best of what is happening internationally, and providing educators with tools that can be used immediately.

Furthermore, the authors' presentation of the framework in this book will help educators to put these materials into context, sharing resources that will avoid reinventing the wheel, but which must be proactively and strategically used with an end-goal in mind, rather than as an *ad hoc* or reactive approach to dealing with accreditation requirements. Used strategically, this publication can be an important guide in reducing the cost and risk of such a transition, which is of particular significance to our colleagues in African higher education institutions.

I look forward to sharing this publication with my African and international colleagues, as we move forward in mainstreaming sustainable development in the higher education sector.

FOREWORD

Professor Walter Leal Filho, BSc, PhD, DSc, DL
Head of the Research and Transfer Centre, Hamburg University of Applied Sciences; Chief Editor, *International Journal of Sustainability in Higher Education*

For the past two decades, I have been fascinated by the development of education for sustainability literature in the higher education sector. From this early environmental education discourse has emerged a strong and independent 'Education for Sustainable Development' field underpinned by a diversity of literature from curriculum development theory, organisational change theory, and sustainable development theory and practice.

Within this field, engineering education has emerged as a focus area, assisted by the community of practice that has formed around the biennial Engineering Education for Sustainable Development (EESD) conferences beginning in 2002, the creation of the Barcelona Declaration on EESD in 2004, and the production of a biennial university survey called the EESD Observatory beginning in 2006. The *International Journal of Sustainability in Higher Education* (*IJSHE*) has collaborated with the previous EESD conferences to publish a special issue journal edition in 2005 on EESD, and to publish key papers from subsequent conferences. Through this ongoing relationship, it is clear that the discourse has evolved from the question of 'what is EESD?' to 'how do we implement EESD?' and 'who is implementing EESD?'.

The 2005 *IJSHE* special issue included a paper on content development by the authors of this publication and I have since had the pleasure of mentoring the TNEP team regarding opportunities for raising awareness about their research and experiences in the process of rapid curriculum renewal for EESD. In particular, I have observed the authors' dedicated efforts in involving a significant representation of the international community of practice, to review and critique this emerging framework for rapid curriculum renewal. As part of the development of this book the *IJSHE* also provided a double-blind peer review which contributed to the development of the resultant manuscript. A strong summary paper of the Time Lag Dilemma and the emerging framework was published in *IJSHE* in April 2009.

As a consumer and distributor of higher education literature, I am excited by the potential for this publication to set a benchmark for other disciplines to follow. The authors have synthesised a wealth of literature and experiences on the 'how' and 'who' questions to date, and present us with a very flexible and practical preliminary framework for educators to consider in their strategic planning for curriculum renewal. Their hard work deserves a warm word of thanks.

I would be very interested to hear about and document colleagues' experiences with the content presented in this publication, in forthcoming issues of the *IJSHE* (walter.leal@haw-hamburg.de).

FOREWORD

Dr Tony Marjoram
Senior Programme Specialist, Former Head of Engineering Sciences, Division of Basic and Engineering Sciences, Natural Sciences Sector, United Nations Educational, Scientific and Cultural Organisation (UNESCO), Paris, France

The Division of Basic and Engineering Sciences of UNESCO has collaborated with The Natural Edge Project since 2003 to produce a number of capacity-building materials for the engineering and science professions. These include *The Natural Advantage of Nations*,[1] *Engineering Sustainable Solutions Program*,[2] and *Whole System Design*.[3] Further to supporting such important content development, I am very pleased to be associated with this new and timely publication, which explores how the higher education sector can play a part in the transition to sustainable development, by undertaking rapid curriculum renewal to embed such content within programs. As the United Nations (UN) draws attention to such issues during the International Decade of Education for Sustainable Development (2005–2014), the authors are to be congratulated on their proactive, collaborative and systemic approach to the important issue of curriculum renewal.

Sustainable engineering and technology are vital in addressing basic human needs and poverty reduction, to bridge the 'knowledge divide' and promote international dialogue and cooperation. Indeed, environmental sustainability is underlined as one of the eight Millennium Development Goals (MDGs), and the Intergovernmental Panel on Climate Change (IPCC) has emphasised the importance of technology in mitigating the impacts of climate change and helping society adapt to changes already locked into place.

Despite this significant need, there is a declining interest and enrolment of young people, especially young women, in engineering, which will have a serious impact on capacity in engineering, and our ability to address the challenges of sustainable social and economic development, poverty reduction and the other MDGs. Although science and engineering have changed the world, they are professionally conservative and slow to change. Rather than incremental change, a transformation of engineering and engineering education is essential if engineering is to play its vital role in assisting to reduce poverty, promote sustainable development and address climate change mitigation and adaptation. There are clear needs to show that science and engineering education and university courses are inherently interesting, and to promote science and engineering as a part of the solution, rather than part of the problem. To promote engineering and attract young people we need to emphasise these issues in teaching curricula and practice. Student interest is already evident, in activities such as the UNESCO-Daimler Mondialogo Engineering Award and work of Engineers Without Borders groups around the world.

In this context, this publication could not be more timely and relevant. While the need for curriculum renewal has been recognised for some time, there is still a need to share information on what this means in practice, and to share pedagogical approaches and curricula that can be mainstreamed for a whole-scale transition to education for sustainability. In this publication, the authors provide us with a practical guide on how such a transition might be strategically planned, within existing cultural and organisational contexts. They also include reference to a number of innovative examples of engineering departments who are responding to rapid changes in knowledge production and application, and changing government, industry and societal demands.

At a time when transformational change to sustainable development is critical, the authors have taken care to ensure that this publication is globally pertinent for a profession that will play a key role in the transition. In particular I commend the publication as a valuable and empowering resource to higher education colleagues in developing countries, who face serious constraints regarding human, financial and institutional resources to develop such curricula and learning/teaching methods.

Notes

1 Hargroves, K. and Smith, M. (2005) *The Natural Advantage of Nations: Business opportunities, innovation and governance in the 21st century*, The Natural Edge Project, Earthscan, London.

2 Smith, M., Hargroves, K., Desha, C. and Palousis, N. (2007) *Engineering Sustainable Solutions Program: Critical literacies portfolio*, The Natural Edge Project, Australia.

3 Stasinopoulos, P., Smith, M., Hargroves, K. and Desha, C. (2008) *Whole System Design: An integrated approach to sustainable engineering*, Earthscan, London, and The Natural Edge Project, Australia.

We work day after day not to finish things, but to make the future better because we will spend the rest of our lives there.
Charles Kettering[1]

A sustainable society into the indefinite future . . . depends totally and absolutely on a vast re-design triggered by an equally vast mind-shift – one mind at a time, one organization at a time, one technology at a time, one building, one company, one university curriculum, one community, one region, one industry at a time, until the entire system of which we are each a part has been transformed into a sustainable system, existing ethically in balance with Earth's natural systems, upon which every living thing utterly depends – even civilization itself.
Ray Anderson, founder and former CEO, Interface Carpets (1934–2011)[2]

Education is a prerequisite for promoting the behavioural changes and providing all citizens with the key competences needed to achieve sustainable development. Success in reversing unsustainable trends will to a large extent depend on high quality education for sustainable development at all levels of education including education on issues such as the sustainable use of energies and transport systems, sustainable consumption and production patterns, health, media competence and responsible global citizenship.
European Union Strategy for Sustainable Development, Luxembourg 2002[3]

. . . what we now need [is] strategies, international understandings and policies that will guide action, correct the biggest market failure the world has ever seen [climate change] and provide a framework for the entrepreneurship and discovery across the whole of business and society, which can show us how to achieve a cleaner, safer, more sustainable pattern of growth and development.
Sir Nicholas Stern[4]

Notes

1 Kettering, C. (n.d.) American engineer and inventor, 1876–1958.
2 Anderson, R. (2005) 'Rethinking development: Local pathways to global wellbeing' Keynote presentation, in proceedings of The Second International Conference on Gross National Happiness, St. Francis Xavier University, Antigonish, Nova Scotia, Canada, 20–23 June.
3 Office for Official Publications of the European Communities (2002) *European Union Strategy for Sustainable Development*, Luxembourg, in Nadolny, A. and Schauer, T. (2007) *The Future of Europe: Sustainable development and economic growth?* Proceedings of the International Symposium, Vienna, 12–13 September 2007.
4 Stern, N. (2009) *A Blueprint for a Safer Planet: How to manage climate change and create a new era of progress and prosperity*, Random House.

OUR ONE-MINUTE PITCH

This book is for colleagues who have seen the need to act, and who are looking for guidance on what to do next. For any colleague who is still considering whether to embark on a journey of curriculum renewal towards education for sustainable development, here is our 'one-minute pitch'.

There is an unprecedented urgency for capacity building for sustainable development. Employers are looking to new graduates with sustainability competency to help them address the challenges and opportunities in the 21st century. We now have a much better understanding of our predicament, with sufficient technology to address the most serious of challenges within the next 2–3 decades. However, there is a gap in capacity to apply these knowledge and skills, and action is too slow. The time has come for the education sector to create this capacity, moving beyond high-level action (declarations and big-picture visions), to transform curricula. Students need to learn the intricacies of how to address these challenges, including technical and enabling knowledge and skills. This transformation needs to be across undergraduate and postgraduate education and continuing professional development. Students will look to academically rigorous and technically challenging courses that they can trust, beyond 'warm and fuzzy' conversations or 'green-wash' content. Professional bodies themselves are also looking to universities for guidance on how to make the transition. If education institutions don't meet this need, then companies will find it elsewhere through in-house/professional development.

With this in mind, the stage is set for an action-oriented, reader-friendly tour of possibilities in curriculum renewal towards sustainable development – let's begin.

PREFACE

Society is increasingly calling for professionals to innovate and problem-solve cost-effective ways to reduce environmental pressures, now being found to pose threats to economies and societies. This poses a significant challenge to the higher education sector, requiring both capacity building for professionals and practitioners who can deal with immediate and short-term issues, along with students who will enter the workforce in the future and assist in dealing with medium- to longer-term issues.

An example of a short-term challenge is to halt the growth of greenhouse gas emissions in the near future. This would then be followed by a medium- to long-term challenge of sustaining reductions in emissions over following decades to reach stabilisation targets. Each challenge requires very different strategies. Tertiary education thus requires curriculum renewal to begin immediately and focus on both undergraduate and postgraduate/professional development programs.

Considering the complexity of capacity building in this field, we have focused on developing processes for doing so over the last decade. Experiences since 2002 have informed the development of a number of curriculum renewal aids for accelerating the process. This has included two awareness-raising textbooks on 'what is' sustainable development (*The National Advantage of Nations*, 2005; *Cents and Sustainability*, 2010), two technology-focused textbooks on 'how to' apply sustainable development principles (*Whole Systems Design*, 2008; *Factor 5*, 2009) and a number of supporting online curriculum resources (*Engineering Sustainable Solutions Program*, 2007; *Energy Transformed*, 2007; *Water Transformed*, 2011).

Further to these publications, this process-focused book focuses on 'how to' build capacity in such knowledge and skills, for the principles and practices to be embedded within daily life, as society adjusts to a low-carbon way of life. In particular, it supports our colleagues in higher education institutions seeking to bring about rapid curriculum renewal for sustainable development. By collating, synthesising and contributing to the body of knowledge on the process of embedding sustainability within higher education, we hope to reduce the barriers to curriculum renewal, and in doing so, help to build momentum for a rapid and large-scale transition. Building on experiences in engineering, we look forward to continuing our inquiry and research within other disciplines to make education for sustainability a reality in coming years.

In summary, this publication:

- Provides a summary of the unprecedented context that our generation of leaders in academia, business and government are living in;
- Highlights what is happening internationally in the education sector and implications for engineering professionals;
- Presents educators with a practical model and approach to move forward, acknowledging challenges and opportunities involved with the transition; and
- Provides professional organisations, accreditation agencies and industry with insight into the world of academia, presenting opportunities for how they can assist the transition to education for sustainable development.

Commentary on our journey of inquiry

This book explores a case account of a sociological phenomenon, namely the need for sustainable development knowledge and skills to be embedded within curricula. In this case, our personal experiences as young engineers and early career academics suggested a shortfall in engineering education for sustainable development and subsequently an urgent need for curriculum renewal in this area.

A review of literature regarding 21st-century challenges (Chapter 1) finds strong evidence of a critical and extraordinary role for all professions to urgently help society address a multitude of emerging issues of sustainable development. The literature review also finds clear evidence of higher education institutions (HEIs) around the world facing increasing pressure from a variety of sources, including professional bodies, industry, government and prospective students, to urgently equip graduates with knowledge and skills to address such challenges. Furthermore, we observe a time lag dilemma for the higher education sector, particularly in engineering education, whereby the timeframe for producing graduates with the required knowledge and skills lags behind the demand for graduates with such knowledge and skills – and indeed the timeframe within which the profession is expected to have acquired this increased capability. This was also evidenced in the findings of several key international surveys over the last decade, as outlined herein.

A variety of catalysts or 'drivers' for accelerated curriculum renewal are identified (Chapter 2), from which we conclude that a focus on engineering education and sustainable development is appropriate, with future potential application to a variety of other disciplines facing similar urgent and challenging circumstances. It is also concluded that although there is evidence of frustration with the current 'slow' process, in the absence of documented discourse about dealing with potential time constraints there has been little discussion of alternative strategies for curriculum renewal in this area. Despite discussion about timing issues existing for more than four decades, there has been little consideration for the speed at which curriculum is constructed and implemented or reviewed. While existing models provide significant guidance on systematic curriculum construction, none consider – either explicitly or implicitly – how to vary the pace at which curriculum renewal may be undertaken.

Through exploring documented cases of curriculum renewal and through personal experiences in various research projects, a number of mechanisms are identified that could be grouped under a number of themes or 'elements' of curriculum renewal (Chapter 3, Chapters 6–9 and Chapter 11), resulting in a curriculum renewal model coupled with an organisational change model, extending the discourse on 'curriculum in context'.

It is also concluded that a number of catalysts play a critical role in ensuring timely curriculum renewal beyond faculty and units within the larger institution (Chapter 4). We discuss an existing schematic for an organisational change model that could be adapted to provide a schematic for the model for rapid curriculum renewal. Not only does this model provide the sense of non-linear dynamism necessary within the higher education industry, it also demonstrates the non-linear behaviour of the elements of curriculum renewal, intertwining in a complex pathway, which is highly dependent on the organisational structure and context, but moving towards the goal of rapid curriculum renewal.

There are also several important strategic considerations to address a number of identified barriers to the process, taking a holistic, non-linear and integrated approach to using the elements (Chapter 5). Ultimately, institutional leadership and support is also critical in ensuring that an institution adopts a process of rapid curriculum renewal, setting and meeting the planned milestones (Chapter 10).

Opportunities for future research

This book is based on the premise that issues of sustainability and the contributing role of the professions are both critical. Furthermore, we believe that the education of professionals to address these issues is a pressing world-wide problem. With the future well-being of society in mind, our approach can be considered by educators world-wide as a potential way forward to achieve rapid curriculum renewal. Furthering our exploration within this book, a number of additional research opportunities are highlighted here:

- Trialling the curriculum renewal model and organisational change model: through action-based research and reflection by others on curriculum renewal experiences.
- Investigating the role of accreditation in driving rapid curriculum renewal: furthering discussion on the role of accreditation as a potential major catalyst for rapid curriculum renewal.
- Investigating supporting government policy mechanisms: considering the potential for national guidance to also contribute to rapid curriculum renewal, including policy mechanisms and other leverage opportunities.
- Further enhancing the theory associated with models: involving consideration of how these complement and challenge existing philosophical constructs of curriculum renewal.

This publication builds on the topic of education for sustainable development within engineering education. However, there are many other professions (for example, including law, business, nursing and medicine) and sectors of society (for example, schools and vocational education) facing similar pressure to incorporate emerging knowledge and skills related to sustainable development. The potential for wide-scale application is also apparent when considering that there are around 60 million teachers in the world spanning kindergarten through to higher education, and the majority have been trained through the higher education system:[1]

- Investigating the applicability of the models to other disciplines and cultural contexts: exploring the concept of rapid curriculum renewal in higher education, transcending boundaries between disciplines and continents.
- Investigating the applicability of the model and helix to K–12 providers: from kindergarten to senior high school (i.e. K–12) education providers, where the professional development of teachers in education for sustainability has been identified as 'the priority of priorities'.[2]

- Investigating the applicability of the models to other further education providers: alongside professional education, technical and vocational education and training providers are also grappling with the significant challenge of embedding sustainability knowledge and skills within their programs, as highlighted by the NSW Department of Education and Training in their 2009 report *Skills for Sustainability*.[3]

Sustaining and building communities of practice

This publication focuses on curriculum that is heavily regulated and which undergoes incremental change as a long-term 'evolutionary' – rather than short-term 'revolutionary' – timescale. Moreover, rapid curriculum renewal is as much about process as it is content related. Hence, it is anticipated that the approaches we discuss here will be useful wherever there is an imperative for urgent change regarding any new knowledge and skills that are complex, not just the 21st-century challenges discussed in this publication. Within this context, a dynamic and responsive curriculum relies on sustaining the enthusiasm of educators exploring curriculum renewal and further building this community amongst our 60 million colleagues globally. The goal of 'education for sustainable development' provides an immediate focus for such efforts that is urgent and challenging for everyone to engage with.

There are so many questions within the realm of timely curriculum renewal that could drive collaboration, inquiry and action – a lifetime's endeavour for many! There is the potential for curriculum-related research initiatives underway to consider the implications of rapid curriculum renewal, investigating how to systematically achieve a rapid process of integration. Any aspect of this publication could be further explored through action-research, expanding the conversation. For example, there is not yet a significant literature that auditing a program will lead to curriculum renewal or changed graduate capabilities. To rigorously demonstrate this could involve a significant longitudinal behaviour change research project comprising a number of schools internationally, including those who have decided to proceed with the transition (i.e. the trial group) and others who have decided not to (i.e. the control group). The trial group could undertake an audit of one or more programs, and then track curriculum modifications and the capabilities of graduates against a pre-determined set of 'graduate attributes' through subsequent audits. We look forward to hearing from colleagues who may be interested in such enquiry.

Notes

1 United Nations Educational Scientific and Cultural Organisation (UNESCO) (2002) *Education for Sustainability – From Rio to Johannesburg: Lessons learnt from a decade of commitment.*

2 UNESCO-UNEP (1990) 'Environmentally educated teachers: The priority of priorities', *Connect* Vol XV, No 1, pp1–3.

3 NSW Board of Vocational Education and Training (2009) *Skills for Sustainability*, 2nd ed, NSW Department of Education and Training, Sydney.

ACKNOWLEDGEMENTS

The education for sustainable development field is still emerging, and we are early career academics in the topic area. To present a rigorous discussion we have relied on the extensive experience and wealth of knowledge within our international network of researchers and practitioners. Drawing on many voices, we have endeavoured to communicate the latest research and opportunities while being pragmatic about the scale of challenges and existing inertia within higher education and academia.

Mentors and collaborating partners are the real champions of this field and have provided wisdom and experience that bring this publication to life. Over the years the team has collaborated with hundreds of colleagues through a range of university partners to create, implement and review a range of curriculum renewal options. Notably, these include investigating graduate attributes with James Cook University and Queensland University of Technology; considering strategy development with Monash University; problem inquiry and sharing at international forums with University of Tokyo, UNESCO, WFEO, the International Symposium on Engineering Education in Ireland; collaborative research with the Australian Government and Engineers Australia; auditing existing courses with James Cook University and Monash University; and content and program development with the University of South Australia, Griffith University, Australian National University, the University of Adelaide and the Queensland University of Technology. Further to this we have attended numerous associated events and forums and have been fortunate to have been mentored by many of the world's leading sustainability educators and action-heroes such as the late Leo Jansen, Stephen Sterling, Debra Rowe, John Fien, Peter Newman, Hunter Lovins, Don Huisingh, Walter Leal Filho, Karel Mulder, Janis Birkeland, Roger Hadgraft and Simon Kemp. We also received blind peer review of the key concepts by 40 colleagues in the field, in collaboration with Walter Leal and his team at the *International Journal of Sustainability in Higher Education*.

We wish to acknowledge the partners that have assisted our efforts to apply the 'Management Helix for the Sustainable Organisation' model to the higher education sector. We thank our colleagues from The Natural Edge Project (TNEP) research group for their support and commitment during this time, without which breakthroughs would not have been possible. In particular Ms Angela Reeve, Ms Omniya el Baghdadi, Ms Annabel Farr, Ms Fiona McKeague, Mr Peter Stasinopoulos and Mr David Sparks, and Professor John Fien (RMIT), Professor Neil Dempster and Professor David Thiel (Griffith University) for their trans-disciplinary mentoring and supervision.

We also extend our gratitude to Monash University, particularly Dr Geoff Rose and Dr Gary Codner, who have been applying the helix model to their curriculum renewal agenda, for access to enquire into their experience. We also thank the more than 150 project and peer review participants from more than 40 countries who have been involved in the international peer review of findings and models described in this book over the last eight years, through international conferences, symposia, workshops and online collaborations.

The following colleagues from around the world are acknowledged for contributions over the years of development: Dr Esat Alpay, Imperial College, UK; Professor Martin Betts, Queensland University of Technology, Australia; Associate Professor Gary Codner, Monash University, Australia; Dr Didac Ferrer-Balas, Universitat Politècnica de Catalunya, Spain; Dr Amanda Graham, MIT, US; Ms Michelle Grant, ETHsustainability, Switzerland; Professor Doug Hargreaves, Queensland University of Technology, Australia; Professor Kwi-Gon Kim, Seoul National University, Korea; Mr David Singleton, Global Infrastructure Business, and Corporate Sustainability, ARUP; Professor Mino Takashi, Tokyo University, and Integrated Research System for Sustainability Science (IR3S), Japan; Professor Wu Zhiqiang, Tongji University, China.

We also thank colleagues for their review assistance with the manuscript text (affiliations correct at the time of review, in alphabetical order):

Dr Azizan Zainal Abidin, Petronas University, Malaysia;
Dr Esat Alpay, Imperial College, UK;
Professor Adisa Azapagic, University of Manchester, UK;
Professor Martin Betts, Queensland University of Technology, Australia;
Dr Carol Boyle, University of Auckland, New Zealand;
Dr Martin Bremer, Monterrey Institute of Technology, Mexico;
Mr Wynn Calder, Director, Association of University Leaders for a Sustainable Future;
Mr Tom Connor, KBR, Australia;
Professor Neil Dempster, Griffith University, Australia;
Ms Elizabeth Ellis, Griffith University Business School, Australia;
Dr Didac Ferrer-Balas, Universitat Politècnica de Catalunya, Spain;
Professor John Fien, Royal Melbourne Institute of Technology, Australia;
Dr Amanda Graham, MIT, USA;
Mr Barry Grear, President, World Federation of Engineering Organisations;
Professor Doug Hargreaves, Queensland University of Technology, Australia;
Professor Jan Harmsen, Shell and University of Groningen, Austria;
Ms Chandler Hatton, Delft University of Technology (masters student), Netherlands;
Professor Don Huisingh, Chief Editor Journal of Cleaner Production; University of Tennessee;
Professor Francisco Lozano-Garc›a, University of Monterrey, Mexico;
Dr Karel Mulder, Delft University of Technology, Netherlands;
Dr James Newell, Rowan University, America;
Dr Euan Nichols, Victoria University, Australia;
Professor Ned Pankhurst, Griffith University, Australia;
Dr Margarita Pavlova, Griffith University, Australia;
Professor Michael Powell, Griffith University, Australia;
Professor Yi Qian, Tsinghua University, China, and Member, Chinese Academy of Engineering;
Ms Milena Ràfols, Polytechnic University of Catalunya, Spain;
Dr Debra Rowe, President, U.S. Partnership for Education for Sustainable Development;
Mr Fabian Sack, Manager Environment and Sustainability, Downer EDI, Australia;

Dr Mariano Savelski, Rowan University, USA;
Dr Luis Serra, University of Zaragoza, Spain;
(the late) Mr Hisham Shabiby, Vice President, World Federation of Engineering Organisations;
Mr David Singleton, Global Infrastructure Business, and Corporate Sustainability, ARUP;
Mr Niek Stutje, Delft University of Technology (masters student);
Associate Professor Magdalena Svanström, Chalmers University of Technology, Sweden;
Professor David Thiel, Griffith University, Australia.
Workshop participants, in particular the 2007 International Conference on Engineering Education and Research, Melbourne; 2007 Australasian Association of Engineering Education Conference, Melbourne, and 2008 Engineering Education for Sustainable Development Conference (EESD08).

The original helix model was co-developed by The Natural Edge Project (including Charlie Hargroves, Cheryl Desha, Peter Stasinopoulos, Michael Smith and Nick Palousis) with Hunter Lovins and the team at Natural Capitalism Solutions, and was informed by Global Academy and the TABATI Group, and MBA candidates at the Presidio School of Management under the supervision of Hunter Lovins.

ACRONYMS

AASHE	Association for the Advancement of Sustainability in Higher Education
ABET	Accreditation Board for Engineering and Technology (US)
ACUPCC	American College and University Presidents Climate Commitment
ADBED	Australian Deans of Built Environment and Design
AGS	Alliance for Global Sustainability
AQF	Australian Quality Framework
BCA	Business Council of Australia
CSIRO	Commonwealth Scientific and Industrial Research Organization
DANS	Disciplinary Associations Network for Sustainability
DCCEE	Department of Climate Change and Energy Efficiency
DESD	Decade of Education for Sustainable Development
EA	Engineers Australia
EE	Energy efficiency
EEAG	Energy Efficiency Advisory Group (Australian)
EEO	Australian 'Energy Efficiency Opportunities' program
EfS	Education for sustainability
EIF	Education Investment Fund (Australian federal government)
EPA	Environment Protection Agency
ESDGC	Education for Sustainable Development and Global Citizenship
FIDIC	International Federation of Consulting Engineers (French)
GDP	Gross domestic product
HEASC	Higher Education Associations Sustainability Consortium
HEEPI	Higher Education Environmental Performance Improvement initiative (UK)
HEFCE	Higher Education Funding Council for England
HEFCW	Higher Education Funding Council for Wales
HEI	Higher Education Institution
HESA	Higher Education Sustainability Act
IAU	International Association of Universities
IEA	International Engineering Alliance

JFS	Japan for Sustainability
LTTS	Long Term Training Strategy for the Development of Energy Efficiency Assessment Skills (Australian federal government)
NAE	National Academy of Engineering (American)
NCCARF	National Climate Change Adaption Research Facility (Australian)
NFEE	National Framework for Energy Efficiency
OECD	Organization for Economic Co-operation and Development
PBL	Problem-based learning
RAE	Royal Academy of Engineering (UK)
RET	Department of Resources, Energy and Tourism (Australian)
RoHS	Restriction of Hazardous Substances directive (European)
SARE	Sustainable Agriculture Research and Education
SudVEL	Sudanese Virtual Engineering Library
SWOT	Strengths, Weaknesses, Opportunities and Threats analysis
TEQSA	Tertiary Education Quality and Standards Authority (Australian)
TNEP	The Natural Edge Project
ULSF	University Leaders for a Sustainable Future
UNECE	United Nations Economic Commission for Europe
UNESCO	United Nations Educational, Scientific and Cultural Organization
VOC	Volatile organic compound
WBCSD	World Business Council for Sustainable Development
WEEE	Waste Electrical and Electronic Equipment directive (European)
WFEO	World Federation of Engineering Organizations
WHO	World Health Organization

INTRODUCTION

Professor Stephen Sterling
Centre for Sustainable Futures, University of Plymouth, UK

Beyond any doubt, there is an unprecedented challenge facing educators today in delivering responsible education for a sustainable tomorrow. I have travelled widely over the last decade in particular, meeting with higher education colleagues internationally to discuss the extent and nature of this challenge, and the need to orient education accordingly.

Over some years I have often commented on the tendency for significant challenges (such as education for sustainability) to be understood and accommodated within the norms of the existing system, rather than changing the system to be congruent with the challenge. It is still the exception rather than the norm for institutions to rethink radically how they will equip graduates with knowledge and skills necessary for life in this century. Perhaps the challenge is just too large: can the higher education sector reinvent its role in society in time?

As my colleague Arjen Wals reflects, 'at present most of our universities are still leading the way in advancing the kind of thinking, teaching and research that . . . accelerates un-sustainability'.[1]

At Plymouth University, an urban university very close to the city centre, we have completed a major initiative through the HEFCE-funded Centre for Sustainable Futures (CSF), which sought to embed sustainability as a key institutional principle (see page 171). In 2005, with a staff of nine, we had the ambitious aim of reaching 3,000 academic staff and 30,000 students in 5 years. We were very fortunate to have a significant budget, with which we provided buyout to 48 staff, and we also created an interdisciplinary network. There have been successes: in 2011 Plymouth won the 'Whole Institutional Change' category in the Green Gown Awards, and it has averaged out as the top green university in the UK since the People and Planet 'Green League' table began in 2007. However, two internal curriculum surveys have indicated that although there is support amongst senior staff virtually across the board, there is still a way to go to ensure all our students are receiving some sustainability education. A number of lessons that we learned are reflected in *Sustainability Education*[2] and *The Future Fit Framework – an introductory guide to teaching and learning for sustainability in HE*,[3] which bring together some of our experiences and those of the Higher Education Academy's work in this area.

I am very pleased to see that this publication builds on experiences from Plymouth University, in addition to a host of other institutions, to develop a grounded, practical and rigorous model for rapid curriculum renewal that does not rely on a super-budget or the energy of one or two champions. Here we have a timely set of tools to imagine and implement stretch goals that can result in transformational change in our institutions, now. Moreover, the inclusive, whole of university approach presented here sets a rapid curriculum renewal agenda for sustainability education with clear responsibilities and tasks for staff throughout the campus. In doing so, the methods shown create a possibility for real creativity and innovation within each institution that takes the model on board. I look forward to tracking the experiences of those who do.

Notes

1 Wals, A. (ed.) (2008) *From Cosmetic Reform to Meaningful Integration: Implementing education for sustainable development in higher education institutes – the state of affairs in six European countries*, DHO, Amsterdam.
2 Jones, P., Selby, D. and Sterling, S. (2010) *Sustainability Education: Perspectives and practice across higher education*, Earthscan, London.
3 Sterling, S. (2012) *The Future Fit Framework – An introductory guide to teaching and learning for sustainability in HE*, Higher Education Academy, York.

INTRODUCTION

Dr Debra Rowe
Detroit Area Green Sector Skills Alliance
President, U.S. Partnership for Education for Sustainable Development
Professor of energy management and renewable energy, Oakland Community College

I want to begin this introduction by honouring the good work that educators all over the world are doing for our future quality of life on this planet. I know we want to do more, and we want everybody to do more, but I really want to honour your efforts so far in creating graduates who can play a part in a truly sustainable society.

I also want to thank Interface Flooring. When we started our higher education sustainability consortium (www.aashe.org/heasc), the former CEO Ray Anderson was our first speaker – the voice of business saying 'we need your graduates to be literate in the sustainability challenge, in order to be able to engage in the solutions'. He had a lot of impact on mainstream higher education leaders who were only just starting to hear about sustainability, taking the conversation from why education for sustainability should happen to how it can happen. Now examples of policies, processes, learning outcomes and so on are freely available through AASHE and the Disciplinary Associations Network for Sustainability.

A prerequisite for rapid curriculum renewal is a rapid shift in mindset, from critical thinking to action skills for sustainable abundance. We simply cannot afford to graduate one more armchair pontificator! This is why I am so excited to be able to help introduce this book to you. It is a multi-pronged approach packed full of ideas and examples for action towards all students and the community becoming environmentally, socially and economically responsible. Every day we make decisions about what we teach, and in doing so we also decide what we don't teach. These decisions can either create more scarcity and suffering, or a future of greater abundance and higher quality of life. These authors are committed to the latter and have an action plan for educators to get there. I think the message of this book goes far beyond engineering and I look forward to future editions exploring application throughout education institutions.

I would also like to take this opportunity to share about a summit that was held in late 2010 in Washington DC, hosted by the US Department of Education, on the importance of sustainability education. The conversations during this event were committed and rich with plans, clearly demonstrating the power of networks and partnerships to bring about unprecedented action for education for sustainability. To go beyond incremental change we need to share culture change efforts, and share ways all parts of society can collaborate with students working on sustainability to create systemic change. As institutions move from asking 'why' to asking 'how', we also need to be giving them manuals like this so that they can get going immediately without wasting time and resources.

I look forward to living the next decade of curriculum renewal within our higher education institutions as we collectively create a shift to sustainable development that ripples through society with our graduates. The leaders of tomorrow who emerge from this transformed education experience will know that our generation of academics made a significant contribution to addressing quality of life on this planet.

INTRODUCTION

Robin W. King
Emeritus Professor, University of South Australia

As ever larger proportions of school leavers and others gain university degrees, the world's population has never been as well credentialed. New graduates will face similar employment and professional challenges as those of earlier generations, and will have new sciences, tools and techniques to tackle them. They will also be practising their professions in increasingly complex contexts in response to the impacts of climate change and other environmental pressures. Future graduates of engineering and many other disciplines will individually and collaboratively have special roles, especially around the realisation of physical infrastructure, products and systems that embrace stronger concepts and properties of sustainability. How the education system in universities can meet this challenge is the subject of this book.

The authors are well known for their groundbreaking work on environmental issues and their intersections with education. In this book, they focus more on the education process itself, and challenge the sector to engage in transformative change, especially in engineering education. This sector is, like its profession, familiar with working with standards and codes. For education, such standards are now commonly expressed in the languages of learning outcomes, national qualification frameworks, and externally applied accreditation and registration processes. Most of the international standards on engineering accreditation include sustainability and other contextual matters in their specifications of required graduate learning outcomes. However, the extent and rate at which sustainability is introduced remain matters of discretion by the education providers. The authors discuss these issues in terms of limiting factors and driving factors.

Climate change threatens widespread and transformative change over several decades, but demands immediate policy change and action. Similarly, the authors' approach to transformative curriculum change to include sustainability is pro-active and strategic, rather than relying on more common processes of incremental improvement. A strong value of the book is its guidance on both principles and implementation of rapid curriculum change, drawing on analogous processes in engineering project design and management. Examples of graduate attribute mapping in the curriculum, learning outcomes tracking, and curriculum change and implementation are drawn from Australian and international universities. These examples will assist other universities to learn from successful practice.

PART 1

A compelling case for rapid curriculum renewal

1

HIGHER EDUCATION IN URGENT AND CHALLENGING TIMES

The higher education sector faces its most significant challenge since emerging in the 12th century: to equip society with knowledge and skills to address unprecedented environmental threats and population pressures. The imminent risk from inaction to reduce greenhouse gas emissions, curb energy demand and adapt to extreme weather patterns and temperature fluctuations means that capacity building is urgently required across all professional disciplines and vocational programs.

In this chapter we briefly overview why these times are 'urgent', considering growing pressures on the environment, growing economic impacts of environmental issues, and growing levels of enforcement (from regulation and policy changes, and professional body and accreditation agency requirements). We also discuss why these times are 'challenging', considering the scale and complexity of efforts required in a short period of time, alongside an increasing pace of technological innovation. The literature suggests that within the next decade there are likely to be abrupt market, regulatory and institutional shifts responding to global challenges, which will require graduates to be equipped with a range of new knowledge and skills.

While many authors have commented on the slow nature of curriculum for the last half-century, there is a lack of literature addressing how the process may be accelerated. Without such strategic guidance it is not surprising that universities and educational institutions around the world are struggling to update curriculum at a pace that matches societal progress. This is creating a time lag dilemma for the higher education sector where the usual or 'standard' timeframe to update curriculum for professional disciplines is too long to meet changing market and regulatory requirements for emerging knowledge and skills.

We conclude that given the current state of affairs, curriculum renewal activities must be accelerated, paying attention to the complexity embedded in producing graduate and postgraduate students within useful timeframes.

As the tertiary education sector transitions to significantly embed sustainability into its offerings over the next decade or so, a range of strategies will be used by higher education institutions. In this chapter we briefly discuss a number of risks and rewards associated with embedding sustainability into the curriculum, and highlight a number of organisations working to assist those who are transitioning their curricula now.

Introduction

With considerable growth and development of higher education over the last century, the effectiveness of preparing professionals to contribute to society would appear to be 'fait accompli'. The higher education sector has risen to the challenge of times of rapid change and upheaval, such as the industrial revolution and the world wars.[1] The first recognised universities grew out of 'cathedral schools' in 12th-century Europe. Devastated by Germanic and Viking invasions, cities demanded trained elites to serve the bureaucracies of the church and fill the emerging professions of the clergy, the law and medicine. The European universities of Oxford and Cambridge actually arose through emulating the successes of the earliest known universities in Paris and Bologna. The British then exported their model of higher education to North American colonies and quickly founded nine colonial colleges before the American Revolution, including Harvard in 1636 and Princeton in 1748.

The industrial revolution, which began in 18th-century Britain, forced universities away from their traditional medieval curricula (which included arts, theology, law and medicine) into a new era of natural, physical and social sciences. Industrial society required the invention of the modern research university and the technical college to teach applied sciences, such as chemistry, biology, engineering and medicine. Towards the end of the 19th century, student numbers had increased all over Europe and dramatically in the United States. During the first half of the 20th century, economic demands influenced the course of university curriculum. For example, in the sciences, as institutions focused on improving their research capacity, the focus shifted to fields that could directly improve industrial production, such as physics and chemistry. By the Second World War in the 1940s, a huge variety of academic disciplines could be found all over the world.

Now at the end of the first decade of the 21st century, higher education is entering urgent and challenging times where compelling evidence suggests that the imperative is now to rapidly and effectively incorporate education for sustainability (EfS) across all education programs. Despite successes in incorporating the digital wave of innovation into programs over the last two decades, signals clearly suggest that higher education has been slow to move to incorporate sustainability, and is generally poorly prepared to do so.[2]

> New forces are transforming higher education at a speed that could not have been foreseen 10 years ago . . . Higher education institutions play a strategic role in finding solutions to today's leading challenges in the fields of health, science, education, renewable energy, water management, food security and the environment . . . We need higher education institutions to train teachers in the conduct of pedagogical research and develop relevant curricula that integrates the values of sustainable development.
>
> Mr Walter Erdelen, Assistant Director-General for Natural Sciences, UNESCO[3]

David Orr, one of the world's leading environmental educators, has argued for decades that the planetary crisis we face is a crisis of education.[4] Sustainability, or sustainable development, poses educators the significant task of renewing programs to provide knowledge and skills in a range of relatively new areas across industry, government and society, in both developed and developing countries. Ian O'Connor, Vice Chancellor of Griffith University, spoke about this challenge at the Green Cross International 2006 Earth Dialogues forum (chaired by President Gorbachev), concluding,

> Higher education is beginning to recognise the need to reflect the reality that humanity is affecting the environment in ways which are historically unprecedented and which are potentially devastating for both natural ecosystems and ourselves. Like the wider community, higher education understands that urgent actions are needed to address these fundamental problems and reverse the trends . . . The urgent challenge for higher education now is to include ecological literacy as a core competency for all graduates, whether they are in law, engineering or business.[5]

A number of studies have since been undertaken in various parts of Europe, the United States and Australia in particular, to understand the state of higher education in providing education for sustainability opportunities for students. Typical of these is the 2007 UK Higher Education Funding Council for England study, which found that sustainable development education was disparate and widely dispersed across higher education institutions.[6] For the most part this comprised 'education *about* sustainable development' including awareness lessons or theoretical discussions, rather than education *for* sustainable development, which increases the capacity of individuals, groups or organisations to act, through developing knowledge and skills.

Such findings are supported by our work in Australia on energy efficiency education over the last six years,[7] where we have found significant mismatches in what industry expects should be taught, what faculty think they are teaching, and what students think they are learning. We have found actual knowledge and skill development to be *ad hoc* and highly dependent on the expertise and interests of individual champions. Often this has no foundation in the overall program design, instead being 'bolt-on' attempts to embed sustainability within the curriculum.

Living in 'urgent' times

As 'Generation X' authors with engineering training, we are self-professed problem solvers and keen to get into the solution space of 'how to' engage in capacity building that makes a difference! However – following our own advice to others in 'whole system thinking' – we realise the importance of first appreciating the full extent and context of the problem. With this in mind, we use the following several pages to reflect on the question, 'what are the issues with 21st-century living that are so urgent to address?'

Growing pressures on the environment

As a result of the impact of the first 200 years of the industrial revolution, the second and third decades of the 21st century are shaping up to be characterised as the time in human history when the impact from our collective activities on our Earth grew to a scale that threatened the very conditions that support life as we know it. Furthermore, we like to think that our grandchildren will look back at this period as a time when there was a swift movement to significantly reduce environmental pressures while strengthening economies around the world.

Since 2002 the work of The Natural Edge Project and its partners has been focused on assisting efforts to achieve such a movement, by contributing to, and succinctly communicating, leading research, case studies, tools and strategies across government, business and civil society (see *The Natural Advantage of Nations*,[8] *Cents and Sustainability*,[9] *Whole Systems Design*,[10] *Energy Transformed*,[11] *Water Transformed*,[12] and *Factor 5*[13]). In this work our team has focused on a selection of key pressures on the environment, namely reducing greenhouse gas emissions, reducing impacts on biodiversity and natural systems, improving freshwater management, reducing waste production, and reducing air pollution. In each of these areas pressures on the environment have reached levels previously unobserved throughout history, and scientists are finding themselves analysing and attempting to quantify projections and predictions into unknown territory, often with an unknown number of variables to take into account. As it is outside the scope of this book, we provide a summary of material presented in *Cents and Sustainability* in Table 1.1 to demonstrate the severity of the impacts of environmental pressures, and refer readers to this work and the many works referenced within.

TABLE 1.1 Examples of environmental pressures set to increase with additional population growth and economic development

Greenhouse gas emissions: Considerations include increasing greenhouse gas concentrations, increasing temperatures, melting ice sheets, and melting permafrost.

- As Dr Rajendra Pachauri, IPCC Chairperson, states: 'The increased evidence of abrupt changes in the climate system, the fact that CO_2 equivalent levels are already at 455ppm, plus the current high rate of annual increases in global greenhouse gas emissions reinforces the IPCC's 4th Assessment finding that humanity has a short window of time to bring about a reduction in global emissions if we wish to limit temperature increase to around 2°C at equilibrium'.[14]
- The 2006 UK Stern Review concluded that within our lifetime there is between a 77% and 99% likelihood (depending on the climate model used) of the global average temperature rising by more than 2°C, with a likely greenhouse gas (GHG) concentration in the atmosphere of 550 parts per million (ppm) or more by around 2100.[15] Modelling by Stern further suggests that to stabilise atmospheric GHG concentrations at or below 550ppm, global emissions must peak between 2020 and 2030 and subsequently decline by 1.5–4% per year.[16]
- Scientists estimate that annual emissions of methane from the thawing of permafrost and wetlands may increase by more than 50% – which would be equivalent to 10–25% of current human-induced greenhouse gas emissions in the atmosphere.[17]
- Current models suggest that if global average warming were sustained for millennia in excess of 1.9–4.6°C relative to pre-industrial values, the Greenland ice sheet may completely disappear, contributing 7m of sea level rise,[18] significantly affecting the 'global conveyor' ocean current which assists in regulating global temperatures.[19]

Biodiversity and natural system resilience: Considerations include current species loss, species committed to extinction, deforestation, desertification, human population dynamics.

- Coral reefs, such as the Great Barrier Reef, are predicted to rapidly bleach if sea surface temperatures increase beyond 1°C above the usual seasonal maximum, and to potentially die beyond increases of 2°C. Coral vulnerability is increased due to increasing acidity from ocean absorption of greenhouse gases, and nutrients from land runoff.[20] The current rate of species loss is predicted to accelerate due to climate change such that by 2050 'between 15 and 37% of the species on the earth might be committed to extinction'.[21]
- It is estimated that some 75% of fisheries worldwide are currently fished at or beyond sustainable levels.[22] A 2003 study concluded that 90% of the large fish in the ocean have disappeared since the middle of the last century.[23]
- In Europe, over 80% of crops are pollinated by insects. However, a decline in bee diversity and abundance threatens the viability of many crops this decade. The collapse in bee populations is linked to habitat loss and disease.[24]
- Globally, natural forests are disappearing at a rate of 13 million hectares a year (ha/year, roughly the size of Greece). Regrowth and commercial plantations replaces approximately 5.6 million hectares, leaving a net loss of 7 million ha/year. It is estimated that the natural forests in Indonesia and Myanmar will be gone within 10 years, and those in the Russian Far East within 20 years.[25]
- The Amazon rainforest is approaching an ecological tipping point, where the last remaining forest is unnaturally dry and vulnerable to fire from lightening strikes.[26] It is predicted that this threat will peak when deforestation exceeds 20–30% of the Amazon; it is currently at 17%.[27] In many countries such as Haiti,

TABLE 1.1 Continued

Madagascar and Malawi, deforestation is resulting in soil loss and a disruption to the hydrological cycle that will result in a collapse of crops in our lifetime.[28]

- In semiarid Africa, human and livestock demands for trees and vegetation are converting large swathes of land to desert; and rapid population growth is exacerbating this trend.[29] Continued degradation has already caused biodiversity and food, water and fibre shortages.[30]

Freshwater extraction: Considerations include global water demand, groundwater depletion, lake water consumption, and effects of climate change.

- Global freshwater use has tripled over the last century, with much of this demand met from aquifers where water is extracted faster than it can be replaced.[31] In Yemen, groundwater resources are depleting so quickly that some towns have access to water only once every 24 days, with the capital Sana'a receiving water once per week. It is estimated that Sana'a will have completely exhausted groundwater sources by 2025.[32]
- Aquifers are being depleted throughout the world causing a significant loss in food production capacity. In Pakistan, it is estimated that the city of Quetta will run out of water by 2016,[33] around the same time the groundwater supplies of the surrounding grain growing region are expected to be exhausted.[34]
- Population pressures have reduced Lake Chapala, the primary water source for Guadalajara, Mexico, to 20% of its volume.[35] In China's Qinhai province, over 2,000 lakes have disappeared over a 20-year period, while in the Hebei province, falling water tables have claimed 969 of the 1,052 lakes which used to exist in the region.[36]
- By 2050, climate change is predicted to exacerbate existing water shortages with areas subject to increasing water stress projected to be twice the size of those with decreasing water stress.[37]

Waste production: Considerations include diminishing landfill space, and diminishing access to finite resources such as oil and precious metals.

- In the United States, at least five states estimate there remains less than 10 years' worth of landfill capacity, after which point waste will need to be trucked into other states and regions.[38] New York City has been trucking waste to landfill sites up to 300 miles (close to 500km) away since 2001.[39]
- In urban areas in China, waste generation is predicted to increase three-fold between 2000 and 2030.[40]
- Assuming annual extraction growth of 2%, US Geological Survey data show economically recoverable reserves of primary lead will run out by 2025, tin by 2027, copper by 2033, iron ore by 2062 and bauxite by 2076.[41] This will lead to a focus on mining urban waste streams and reprocessing vast existing waste sites.

Air pollution: Considerations include (in addition to greenhouse gas emissions) premature deaths and acid rain.

- The WHO estimates there are over 400,000 premature deaths each year due to outdoor air pollution and a further 1.6 million due to indoor air pollution.[42]
- Based on current trends, the OECD projects that the number of deaths in 2010 due to airborne particulate matter will double from approximately 1.5 million to just over 3 million by 2050.[43]
- Climate change may exacerbate existing air pollution issues through increased ozone and VOC formation, increased frequency of forest fires, and the potential formation of inversion zones trapping pollutants at ground levels.[44]
- In China, acid rain affects just under one-third of the country and around half of the cities and counties being monitored.[45] In some cities, all precipitation is in the form of acid rain.[46] NOx and SOx emissions are generated primarily in China's coal-fired power plants and from burning oil, and recent projections suggest that annual coal consumption will reach 3.8 billion tonnes by 2015, an increase of 800 million tonnes compared to 2009.[47] Further, the increase in greenhouse gas emissions from China's coal use is predicted to exceed that of all industrialized countries combined by 2031, surpassing by five times the reduction in such emissions that the Kyoto Protocol seeks.[48]

Source: Summarised from Smith, M., Hargroves, K., and Desha, C. (2010)[49] with references noted within table.

Growing financial impacts on economies

> The 'business as usual' model, where profits come before sustainability, is absolutely finished. We now have a window of 10 to 15 years (up to 2017–2022) to adopt a sustainable approach before we reach a global 'tipping point' – the point at which mankind loses the ability to command growth and development.[50]
>
> Jonathon Porritt, Founding Director, Forum for the Future, Chairman of the UK Sustainable Development Commission, and author of *Capitalism as if the World Matters*

As Jonathon Porritt so powerfully reflects, mankind is running the risk of losing the ability to 'command growth and development' without regard for natural systems. What makes the early 21st century such a time of urgency is the fact that not only is the scale of environmental pressure creating impacts that will threaten the Earth's ability to sustain the conditions we have grown accustomed to, the change in these conditions will have a meaningful impact on economies around the world. As can be seen from Figure 1.1 the trends in GDP are now being replicated in the growth of direct and indirect costs related to environmental damages.

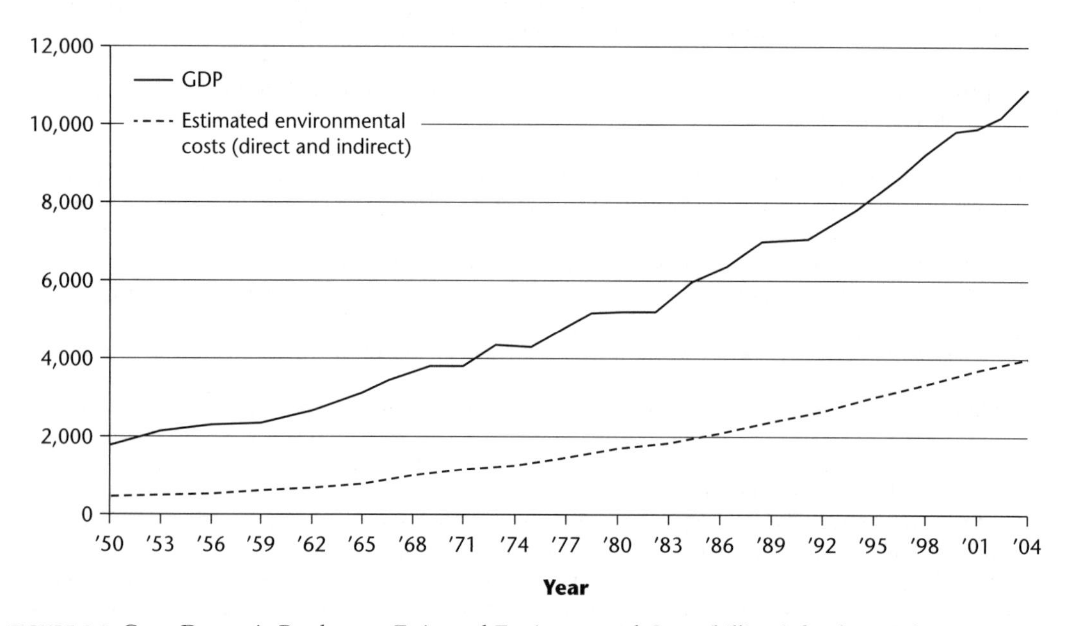

FIGURE 1.1 Gross Domestic Product vs. Estimated Environmental Costs (billions) for the US from 1950–2004

Source: Data reinterpreted by K. Hargroves from J. Talberth *et al.* (2006)[51], and presented in Smith, M., Hargroves, C. and Desha, C. (2010)[52]

This coupling of costs related to environmental pressures and economic growth is now evident across each of the key areas listed above. For brevity, we provide a sample of cases discussed in detail in *Cents and Sustainability*:[53]

- Greenhouse gas emissions: In 2006 a study on the economics of climate change estimated that each year on average the cost to the global economy of not acting to reduce greenhouse gas emissions could be in the order of 5–20% of GDP, compared to an estimated 1% cost of acting to stabilise emissions.[54]
- Biodiversity: In 2006 a study on biological diversity concluded that, 'The intensification of fishing has led to a decline of large fish. In the North Atlantic, their numbers have declined by 66% in the last 50 years,' with this having economic impacts on fisheries globally .[55]
- Water consumption: In 2004 a study on the millennium development goal of halving the population without access to water and sanitation by 2015 estimated that this would cost in the order of US$10 billion annually, and the cost of not achieving it would be in the order of US$130 billion annually.[56]
- Waste production: In 2001 leading environmental business advocate Amory Lovins stated that: 'It is extremely profitable to wring out waste, even today when nature is valued at approximately zero, because there is so much waste – quite an astonishing amount after several centuries of market capitalism.'[57]
- Air pollution: In 1996 a study found that the economic losses due to health costs of air pollution in India in 1995 was slightly over the amount of growth of GDP for that year, meaning the growth in GDP was invested in addressing health costs.[58] In 2006 a study on agricultural economics estimated that reductions in crop yields from tropospheric ozone in Europe was in the order of €4.4–9.3 billion/year.[59]

Growing levels of enforcement

History clearly shows that humanity takes time to acknowledge, accept and then deal with issues that have widespread and significant ramifications on daily life.[60] Over the last two decades there have been a number of declarations and action plans developed to encourage and assist the higher education sector to incorporate the imperative to reduce environmental pressures, as shown in Table 1.2.

> Universities educate most of the people who develop and manage society's institutions. For this reason, universities bear profound responsibilities to increase the awareness, knowledge, technologies and tools to create an environmentally sustainable future.[61]
>
> Talloires Declaration, 1990

According to the Copernicus Alliance in 2012

> Perhaps the greatest challenge of all is to reorient the higher education curriculum so that it aligns with sustainable development. This requires not just the inclusion of relevant subject matter and the pursuit of inter- and transdisciplinary approaches, but also the development of education for sustainable development competences of university and college educators as well as learners. Competences associated with: systemic thinking; critical reflective thinking; futures engagement and values clarification; the ability to deal with complex and contradictory situations; the capacity to work in partnership in order to facilitate transformative actions towards sustainability are vital . . . The curriculum gate-keepers, professional bodies, government agencies, student groups and academic development bodies as well as teaching colleagues have a key role to play to achieve this ambition.[62]

TABLE 1.2 Examples of declarations and action plans promoting education for sustainability

Date	*Declaration*	*Brief Description*
1990	The Talloires Declaration	A ten-point action plan for colleges and universities committed to promoting education for sustainability and environmental literacy in teaching, research, operations and outreach at colleges and universities.[63]
1992	Agenda 21	Chapter 36 articulated the need for education to play a key role in addressing the challenge of sustainable development.[64] This was subsequently acknowledged in a range of documentation around the world.[65]
1997	Thessaloniki Declaration	A declaration signed by 83 countries on the need for public awareness and education for sustainability.[66]
1998	World Declaration Higher Education in the 21st Century	A declaration that articulated the need for a critical mass of skilled and educated people to ensure sustainable development.[67]
2000	United Nations Earth Charter	Provided a statement of ethics and values for a sustainable future, including the need for education for sustainability.[68]
2001	Lüneburg Declaration	Endorsement of Agenda 21 and numerous other declarations around the role of higher education in education for sustainability in preparation for the 2002 World Summit on Sustainable Development.[69]
2002	Ubuntu Declaration	Declaration for all levels of education, highlighting the need for education in science and technology for sustainable development.[70]
2002	Decade of Education for Sustainable Development (DESD) 2005–2014	Led by Japan, the DESD created a global platform for dialogue in the higher education sector. Since the declaration there has been a rapid growth of education for sustainability literature around the role of universities in education, research, policy formation and information exchange necessary to make sustainable development possible.[71]
2009	G8 University Summit Declaration	Declaration on research and education for sustainable and responsible development, locally and globally.[72]
2009	World Conference on Higher Education Communiqué	A detailed account of what should be occurring in higher education institutions, and a call for action for member states and UNESCO.[73]
2010	AASHE Call to Action	A call to action for sustainability curriculum in higher education, by the Association for the Advancement of Sustainability in Higher Education, based in America.[74]
2012	People's Sustainability Treaty on Higher Education	Drafted by representatives from 25 higher education agencies, organisations, associations and student groups to create a consolidated platform for cooperation beyond the Rio+20 event in June 2012.[75]

Source: References noted within.

Despite such declarations that highlight the significant impacts that our global society is having on the environment, and the ramifications this in turn will have on global economies, there is very little actual commitment to act on a meaningful scale. As with any new stage of development there are those who are first to act, and who take the early risks and position themselves well for the future. These leading efforts then inspire others to follow and demonstrate what can be achieved with dedicated effort. Again rather than outlining these leading efforts herein we provide a sample as presented in *Cents and Sustainability*:[76]

- Greenhouse gas emissions: The UK Government's 'Code for Sustainable Homes' is the first national legislation to set minimum standards for energy performance in new homes, calling for reductions in energy use compared to 2006 standards of 25% by 2010, 44% by 2013, and 100% (zero emissions) by 2016.[77]
- Biodiversity: When the Korean War ended, South Korea was largely deforested. Since the early 1960s the government has initiated programs to achieve some 6 million hectares of tree planting, nearly 65% of the country.[78]
- Water consumption: In Bogor, Indonesia, the water tariff was increased from US$0.15 to US$0.42 per cubic metre, resulting in households decreased demand by 30%. In São Paulo, when effluent charges for industry were introduced, three industries decreased their water consumption by 40–60%.[79]
- Waste production: In 2006 the European Union released its 'Restriction of Hazardous Substances'[80] (RoHS) directive that then triggered an international response with the percentage of RoHS-compliant manufacturers rising from 51% to over 93% in nine months,[81] and aligned policies were introduced in China in 2007[82] and in South Korea in 2008.[83]
- Atmospheric pollution: A succession of agreements by European countries has resulted in a decrease in sulphur and other air emissions. The 1983 'Convention of Long Range Trans-boundary Air Pollution' set a target for emission reduction of 30% compared to 1980 levels.[84] The Convention was updated twice, and was followed by the 1994 'UNECE Second Sulphur Protocol', which set a target for emission reduction of 50% by 2000, 70% by 2005, and 80% by 2010.[85] During 1980–1998, European sulphur dioxide emissions decreased by over 70%, while GDP grew by 44%.[86]

These and an increasing number of examples of leadership from governments are now creating real precedent for industry, business and society to act to reduce impacts on the environment.[87] Considering the economic impacts already being felt from the growing levels of environmental pressures, it is clear that the current low levels of compliance on environmental performance required by governments will rapidly increase in the near future, shown stylistically to begin at a hypothetical 'Time (t)' in Figure 1.2. Factors that will influence the timing of 'Time (t)' will include the level of perceived economic risk from environmental damage and potential collapse of ecological systems such as bee communities required for wide-scale pollination, fish stocks, storm surges and sea-level rises, increased natural disasters, and so on. Following 'Time (t)' the level of environmental performance of an organisation, business or education institution will dictate the pace at which action is taken to comply with enforcement. When requirements for change begin to ramp up in the near future, those institutions that have maintained 'compliance' or lower will have a very steep curve requiring significant action, while those that have improved performance in anticipation of the transition will have a stronger strategic position.

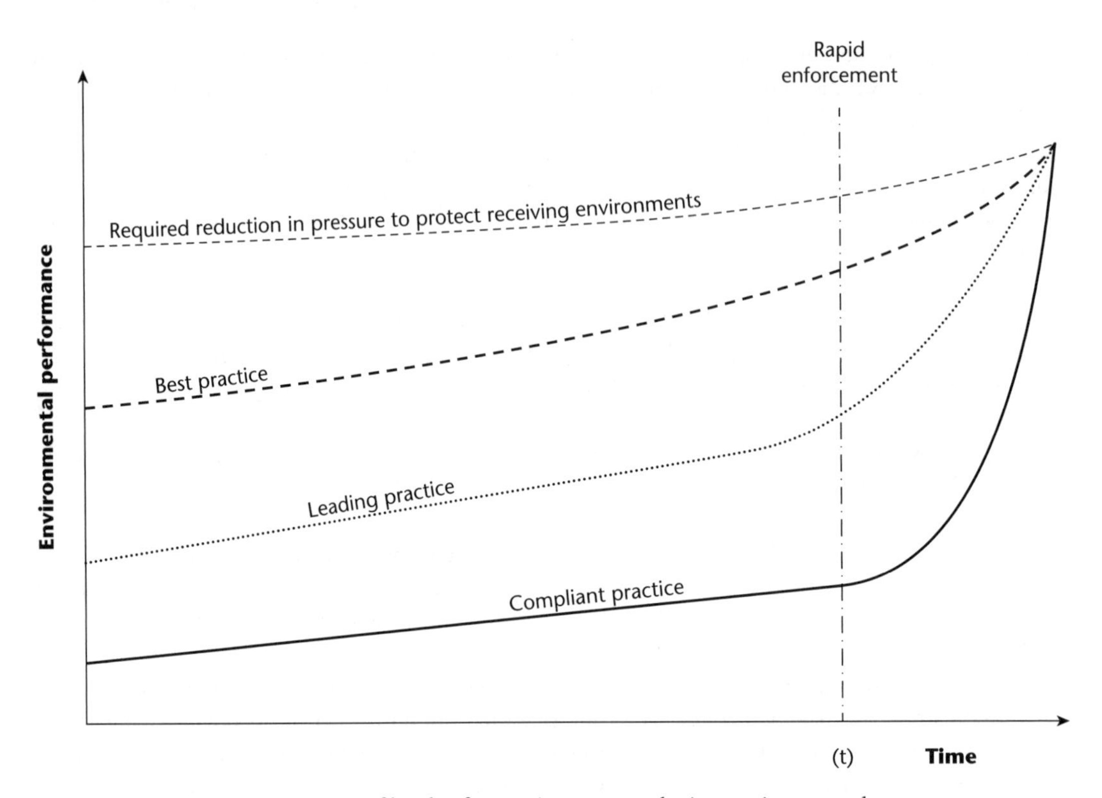

FIGURE 1.2 Stylistic representation of levels of commitment to reducing environmental pressures

Source: Smith, Hargroves and Desha (2010)[88]

> In its sixty years of existence, the IAU [International Association of Universities] has witnessed significant changes that are shaping higher education as well as the increasing pressure placed on higher education systems and institutions to change so that they meet national development objectives and individual aspirations. This pressure has probably never been as great as today . . . Higher education institutions retain their mission to educate, to train, to undertake research, and to serve their communities but are asked to do so in a rapidly changing environment . . . Higher education institutions are asked to equip increasing numbers of learners with the knowledge, skills, and critical thinking that will ensure their employability and respond to national sustainable development objectives . . . A very tall order – although I mentioned only some of the challenges that higher education institutions are facing – which attests to the vital contribution of higher education and its institutions to sustainable national development.[89]

Within the higher education sector, the anticipated rapid increase in compliance requirements will be evident in a range of ways, such as regulatory and policy changes to enforce improved environmental performance in industry and business practices requiring changes to graduate attributes; professional body and accreditation agency requirements for specific graduate attributes to be included in education programs; funding agencies requiring incorporation of related topics in research grant and capital funding applications; and a significant increase in demand from potential employers for graduates with associated graduate attributes.

Regulation and policy changes

Changes to regulations and policy will impact universities in a range of different ways including changes to research and teaching policies and through direct regulation of industry, business and organisational practices. Governments are increasingly modifying selection criteria for research funding to include clear and increasingly stringent requirements to demonstrate the contribution of the research proposal to assisting society to reduce environmental pressures. An example of this is the Australian Government's four National Research Priorities including 'an environmentally sustainable Australia' and 'frontier technologies for building and transforming Australian industries'. In relation to teaching policy, governments are likely to link federal funding for higher education institutions to their ability to address education for sustainability and to integrate associated knowledge and skills into curricula (in a similar manner to the way in which institutions currently track integration of priority areas such as indigenous knowledge and research-led teaching).

There are also a growing number of examples of increasingly stringent environmental reporting and performance requirements on business, such as the Australian Energy Efficiency Opportunities (EEO) program. The program was launched in July 2006 and required businesses that used more than 0.5 PJ (139,000 MWh, equivalent to the energy use of around 10,000 households) of energy per year – some 220 businesses representing around 45% of national energy demand – to undertake an energy efficiency assessment and report publicly on opportunities with a payback period of up to 4 years.[90] Extending this, the Victoria State Government was the first state to require all EPA license holders using more than 0.1 PJ (27,800 MWh per year) to implement opportunities with a payback period of up to 3 years in order to retain their licence.[91]

By imposing such public reporting, businesses were forced to undertake an internal process to identify energy efficiency opportunities knowing that the results would be publicly scrutinised by customers and shareholders. The value of the results was that managers and shareholders could see the range of potentially profitable activities that could be undertaken in a 4-year period to reduce energy demand. A 2008 progress report found that 'the 199 corporations reported that assessments had identified over 7,000 opportunities to improve energy efficiency with the potential to save 62.5 PJ of energy with a better than four year payback (the equivalent of 5.7 million tonnes of CO2-e) and $626 million in net annual financial savings'.[92] A significant finding of the program was that there was a lack of capacity in industry and business to understand the program requirements and report on energy efficiency opportunities. The program found that 'Almost 40% of survey participants did not consider the requirements of the EEO Legislation and associated reporting easy to understand'.[93]

Following this finding the government initiated an investigation into the development of a 'Long Term Training Strategy for the Development of Energy Efficiency Assessment Skills', across the energy-intensive industries, energy service providers and universities.[94] The findings of this study have been used to catalyse a number of other exploratory studies in vocational education and training (national context) and higher education (engineering and business), across departments including 'Resources, Energy and Tourism', 'Industry, Innovation, Climate Change Science, Research and Tertiary Education', and 'Sustainability, Environment, Water, Population and Communities'. It also informed the development of a National Energy Efficiency Advisory Group on engineering curriculum renewal (described further in Chapter 2), which has worked with the Institution of Engineers Australia to explore emerging expectations for undergraduate engineering education in the area of energy efficiency knowledge and skill development (led by the authors).

Reflecting on the original EEO program and its ambitions, it is clear that the program has had a direct impact on the tertiary education sector in the topic area of energy efficiency education. In this example, regulation led to uncovering gaps in knowledge and skills, which catalysed the exploration of issues and subsequent federal interaction with professional organisations to address remedial action in the education sector. Budgets were also created within four federal departments for structural adjustment within the tertiary education sector around energy efficiency. Several programs then targeted capacity-building initiatives that directly involved collaboration between educational institutions, professional organisations and industry. In this particular 'Time (t)' transition, the half a dozen or so Australian higher education institutions with expertise and existing curriculum in energy efficiency positioned themselves well to receive funding and recognition, manoeuvring into a post-t leadership position in the sector.

In 2007, The Natural Edge Project collaborated with CSIRO and Griffith University to develop a comprehensive free access online resource to assist professionals and students to build capacity in energy efficiency – 'Energy Transformed: Sustainable Energy Solutions for Climate Change Mitigation', containing 30 lectures of peer reviewed materials.[95]

Professional body and accreditation agency requirements

Within regulated disciplines such as engineering, business and law, program accreditation is a strong driver of change, setting a review period of 3–6 years for universities to reflect on and demonstrate how they have addressed accreditation requirements in their program/s. Accreditation requirements are influenced by a range of factors including industry requirements, student demands, government policy, the regulatory environment and globalisation.

In the absence of a clear mandate for government to develop legislation to significantly reduce environmental pressures, and current legislative compliance levels being well below what is required (to mitigate climate change, etc.), government and society are increasingly looking to accreditation agencies to take a leadership role. In such a role, accreditation agencies would not only continue to ensure that programs meet compliance, they would also force higher education institutions to go beyond compliance and update programs to include an outcome-based approach to education for sustainability through an enhanced accreditation process. This aspirational role for accreditation agencies goes beyond being the 'professional police' that enforce compliance, to one that both ensures quality of graduates and actively contributes to shaping the future of the profession and its contribution to society.

In the past, accreditation agencies and professional bodies have been largely focused on ensuring compliance with government and industry regulations and expectations and were sound in the knowledge that this was the role society expected them to play. For example, in Australia a 2008 Review of Higher Education by the Australian Deans of Built Environment and Design (ADBED) concluded that accreditation was focused on compliance rather than innovation. Graduate outcomes desired by accreditation panels were those most needed for 'work-ready' graduates (who cater to current employer needs) rather than looking ahead to future expectations.[96] In the UK, a 2008 study by the Higher Education Funding Council found similar barriers to implementation, but also the added barrier of professional bodies themselves, whose perceived conservatism acted as a barrier.[97]

For universities to maintain their role in the formation of leaders for the emerging Australia, its economies and businesses, the accreditation processes need to maintain a focus on innovation and leadership rather than 'training for work'.[98]

However, there are clear signs of changing attitudes regarding the role of accreditation, with a number of professional engineering institutions already embedding sustainability language into codes of ethics and graduate competency statements.

For example, the Royal Academy of Engineering (RAE) argues that 'the accreditation process for university engineering courses should be proactive in driving the development and updating of course content, rather than being a passive auditing exercise'.[99] Furthermore, it promotes sustainable development concepts through a published set of twelve 'Guiding Principles' for engineering for sustainable development in a document that also provides examples and applications for curriculum implementation.[100] It has sponsored a visiting professors scheme in the UK since 1998 to embed the topic of engineering for sustainable development into engineering courses, rather than creating a separate subject. The importance of accreditation as a driver for curriculum renewal is also reflected in countries such as Australia and the United States. In Australia, reviews of the Stage 1 (graduate) and Stage 2 (professional) Competency Standards that underpin program accreditation catalysed the embedding of sustainability elements throughout the competency statements. In America, the Accreditation Board for Engineering and Technology (ABET) has also embedded language regarding the design of systems, components or processes to meet desired needs within realistic constraints including sustainability considerations, supported by documents such as the Society for Civil Engineers' code of ethics, which embeds sustainability into its tenets of practice.

In 2010 we ran a workshop on engineering education and accreditation in Ireland with our Cork University colleagues, as part of the International Symposium on Engineering Education.[101] The focus was in ground-truthing an observed correlation between levels of sustainability-related curriculum renewal across certain regions and countries globally and the introduction of sustainability concepts within accreditation documentation of corresponding regional professional bodies. Findings of this workshop reinforced literature observations, showing at best a wide range of opinions and understandings and a general disparity among what key sustainability themes for engineering curricula should be among academics and practitioners. It did highlight several priority themes to consider regardless of discipline.

At the 2010 International Symposium on Engineering Education workshop in Ireland, education about 'resources' was identified as the most common priority area for focus over the next 5 years. This was followed by renewable fuel sources, life cycle analysis/management and water. Other priority themes for embedding regardless of discipline included ethics and responsibility, design/systems thinking, and thermodynamics.[102]

Within this context, it is clear that the opportunity provided by accreditation processes is increasingly important, developing criteria and guidance for embedding sustainability into the curriculum. Moving forwards, the role becomes one of ensuring that the intentions of such criteria are embedded within the programs. This will require addressing a range of barriers to overcome challenges associated with understanding ways to meet new accreditation criteria, being able to evaluate sustainability-related competencies during accreditation reviews, and empowering accreditation panel members who are often volunteers with their own time and resourcing constraints.

Living in 'challenging' times

Following an appreciation of what it means to live in urgent times, it could be argued that this is fine, so long as we can address such issues quickly. However, in a number of global examples such as available fish stocks, access to fossil fuel and changing climate patterns, we can see that this has not been possible. With this in mind, we use the following several pages to reflect on the question, 'what are the issues with 21st-century living that are so challenging to address?'

Economy-scale efforts required in a short period of time

As the majority of the activity in the economy exerts some form of environmental pressure, economy-wide efforts will be required to reduce such pressure on a meaningful scale. Further, such efforts need to ensure that economic performance is maintained, referred to as the challenge to 'decouple economic growth from environmental pressure'.[103] This challenge, represented in Figure 1.3, calls for the growth of environmental pressures to be reduced 'relative' to economic growth and where possible completely

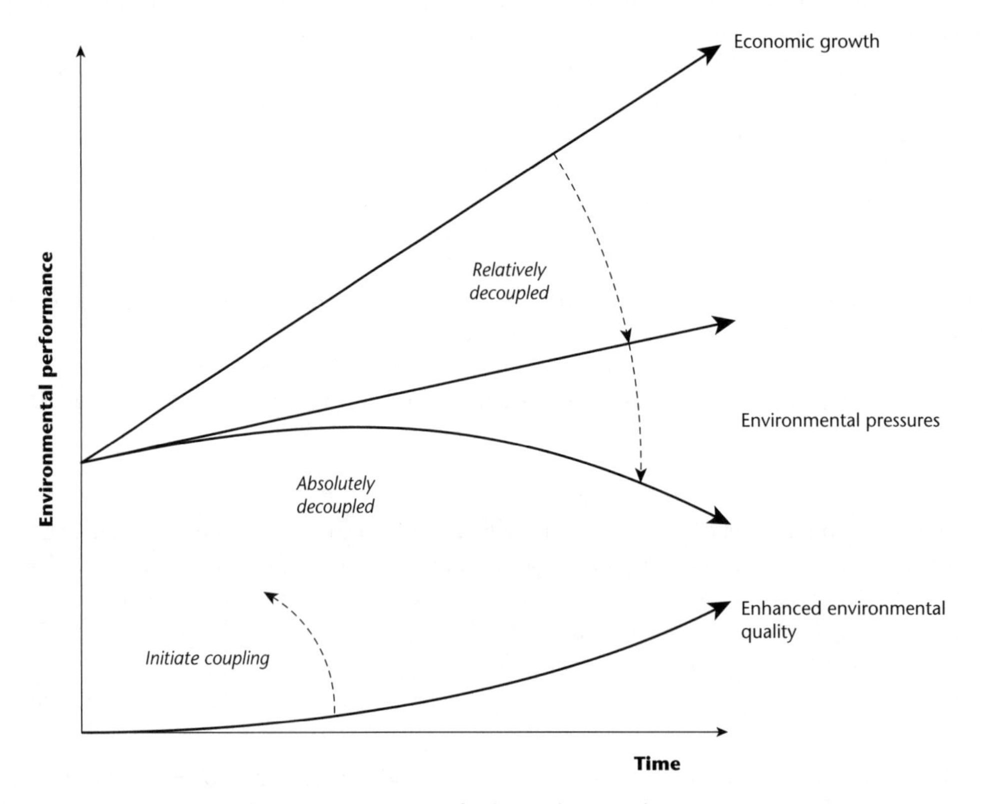

FIGURE 1.3 Conceptual and stylised representation of a decoupling graph

Source: Developed by Karlson Hargroves, Peter Stasinopoulos, Cheryl Desha and Michael Smith, in Smith, Hargroves and Desha (2010)[104]

– or 'absolutely' – decoupled from improved economic performance. Furthermore, positive environmental impacts (for example, reforestation, aquifer recharge, etc.) would be 'recoupled' to economic performance so that as development proceeds, environmental systems are restored.

> Action to achieve decoupling across entire economies will call for significant reorientation of systems, legislation, standards practices, etc. and may pose the most significant challenge to the human race in its history.

In the following paragraphs we provide a summary of the example of reducing environmental pressures, as detailed in *Cents and Sustainability*.[105]

It is now well established that absolute decoupling is required to achieve a stabilisation of atmospheric concentrations of greenhouse gases in the order or 450–550 parts per million (ppm) by 2050, to avoid dangerous climate change.[106] There is a range of scenarios for achieving this goal that are affected by the rate at which emissions are reduced over time. In each scenario, the current growth in emissions needs to be stopped to create the peak of the absolute decoupling curve requiring a focus on short-term performance, and the levels of emissions across entire economies need to be gradually reduced each year over some 30 to 50 years, requiring a medium- to long-term strategy. The level of sustained reduction is dictated by the timing and height of the peak with Figure 1.4 showing that peak in say

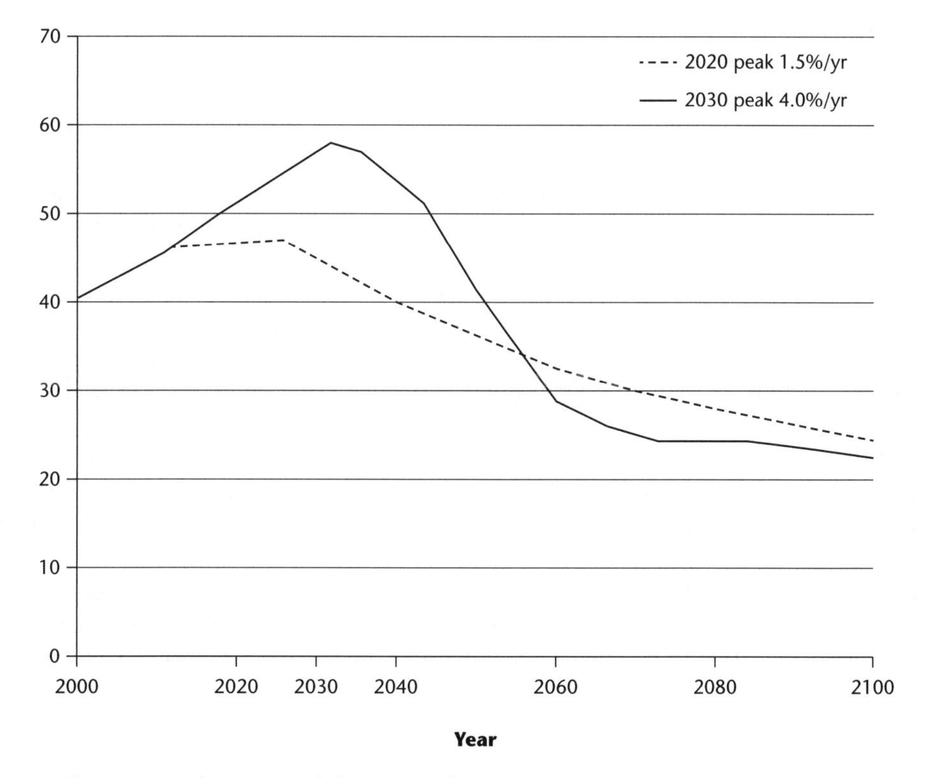

FIGURE 1.4 Illustrative pathways to stabilising greenhouse gas emissions at 550ppm CO_2e

Source: Based on data from Stern, N. (2006)[107]

2020 will result in a lower annual reduction target than a peak in 2030 with both curves achieving 550ppm.

Hence the challenge for economies around the world is to rapidly achieve a low peaking in emissions, around 2020, to then allow a more manageable annual reduction target. The challenge of the higher education sector is that the timeframe to achieve peaking does not allow for the required knowledge and skills,[108] which are largely yet to be incorporated into undergraduate programs, to be developed over the standard curriculum renewal timeline, meaning that it will be largely a postgraduate and professional development challenge. Further, in order to prepare undergraduates to contribute to society achieving gradual sustained reductions after the peaking is achieved, the standard curriculum renewal process will need to be improved and accelerated.

In practice this calls for a dual focus, both on engaging with current practitioners and decision makers around knowledge and skills required to peak greenhouse gas emissions (such as postgraduate certificates, diplomas and Master's programs, along with professional development seminars and short courses), and also focusing on undergraduate programs to develop knowledge and skills required to both continue to maintain the peaking and to then achieve gradual sustained reductions balanced across each sector. In order to achieve absolute decoupling a short-term/long-term approach will be required for each of the major environmental pressures, such as:

- Greenhouse gas emissions: In the short term, highly energy-inefficient processes and appliances can be improved to continue to deliver products and services while using significantly less energy, in many cases as much as 80% less as outlined in *Factor 5*.[109] In preparation for long-term sustained reductions in emissions, low-carbon energy generation technologies need to be innovated, commercialised and brought to scale.
- Biodiversity: In the short term, significant reductions to species loss are required with as much as 40% of species being already lost between 1970 and 2000.[110] In preparation for long-term sustained reductions in pressure on biodiversity and natural systems, a range of approaches to deforestation, fisheries management and control of invasive species need to be developed and implemented.
- Water consumption: In the short term, significant reductions to freshwater withdrawal considering that groundwater extraction rates are exceeding replenishment rates by 25% in China and over 50% in parts of northwest India.[111] In preparation for long-term sustained reductions in water consumption, a range of forestry, agriculture and natural resource management strategies and practices need to be developed, trialled and brought to scale, such as advanced deficit irrigation strategies,[112] holistic resource management methods,[113] and water-sensitive urban design.[114]
- Waste production: In the short term, significant reductions to waste generation are required considering that since 1980 the levels of annual global resource extraction have increased by 36% and are expected to grow to 80 billion tons in 2020.[115] In preparation for long-term sustained reductions in waste generation, a range of design, manufacturing and recycling processes are needed to underpin structural adjustments in a range of industries.
- Atmospheric pollution: In the short term, significant reductions to air pollution are required considering that in 1999 some 10,000 people died prematurely in Delhi due to air pollution, equivalent to one death every 52 minutes.[116] In preparation for long-term sustained reductions in air pollution, a range of new processes and methods are required to reduce emissions of sulphur dioxide, nitrous oxide, lead and particulate matter.

Such dual-track approaches present a significant challenge to the higher education sector, as graduates and professionals need to be up-skilled in areas to contribute to both agendas. As highlighted in a United

Nations Environment Program report on working in a low-carbon world, 'companies in the fledgling green economy are struggling to find workers with the skills needed to perform the work that needs to be done. Indeed, there are signs that shortages of skilled labor could put the brakes on green expansion . . . There is thus a need to put appropriate education and training arrangements in place.'[117]

The pace of technological innovation is increasing

As outlined in *The Natural Advantage of Nations* in 2005, following the industrial revolution society has experienced a series of major waves of technical innovation, with the sixth and current wave underway, which is providing 'a critical mass of enabling eco-innovations making integrated approaches to sustainable development economically viable'.[118]

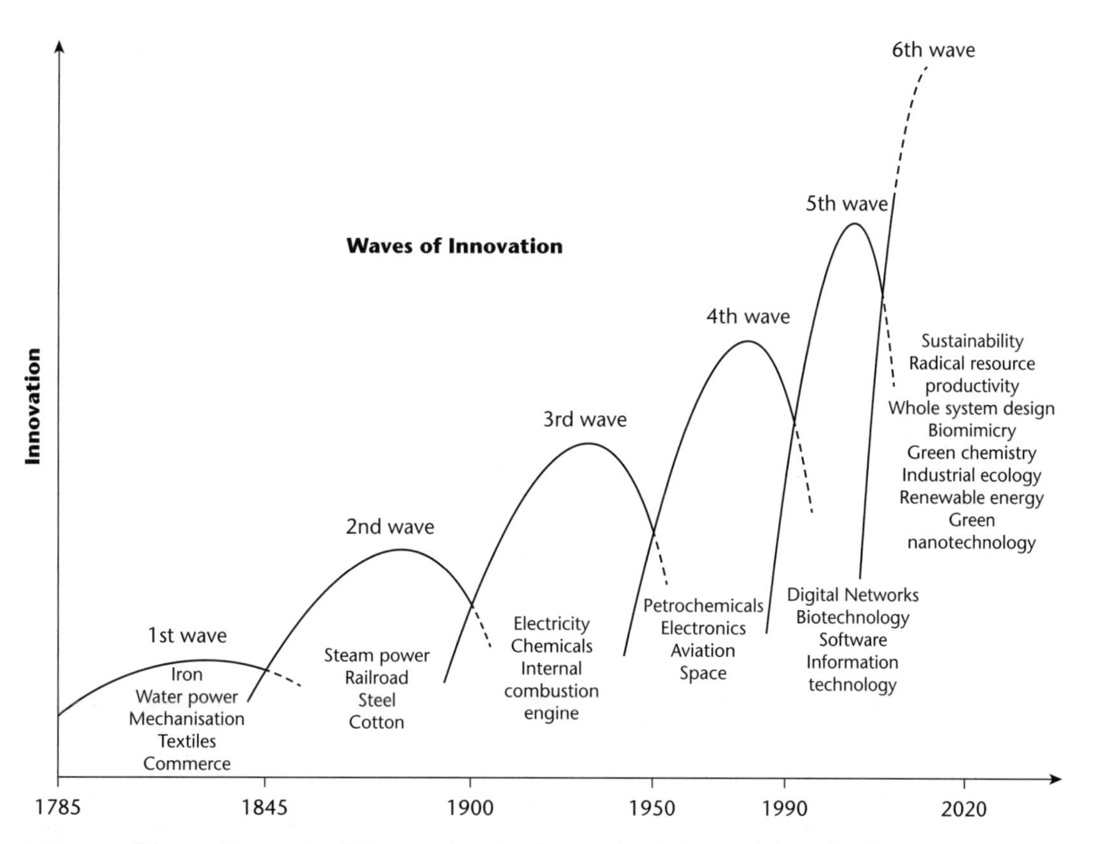

FIGURE 1.5 Waves of Innovation Diagram, showing the associated characteristic technologies

Source: The Natural Edge Project, in Hargroves and Smith (2005)[119]

As Figure 1.5 suggests, the level of innovation has been increasing in each subsequent wave and the timeframe over which the innovations are conceived, trialled, adopted and then form the basis of the next wave, is getting shorter for each new wave. By the early 20th century the world had most of the scientific understanding, enabling technologies and methodologies needed to underpin a number of significant developmental feats. For example, advances in mobility led to trains, cars and planes moving people at a pace and over distances scarcely imaginable when the century began.[120] Air transport connected the world and continued to expand into the 21st century as one of the fastest-rising transport modes – with an 80% increase in kilometres flown between 1990 and 2003.[121] As a result of agricultural innovation and the use of pesticides and inorganic fertilisers, the world grain harvest has quadrupled, and with continued advances in chemistry, global chemical production is projected to increase by 85% by 2020.[122] Humans now have unprecedented access to raw materials and processed goods from around the world, with shipping alone rising from 4 billion tons in 1990 to 7.1 billion tons total goods loaded in 2005.[123]

With this perspective in mind, Table 1.3 outlines how each new wave of innovation calls for the world's economies to be 'upgraded' resulting in significant changes across industry, governance and, in due course, the higher education system. For example, the shift to electrification during the third wave called for education and capacity building in many new areas across the professions, such as engineers and designers learning how to generate and distribute electricity and manufacture electrical fittings and appliances; economists and policy makers learning how to set energy prices; lawyers learning how to consider the new liabilities and risks to infrastructure and equipment from electricity.

TABLE 1.3 Examples of significant capacity-building requirements associated with subsequent waves of innovation

First Wave: Iron, water power, mechanisation, textiles, commerce

- *Canal Design and Construction*: Before the invention of the steam engine, canals proved a vital part of Europe's freight transport network, with canal design and construction a core part of engineering education.[124] At the time, water transport was able to carry bulk goods and freight at a significantly lower cost than was possible on land, with horse power deployed to tow barges on rivers or canals and nearly 2,000km of navigable river by the mid 1700s.[125] However, from the 1830s canal building gradually diminished with the rise of the coal-powered steam railways soon superseding the need for canals.
- *The iron bridge*: The development of a method to economically smelt iron in large quantities in 1709[126] and the production of wrought iron in 1783[127] led to a rapid rise in the popularity of iron bridges. During this time, engineers learnt much about the material properties of iron and its behaviour in compression and tension. However, iron bridges suffered some of the most catastrophic failures in bridge history, including the Tay Bridge Disaster in Scotland where the supporting wrought iron girders collapsed in high winds, killing 75 people.[128] The arrival of economically competitive steel as part of the second wave – a far superior bridge building material – led to the sudden demise of building iron bridges. Engineers turned their attention to steel arches, steel trusses and wire suspension bridges.

Second Wave: Steam power, railroad, steel, cotton

- *Kerosene lamps*: Kerosene lamps became popular in the mid 1850s, when oil wells drilled in America caused the price of oil to fall sharply. As engineers and scientists realised the potential of this light source, nearly 100 patents were granted to fund research and improvements to kerosene-burning lamps.[129] The most promising development to emerge from this research was the incandescent mantle. Unfortunately, the patent for this innovation arrived just as major cities were beginning the switch to the electric light bulb. Although kerosene lamps are still used today, the uptake of electricity led to the demise of research and development into this once promising innovation.
- *The steam locomotive*: Steam technology revolutionised land transport and sparked some of the largest industrial development in human history. Experimentation with steam-powered engines began in 1765, but it wasn't until 20 years later that a Welsh ironworks commissioned a steam locomotive.[130] Steam locomotives subsequently transformed society;[131] however, it was inherently inefficient, expensive to build and maintenance-heavy. This changed with the introduction of the diesel engine, most notably the Pioneer Zephyr by General Motors in 1934.[132] The transition to diesel had an enormous impact on the railways,[133] and by the 1950s it became difficult to find steam locomotives in operation even in America.[134]

TABLE 1.3 Continued

Third Wave: Electricity, chemicals, internal combustion engine

- *Printing and photography*: A retired French military officer took the first photograph of the view outside his workroom in 1826, using recent developments in photochemistry to develop a 'heliograph' – a pewter plate coated with bitumen. The bitumen hardened when exposed to light, resulting in a faint positive image after an 8-hour exposure.[135] Further developments from the 1850s by chemical and process engineers resulted in developments from a plate to a film technology, as well as the development of Polaroid and colour films. These developments went hand in hand with a rapidly increasing knowledge of chemistry and optical physics.[136] Today this technology is almost completely superseded by digital and electronic processes, where cameras have more in common with television and capture live images by converting light into electrical impulses.
- *The LeBlanc soda process*: The LeBlanc soda process refers to the 19th-century production of soda ash, caustic soda and chlorine. The application of these chemicals is extremely diverse, and includes the productions of soap and detergents, fibres and plastics, glass, petrochemicals, pulp and paper, fertilisers and explosives.[137] The process was developed by Nicolas LeBlanc, and later a family of iron founders. Subsequently a Belgian chemist developed a more direct process using ammonia (the 'Solvay process'), which reduced the price of soda ash almost one-third. At first, the Solvay process had difficulty competing with the well-established LeBlanc industry; however, by 1915 the new ammonia-soda process had completely displaced the LeBlanc.[138]

Fourth Wave: Petrochemical, electronics, aviation, space

- *Printing press*: One of the early major breakthroughs in distributing the printed word was the system of moveable type, coupled with the printing press, developed in the 14th century by inventor Gutenberg.[139] The replacement of the hand-operated Gutenberg-style press occurred in the 19th century with the introduction of steam powered rotary presses.[140] Offset printing then caused another revolution in modern commercial printing technology,[141] followed by digital technology and the computer which changed the very mechanisms of printing.[142] Many printing technologies from the fourth – and fifth – waves are now close to being obsolete, such as typewriters, daisy wheel printers and dot matrix printers.[143] These innovations have had significant implications for design, from the way that services are provided to clients, through to the design and manufacture of the machinery.
- *Communications technology*: Communications technology had its beginnings in signalling systems for the emerging British railway lines.[144] Subsequent research and innovation contributed to an impressive progression of technologies, including the development of Morse code (1843) and the electric telephone (1876).[145] Two scientists (Faraday and Maxwell) transformed the basis of communications technology forever with their research into electromagnetism and electromagnetic wave theory, and in 1901 the first wireless message was transmitted across the Atlantic.[146] The emergence of the electronic calculator during the Second World War marked the beginning of a new age of communications technology and education platforms.[147] These developments also sparked major changes in education, from the rise of long-distance education (via radio and telephone) to basic lecture recording and information sharing.

Fifth Wave: Digital networks, biotechnology, software information technology

- *Information technology*: Information technology innovations occurred in parallel with computing power. For engineers and teachers of engineering, this had huge technical implications. For example, prior to the era of personal computers, engineering students spent large amounts of class time learning technical drawing skills, using technology such as slide rules and drawing tables, and handbooks with tables of values. The release of drawing programs such as AutoCAD in 1982 at a trade show in Las Vegas signalled the beginning of a new era in engineering drawing and drafting and the rapid industry take-up led to changes in engineering coursework.[148] Other software such as MODFLOW has also significantly changed the scope of modelling and programming that is taught. Software advances have created jobs in consulting, design and marketing, and graduates now require different personal and professional competencies. Most engineering programs have added to or even redesigned theoretical coursework to incorporate the new engineering design skills, with courses in technology studies and programming.[149]

TABLE 1.3 Continued

Sixth Wave: Sustainability, radical resource productivity, whole system design, biomimicry, green chemistry, industrial ecology, renewable energy, green nanotechnology

Engineering programs already or in the future will incorporate these concepts and successful built environment professionals will incorporate them throughout the life of a project. For example:

- *Energy efficiency in buildings*: Buildings have continually evolved to address social needs. The arrival of the skyscraper a century ago, for instance, was due to the scarcity of space in congested American cities.[150] The face of real estate changed, enabling extraordinary accommodation of people in a contained footprint. Today, the building industry is entering yet another era of change, with a focus on minimising the energy, carbon and environmental footprint of residential and commercial buildings. Forty per cent of global greenhouse gases is attributed to the building sector, along with 12% of global water use, and significant material flows.[151] Transformative technologies could hold the answer to curbing the challenges of greenhouse gas emissions. Energy efficiency has been the centre of engineering–economics discussions with regard to the extent of possible cost-effective savings, from 10–30% in the mid 1970s to 50–80% in the mid 1980s. By the mid 1990s, practitioners were achieving 90–99% improvements in some situations.[152] Factor Four has previously argued, later supported by the IPCC 4th Assessment's Mitigation Working Group findings, that 75% enhanced improvement in energy efficiency could be made in building design.[153] Today, designers, developers and owners are scouting for ways to reduce environmental impacts and operating costs of buildings, as well as enhancing their functionality and appeal to residents.[154]
- *Project management*: Project management is an ancient profession, evident in many of the ancient civilisations such as Ancient Greece and Egypt.[155] Modern project management is a product of detailed examinations of 'work study' that was completed in the United States at the end of the 19th century that evolved into industrial administration, organisation and method, and managerial techniques. These techniques were crucial in converting the small workshops and cottage industries of the 19th century into the giant engineering establishments of the 20th century, with their mass-production and assembly-line techniques.[156] In the 21st century, this process is undergoing another transition from a construction paradigm that was all about more and bigger, to 'less is more' and streamlined resource use.

  The 6th wave is about upgrading to a lifestyle that improves environmental circumstances (i.e. decoupling economic growth from environmental pressure). Courses already or in the future will incorporate concepts such as resource productivity and whole system design. The most successful built environment professionals will incorporate these sustainably concepts throughout the life of a project as they are uniquely placed to ensure that sustainability can be practised throughout the construction industry.

Source: References noted within table.

The curriculum renewal process must be improved and accelerated

Shaping the Education of Tomorrow – Report on the UN Decade of Education for Sustainable Development:

> The boundaries between schools, universities, communities and the private sector are blurring as a result of a number of trends, including the call for lifelong learning; globalization; information and communication technology (ICT)-mediated (social) networking education; the call for relevance in higher education and education in general; and the private sector's growing interest in human resource development . . . This new dynamic provides a source of energy and creativity in education, teaching and learning, which itself provides a powerful entry point for education for sustainable development.[157]

Building on the previous waves, the fifth wave of innovation provided a new technological platform and numerous tools for enhancing communications, computation, design, drafting, and data analysis and storage, allowing operations to be significantly improved; however, the associated environmental pressures from the accelerated development were largely ignored.[158] With the significant environmental impacts outlined above, the sixth wave is focused on innovations that both build on the previous waves and deliver significantly lower environmental pressures.

In essence, while the fifth wave was driven by the economic opportunity of reducing transaction costs and enhancing communications, the sixth wave is driven by the economic risk of failing to reduce environmental pressures from the previous waves.

Each wave has called for significant updating of education and capacity building programs, as shown stylistically in Figure 1.6. Following the emergence of each wave, in general, education programs undertook a curriculum renewal transition to renew courses with the new innovations and apply them to what employers needed from graduates. As the level of innovation in knowledge and skills has progressively increased with each wave, this has called for increasingly larger-scale efforts – and shorter time constraints – to achieve associated curriculum renewal.

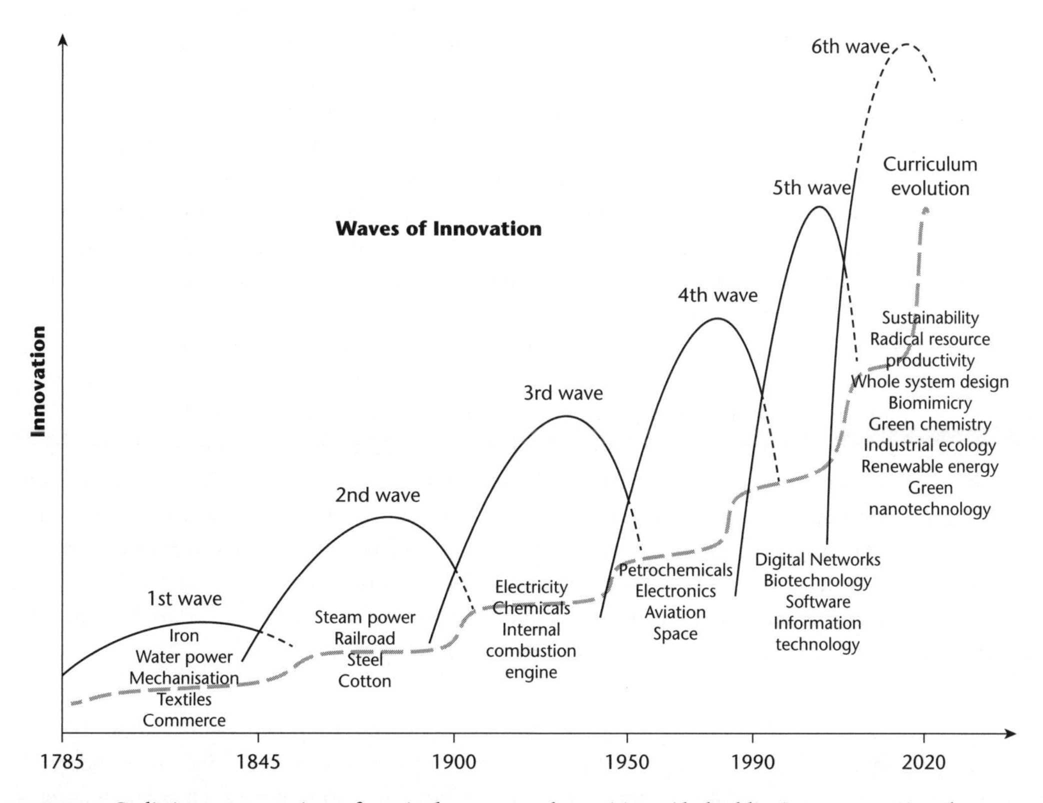

FIGURE 1.6 Stylistic representation of curriculum renewal transitions (dashed line) accompanying the waves of innovation since the industrial revolution

Source: Desha and Hargroves (2011), adapted from Hargroves and Smith (2005)[159]

As mentioned earlier in this chapter, although the higher education sector is still predominantly operating in an environment of low pressure to renew courses for sustainability, governments around the world have been increasingly vocal about action towards curriculum renewal around this sixth wave, providing early signals of a changing requirements. For example, in 2001 the South African National Quality Framework emphasised environmental education for a wide range of education institutions including higher education.[160] In New Zealand, the 2002 Tertiary Education Strategy included sustainability as one of six national development objectives.[161] In the same year in the UK, the government's Sustainable Development Education Panel required all UK further and higher education institutions to have faculty fully trained in sustainability and providing relevant learning opportunities to students by 2010.[162]

Looking forward, given the time imperative to achieve significant reductions across a number of environmental pressures, the current curriculum renewal challenge may be the greatest the modern education system has ever faced.

Given that a typical (or 'standard') process of undergraduate curriculum renewal for an accredited program – including, for example, engineering, architecture, planning, law, business or education – may take 3–4 accreditation cycles (of approximately 5-year intervals), the time to fully integrate a substantial new set of knowledge and skills within all year levels of a degree will be in the order of 15–20 years. Further, as the average pathway for a graduate is approximately 2–4 years from enrolment to graduation, followed by 3–5 years of on-the-job graduate development, if institutions take the typical approach to fully renew such bachelor programs, this will result in a time lag of around 20–28 years; hence it will be some 2–3 decades before students graduating from fully integrated programs will be in decision-making positions using current methods. For postgraduate students the time lag will be shorter as students may already be practising in their field and will return to positions of influence; however, accounting for the time to renew programs, the time lag is in the order of 8–12 years, depending on the pace and effectiveness of curriculum renewal efforts.

Looking across the higher education sector, a number of disciplines have examples of timely and program-wide curriculum renewal, from law (embedding skills training) to business (embedding corporate social responsibility), nursing (embedding evidence-based practice) and medicine (embedding technological advances).[163] Although each discipline can point to leading efforts in particular programs, these efforts are isolated and largely *ad hoc*. Indeed, there appears to be a need to improve curriculum renewal processes across many disciplines of study. Broadly speaking, the standard timeframes to renew undergraduate and postgraduate programs are well beyond the timeframes needed to significantly reduce a range of environmental pressures as outlined previously. We refer to this as a 'time lag dilemma' where the usual or 'standard' timeframe to update curriculum for professional disciplines is too long to meet changing market and regulatory requirements for emerging knowledge and skills.

A 'time lag dilemma' for the higher education sector exists where the usual or 'standard' timeframe to update curriculum for professional disciplines is too long to meet changing market and regulatory requirements for emerging knowledge and skills.

In hindsight, if institutions had acted on previous calls for capacity building related to sustainability, such as in *Our Common Future* in 1987, then the standard processes may have been sufficient over the subsequent 20–30 years. However, this window for such a response has well and truly closed. Hence the urgency and complexity of the challenge to reduce environmental pressures, combined with the scope of associated knowledge and skills required, calls for both an improvement in, and acceleration of, the standard approach to curriculum renewal across higher education. Whether in under- or postgraduate education, curriculum renewal towards education for sustainable development requires immediate attention.

How can the higher education sector respond?

Given that at some time in the next decade there is likely to be an increase in enforcement related to reducing environmental pressures, as with all organisations, higher education institutions will have a choice as to whether they move early or wait until enforcement ('Time (t)' as indicated in Figure 1.2). This decision will affect the level of risk and reward for the institution, with a low commitment delivering high future risks and low future rewards, and a high commitment positioning institutions to capture future rewards and avoid risks, as illustrated in Figure 1.7. Furthermore, the benefits curve may also be affected as the supply of graduates with sustainability knowledge and skills subsequently catches up with employer demand, flattening over time. As more institutions develop graduates with desired traits, a department's efforts in curriculum renewal may actually just be keeping up, rather than creating market niche.

Institutional risks of inaction include, for instance, falling student numbers, increasing accreditation difficulties, ineligibility for research grants, and poaching of key faculty. Rewards for action include, for example, attracting the best students and staff, staying ahead of accreditation requirements, attracting research funding, securing key academic appointments and industry funding.

Those institutions with a high commitment will have access to greater rewards before and after enforcement, whereas those following wrestle for reduced rewards, referred to as 'first mover advantage'. Further, those who maintain a low commitment will see risks increasing before enforcement as efforts to reduce environmental pressures ramp up, and after enforcement as enforcement efforts become more stringent (see Figure 1.7).

Consider an example of introducing a carbon-trading scheme. If the cost of petrol rises significantly, large companies currently producing high levels of emissions will likely require carbon-related competencies in their recruitment strategies. This includes rapid innovation to address the manufacture and supply of goods and services; mechanical and electrical engineers will be expected to design more efficient processes, equipment and vehicles; and civil engineers will be expected to design more efficient transport systems and infrastructure. In the face of such rapid employer shifts in demand, education departments that are unprepared could face increasing accreditation difficulties, falling student numbers, with the potential for faculty loss and restricted research opportunities even before the period of enforcement. In addition, their graduates will be competing with others who are better equipped.

This situation presents significant cause for universities and departments to rethink their strategies related to curriculum renewal, to minimise the current and future risks and position themselves to

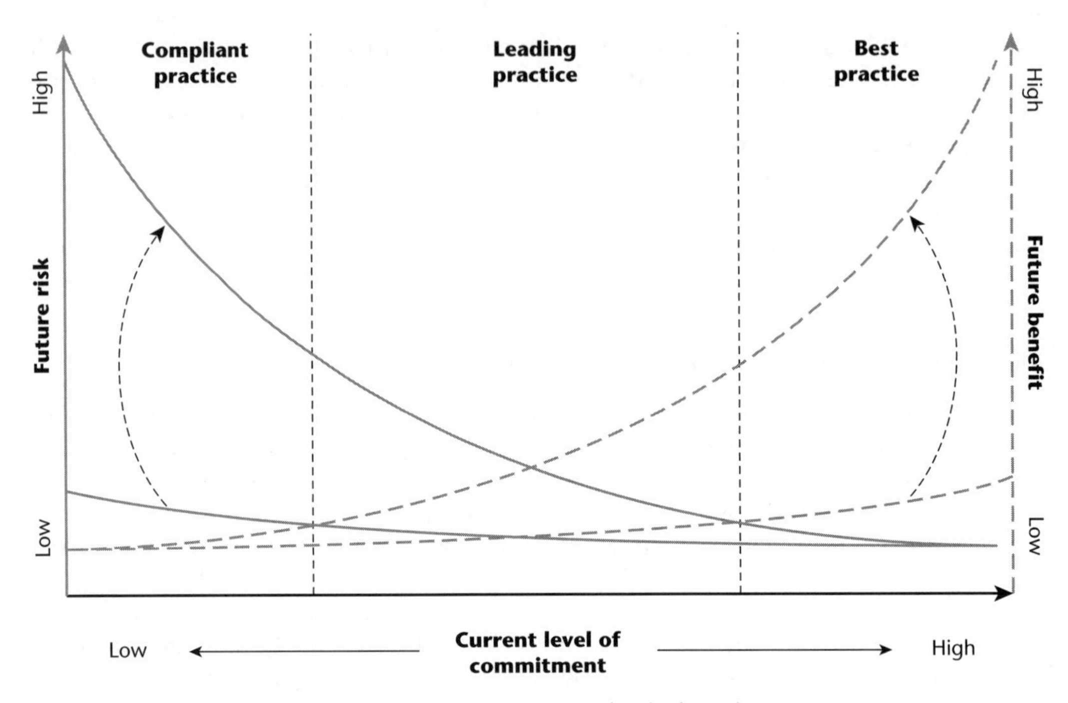

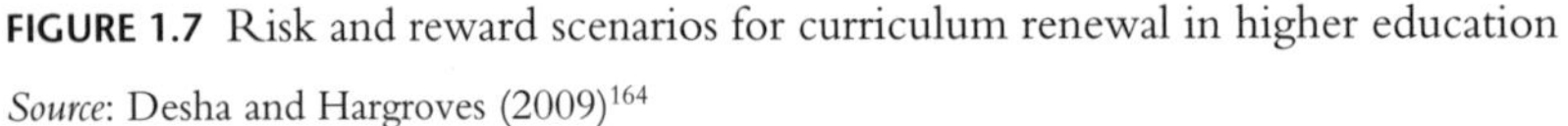
FIGURE 1.7 Risk and reward scenarios for curriculum renewal in higher education

Source: Desha and Hargroves (2009)[164]

capture the growing rewards. In short, over the coming years, departments who do not transition their programs to incorporate sustainability are likely to find it increasingly difficult to operate. Furthermore, their traditional roles as professional education providers may be usurped by private training providers capturing niche opportunities in capacity building, along with firms and government departments developing in-house capacity building programs.

For faculty who are personally committed to sustainability, there are two trends: continuing to influence at the level of individual courses; or leaving to join an institution with a stronger level of commitment, working on systemic integration.

In the UK, the Higher Education Funding Council (HEFC) has funded a number of Leadership, Governance and Management (LGM) projects over the last several years totalling £2.1 million, ranging from developing future leaders to learning in future environments.[165]

In 2012 the UK Higher Education Academy held its first policy 'think tank' to investigate how the higher education sector can contribute to 'greening the economy'. The think tank:[166]

- acknowledged the significance of the emerging green economy policy to the role of universities in furthering the education of their graduates and in preparing them for active and participative socioeconomic roles;
- stressed that the purpose of universities in educating graduates includes more than just the green economy;
- found the UK Government's definition of a green economy limited in not emphasising planetary boundaries and resource scarcity;
- concluded that the green economy (by any definition) is not yet a strategic issue for many universities, in part because of the lack of clarity within many of the national policy documents.

When an institution or department commits to education for sustainability, one of the first considerations is the timeframe in which curriculum renewal can be undertaken. We could imagine an envelope of opportunity with a minimum and maximum timeframe for transitioning the curriculum as shown in Figure 1.8. First, institutions could wait until enforcement (i.e. adopting 'standard curriculum renewal' or 'SCR' processes) and then move rapidly (i.e. through 'rapid curriculum renewal' or 'RCR' processes) to comply along with the rest of the sector – shown as the 'post-t transition'. Alternatively, institutions could move rapidly ahead of future compliance and capture the associated

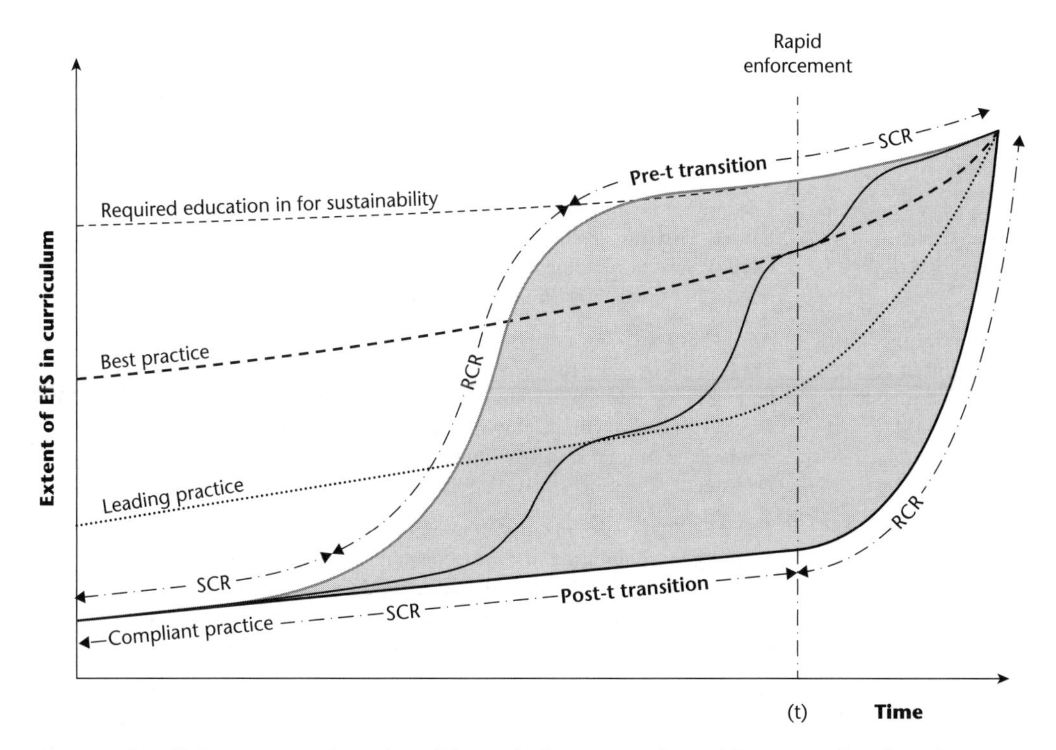

FIGURE 1.8 A stylistic representation of possible curriculum renewal transition curves for education institutions ('SCR' Standard Curriculum Renewal; 'RCR' Rapid Curriculum Renewal)

Source: Desha and Hargroves (2012)[167]

benefits, shown as the 'pre-t transition'. Within the shaded area inside the upper and lower bounds of this envelope, there are a number of possible transitions, including a staged stepping up from 'compliance' to 'leading practice' to 'best practice', as shown in the figure.

A growing number of organisations, alliances and networks have emerged over the last decade, committed to integrating sustainability into the curriculum, as highlighted in Table 1.4.

There are a number of emerging non-profit partnerships that are working to facilitate capacity building for sustainability, extending beyond higher education institutions into professional associations, industry and government. For example, 'Second Nature' is a US non-profit organisation that since 1993 has worked with more than 4,000 faculties and administrators at more than 500 colleges and universities to help incorporate the principles of sustainability in higher education programs. Led by one of the world's leading education for sustainability experts, Dr Anthony Cortese, the organisation's successes include advancing networks at the state, regional and national levels, and conducting a multi-

TABLE 1.4 Examples of university alliances promoting education for sustainability

Alliance	*Brief description*
University Leaders for a Sustainable Future (ULSF)	Since 1992, ULSF has served as the secretariat for signatories of the Talloires Declaration, a ten-point action plan committing institutions to sustainability and environmental literacy in teaching and practice. Over 350 university presidents and chancellors in more than 40 countries have joined by signing the declaration.
Higher Education Partnership for Sustainability (HEPS) programme	One of the earlier university alliance initiatives was a three-year UK partnership (2001–2003) of 18 higher education institutions committed to sustainability supported by the funding councils of England, Northern Ireland, Scotland and Wales. Co-ordinated by Forum for the Future, the partnership worked to generate transferable tools, guidance and inspiration to demonstrate the potential for integrating sustainability in the higher education sector.[168]
Global Higher Education for Sustainability Partnership (GHESP)	Comprising the International Association of Universities (IAU), the University Leaders for a Sustainable Future (ULSF), Copernicus-Campus and UNESCO, GHESP aims to mobilise higher education institutions to support EfS, focusing on responding to Chapter 36 of Agenda 21 regarding the role of education.
Association for the Advancement of Sustainability in Higher Education (AASHE)	AASHE is a member organisation of colleges and universities in the United States and Canada working to create a sustainable future. The *AASHE Bulletin* is the leading news source for campus sustainability in North America, and the *AASHE Digest* is an annual compilation of bulletin items. AASHE has developed a standardised campus sustainability rating system called STARS (Sustainability Assessment, Tracking and Rating System), launched in 2009.
American College and University Presidents Climate Commitment (ACUPCC)	The ACUPCC is an initiative of presidents and chancellors to address global warming by committing to climate-neutral campuses and by providing the education and research to enable society to do the same. Nearly 600 US college and university presidents have signed the commitment and are publicly reporting progress, including greenhouse gas emission reports and climate action plans.
Higher Education Associations Sustainability Consortium (HEASC)	HEASC is an informal network of higher education associations with a commitment to advancing sustainability within their constituencies and within the system of higher education itself. This includes developing in-depth capability to address sustainability issues.

Source: References noted within the table.

million dollar, 10-year advocacy and outreach effort that was instrumental in launching and maintaining momentum for education for sustainability within higher education in the United States, through AASHE and the HEASC (see Table 1.4).

The US Partnership on Education for Sustainable Development was formed to leverage the UN Decade to foster education for sustainability in the US.[169] Led by another of the world's education for sustainability leaders, Dr Debra Rowe, it comprises individuals, organisations and institutions with a vision of sustainable development being fully integrated into education and learning in the country. One of its actions has been to initiate and sponsor the Disciplinary Associations Network for Sustainability (DANS), an informal network of professional associations working on professional development, public education, curricula, standards and tenure requirements to reflect sustainability, and legislative briefings on what higher education can bring to sustainability-related policies.[170]

The choice and strategy for transitioning the curriculum will depend on a number of factors that are usually part of institutional risk management and business planning; this is not specific to the education for sustainability and we have not found the silver-bullet for moving forward. However, in understanding the 'what's so' of the sixth wave of innovation, institutions gain access to positioning themselves in the emerging education marketplace.

In this sixth-wave transition, it is important to contribute graduates who can lead, but not be too far ahead of the reality, at their time of graduation. The balance of 'old' and 'new' needs to be carefully managed to consider to the need to reduce environmental pressures, the needs of society and employer demands, as indicated in Figure 1.9. As there is a large amount of embedded infrastructure (for example roads, bridges, coal-fired power stations, electricity grids, etc.) to be managed, maintained and transitioned, requiring 'old industry' education, integrating 'new industry' content too quickly could be problematic if graduates don't have the skills that the employment market needs at the time that they graduate.

At the level of the institution, targeted effort will be required to incorporate sustainability into existing operational frameworks across the breadth of the institution or department – including governance and management, curriculum design and innovation, operations and facilities, marketing, human relations and stakeholder relations (see Chapter 4). Departments will also need to direct efforts to support the transition, including increasing internal professional capacity, and addressing knowledge gaps to deliver the required curriculum. It will also need to promote such opportunities to potential students, and anticipate shifts in student enrolment.

Throughout government, industry and the higher education sector itself, there are persistent and growing calls for increased capacity building towards sustainability. There is also an emerging awareness of the complexity of this challenge, and the urgent need for curriculum renewal for institutions to meet employer demand for graduates over the coming years. In the current working environment where market and regulatory enforcement of education for sustainability is present but highly variable, we have asked ourselves, 'is it possible for the education sector to engage in transformative curriculum renewal within the decade?' In the following chapter we focus on one of the major professional disciplines in higher education – engineering – to consider the current capacity for such curriculum renewal, and opportunities for educators to engage.

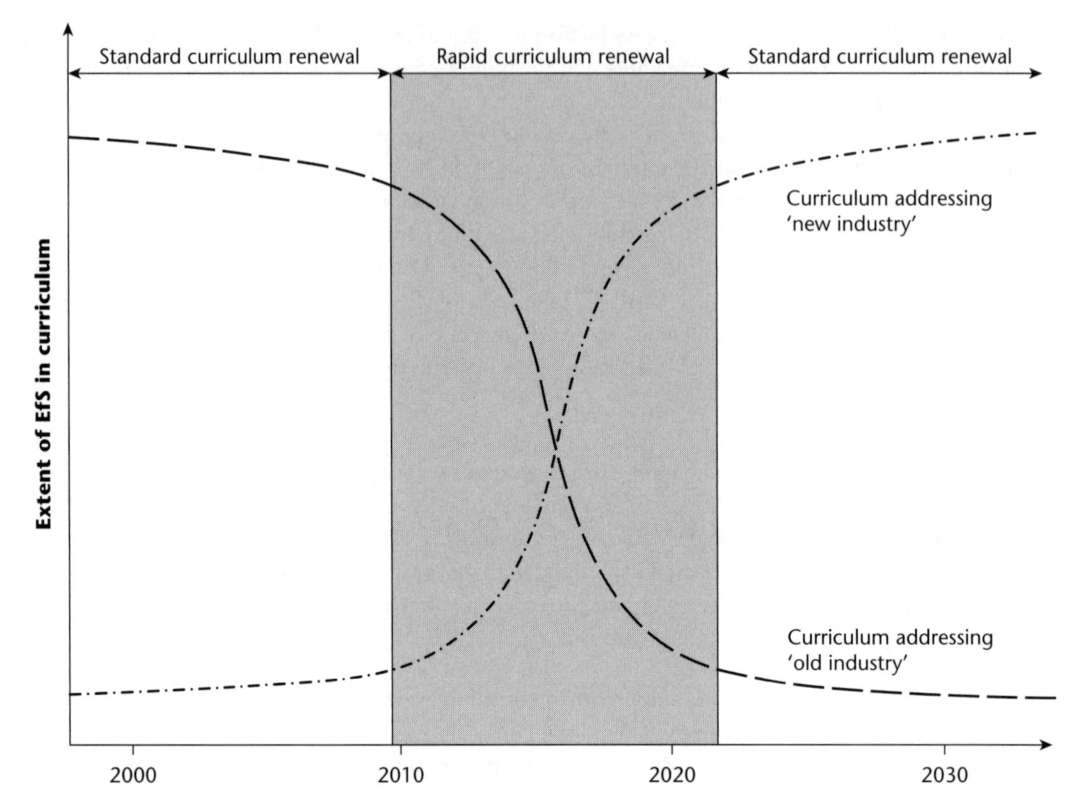

FIGURE 1.9 An illustrative curriculum transition curve, showing a period of rapid curriculum renewal from 'old' to 'new' industry

Source: Desha and Hargroves (2011)[171]

2

DRIVERS AND BARRIERS TO EDUCATION FOR SUSTAINABILITY

The education of professionals will need to be significantly renewed in the coming decade, to align with requirements to respond to a growing range of environmental, social and economic challenges. However, despite awareness for more than 20 years of the need for curriculum renewal to deliver 'education for sustainability', on the whole there has been a slow response, with only a few quick to respond. Although there are signs of change, progress is typically limited to particular examples rather than being mainstreamed across the sectors. In this chapter, we outline a series of drivers and barriers to mainstreaming education for sustainability. Herein we draw on the engineering disicpline to provide examples and context for the theory and models presented.

The chapter begins with a discussion of what educators 'really' want, drawing on the findings of a series of surveys on energy efficiency education involving over 80 per cent of Australian universities, informed by over 70 international leaders in education for sustainability. We reflect on the 'timing' aspect, which is critical to achieve systematic curriculum renewal, particularly considering that pressure to respond to sustainability will increase in the future, as discussed in Chapter 1. As such, educators will require a timely, whole of curriculum approach to renewal, with online open-source content that can be readily integrated into existing curricula, and which enables educators to produce graduates to meet the changing demand. Meeting these needs in time will require government and industry support for developing 'ubiquitous' resources readily transferrable between undergraduate, postgraduate and in-practice training.

Considering the current context for learning and practising engineering, there are a number of key factors constraining change, such as:

- A shortage of engineering graduates
- Short-termism in the higher education sector missing important trends
- Lack of faculty competencies in emerging areas of demand
- Persistent 'traditional economy' industry practices
- Perceived threats to employability and position if not delivering old materials
- Growing disconnect between engineering, science, economics and policy
- Lack of convenient access to emerging and rigorously peer-reviewed content

- Lack of access to information and resources in foreign languages
- Lack of strong requirements for change in curriculum from accreditation bodies

In spite of the power of these constraints, we also observe a number of key factors that are beginning to drive curriculum renewal to include sustainable development:

- Shifting requirements by employers and governments
- Increasing student demand for the subject
- Increasing level of availability of research funding related to sustainable development
- Increasing focus in declarations and conference topics
- Increasing professional advocacy
- Shifting codes of practice
- Shifting accreditation requirements for graduates

Key to achieving education for sustainability is understanding that there are significant challenges to overcome in achieving curriculum renewal of any type, let alone with content that is rapidly emerging and wide reaching in scope. A number of deeply entrenched organisational, pedagogical and cultural factors affect the ability of education providers to respond to changing needs. Within this context, the stage is set for Chapter 3, where we embark on a conversation about how curriculum renewal can be improved, to embed emergent knowledge and skills in a timely manner.

Why a focus on engineering education and sustainable development?

> It is up to engineers to consider sustainability in every project we design and construct and every product that is made. Sustainability is now a fundamental responsibility that we must carry every day . . . I call on all engineers to assume this responsibility – we must quickly learn about sustainability and adapt it into every aspect of our practice – for without engineers' 100 per cent participation, sustainability will just not happen and severe climate change is inevitable.[1]
>
> Doug Jones, President 2004, Institution of Engineers Australia

The curriculum renewal conversation in this book is applicable to all disciplines. It has been designed this way as the challenge to provide society with education for sustainability will affect the full spectrum of programs in higher education. With this in mind, we have focused our conversation on the engineering discipline as a primary example throughout, given our background training as engineers in built environment programs. Furthermore, beyond our personal preferences, engineers are fundamental to any developed or developing nation, designing, constructing and delivering products and services that are used in homes and workplaces, and will be critical to achieving sustainable development.

The engineering profession has also played a key role in activities that have brought about the range of significant environmental pressures discussed previously, probably the most of any profession. As UNESCO concluded in its first report on engineering in 2010:[2]

> Engineering is one of the most important activities in the context of climate change mitigation and adaptation . . . Many countries have already introduced policies and initiatives for climate change mitigation and adaptation . . . this will be one of the areas of greatest demand and challenge that engineering has ever faced.

Hence, the need for engineers with knowledge and skills to address the challenges facing society today and tomorrow is crucial. The report goes on to say that:

> One of the first challenges is to make sure that there will be enough appropriately qualified and experienced engineers to meet this demand – this will require the development of new courses, training materials and systems of accreditation.

In response to the need to significantly reduce environmental pressures while maintaining economic growth, the role of engineering will experience the most dramatic shift in knowledge and skills since the start of the industrial revolution.[3] As Australian engineer and 2009 WFEO president and former president of Engineers Australia Barry Grear (AO) questions,

> What aspirational role will engineers play in that radically transformed world?' . . . An ever-increasing global population that continues to shift to urban areas will require widespread adoption of sustainability. Demands for energy, drinking water, clean air, safe waste disposal, and transportation will drive environmental protection [alongside] infrastructure development.[4]

According to the World Federation of Engineering Organisations (WFEO, representing 15 million engineers from more than 90 nations), engineers have 'an important role in planning and building projects that preserve natural resources, are cost-efficient and support human and natural environments'.[5]

Until the mid 20th century, engineers generally focused on making products well, making them quickly, and making them inexpensively. In the 21st century this focus has expanded to include making products safely *and* in a manner that is environmentally friendly *and* socially responsible.[6] Furthermore, engineers are now expected to quickly find solutions to a range of emerging development challenges such as the need to reduce greenhouse gas emissions to mitigate climate change, and to assist humanity to adapt to climate change impacts, such as changing weather conditions, rising sea levels and stronger and more frequent natural disasters.

In the coming decades engineers will be expected to retrofit, redesign and innovate new products and services that meet rapidly increasing environmental and social criteria. This will need to be done while also dealing with the reducing availability and increasing costs of conventional inputs such as oil and some metals, higher standards on waste and pollution, all the while being cost effective. Hence, while fundamentals knowledge and skills (including statics, dynamics, thermodynamics and fluid mechanics) will continue to form the basis of engineering education, new knowledge and skills associated with sustainable development will be required to be integrated into programs (including low carbon electricity production, resource efficiency, green chemistry, green buildings, industrial ecology, fuel cells and advanced water management).

This shift in knowledge and skills will be across all disciplines, requiring a wealth of new materials and research to inform such efforts. The team from The Natural Edge Project has researched such advances for over a decade now, mentored by many of the world's leading experts. We have identified a number of critical 'threshold concepts' that have been the focus of our previous four books. Such concepts are critical to successfully grappling with the complexity and urgency of sustainable development challenges. These include, for example:

- 'A natural advantage of nations', which focuses on nations using sustainable development as the basis for their economic and social development to achieve a range of benefits across all levels of government and across every part of the economy.[7]
- 'Decoupling economic growth from environmental pressure', which presents a range of new opportunities for wealth creation and improved well-being across all sectors of the economy whilst significantly reducing negative environmental impacts.[8]
- 'Whole system design', which harnesses traditional formulae and design methods through a holistic approach to deliver more sustainable solutions.[9]
- 'Factor 5', which focuses on identifying and implementing options to achieve in the order of 80% improvements in productivity, rather than the typical incremental improvements, in the use of a range of resources such as energy, water and materials.[10]

Once learned, these concepts cannot be 'unlearned', providing a transformational new lens through which professionals can consider design problems and apply technical knowledge and skills.[11]

> Issues of sustainable development, poverty reduction and climate change are fundamentally engineering issues. We have to learn to broaden our design brief beyond the traditional objectives of schedule, cost and conventional scope.
>
> John Boyd, President FIDIC 2007–2009[12]

This is not the first time that new knowledge and skills have been incorporated into already 'full' programs due to shifts in society. As we detailed in Chapter 1, there have been a series of major technological innovation waves since the early stages of the industrial revolution, each resulting in a reorientation of practice, policy and education.

In each case, higher and vocational education was renewed with previously taught knowledge and skills being updated or replaced. Typically these waves have been driven by the motivation to improve economic growth and capture market advantage. What makes the sixth wave unique is that it is also motivated by the imperative to innovate new technologies and processes to address environmental pressures that have arisen from previous waves. Furthermore, the sixth wave focuses on reducing these pressures swiftly, within timeframes in the order of one to two decades, which presents a significant challenge for issues such as reducing greenhouse gas and particulate emissions, as little has been done since the early warnings.

This sixth wave is inspiring action in various parts of the education sector to review the current level of coverage of such topics in programs. However, the extent of action is typically sporadic and short-lived, as outlined in Table 2.1. Colleagues around the world have been tracking progress through a number of surveys on sustainable development and sub-topic areas since the late 1990s, with a common finding that the level of coverage of knowledge and skills for sustainable development in higher education programs appears to be quite low.

For example, in Australia in 2007, the National Framework for Energy Efficiency (NFEE) funded The Natural Edge Project to undertake a national survey of universities on the 'State of Energy Efficiency Education in Australia'.[13] The comprehensive survey received responses from 27 of the 32 universities offering engineering education in the country and found that although there was clearly a desire to integrate energy efficiency content among faculty, there was a substantial shortfall in the inclusion of theory, knowledge, application and assessment in engineering education. Even mainstream topics such as '*carbon dioxide and other greenhouse gas emissions from energy generation*' and '*the link between greenhouse gas emissions and global temperature change*' were only covered in detail by up to a third of surveyed courses, and mentioned by less than half. Moreover, student survey results indicated only a low to moderate appreciation of how energy efficiency might be directly related to their future careers.[14]

These findings were confirmed in subsequent investigations in 2009,[15] 2010,[16] 2011[17] and 2012[18] involving TNEP for NFEE and the Federal Department of Resources, Energy and Tourism (RET). The results of these surveys suggest that in general the curriculum renewal process to date has been slow and *ad hoc*, despite clear signs of the imperative for education for sustainability being seen internationally, across government, industry and academia.

TABLE 2.1 Summary of key surveys on the state of education for sustainability

Year	*Survey*	*Brief description*
1998	World Engineering Partnership for Sustainable Development	Questionnaire circulated to national members of WFEO to provide an improved benchmark. *Conclusion:* No strong or consistent approach to environment and sustainable development in engineering education. On a country average, not much more than 10% of time in 10% of courses is devoted to these aspects.[19]
2000–2002	University of Surrey (UK) and University of Melbourne (Australia)	Survey of a sample of international engineering students on their level of knowledge and understanding of sustainable development.[20] *Conclusion:* (21 respondents from 40 invitees) The level of sustainable development knowledge is not satisfactory, and significant knowledge gaps exist within the curriculum.[21]
2002	Royal Melbourne Institute of Technology	Twenty-one Australian universities invited to participate in a survey on the status of EfS in these institutions. *Conclusion:* (from a quarter of invitees) Few universities are engaged in such education for a wide range of their students. In some universities more students of particular disciplines are gaining exposure. However, there are clear barriers to the introduction and expansion of sustainability education.[22]
2006	Chalmers University of Technology, Delft Technical University, Technical University of Catalonia, Alliance for Global Sustainability	The Observatory assessed the status of EfS in European higher education, benchmarking 51 European universities (survey), against examples from outside Europe. *Conclusion:* In 2006 there was no European university that showed sufficient progress in EfS to be considered an inspiration.[23]

TABLE 2.1 Continued

2007	Forum for the Future's Engineers of the 21st Century Programme	Just under 500 young engineers who had graduated between 1997 and 2005 were surveyed regarding sustainability literacy.[24] *Conclusion:* 40% perceived their university lecturers had inadequate knowledge of sustainability and 30% perceived their lecturers had a positive to passionate attitude about EfS.
2007	National Framework for Energy Efficiency (NFEE, undertaken by TNEP hosted by Griffith University)	National survey of universities teaching engineering education on the state of engineering education in Australia, within the sub-topic of energy efficiency education (82% response rate from 32 universities).[25] *Conclusion:* The state of education for energy efficiency in Australian engineering education is currently highly variable and *ad hoc* across universities and engineering disciplines. Key issues for educators included perceived course overload and lack of time for professional development or to prepare new content.
2007–2008	US Centre for Sustainable Engineering	Benchmarking survey on the extent of sustainable engineering education within 1,368 engineering departments (or the equivalent), with just over one-fifth of the invited 364 American universities and colleges participating.[26] *Conclusion:* The engineering education community is now at a critical juncture. To date there has been a significant level of 'grass-roots' activities but little structure or organisation. The next step will be for engineering accreditation bodies to think critically about what should or should not be included.[27]
2008	Chalmers University of Technology, Delft Technical University, Technical University of Catalonia, Alliance for Global Sustainability	Second survey by The Observatory[28] initiative. Of the 57 universities participating in the 2008 survey, most had not participated in the 2006 survey, making it difficult to directly compare the results of the successive reports. *Conclusion:* A growing number of institutions from European countries are actively engaged in sustainability activities.
2011	Department of Resources, Energy and Tourism (Undertaken by TNEP hosted by the University of Adelaide)	National repeat survey (following 2007 NFEE Survey) of universities teaching engineering education on the state of engineering education for energy efficiency.[29] *Conclusion*: Despite growing awareness of the importance of energy efficiency in both industry and academia, the current depth and breadth of content in courses does not reflect this.

Source: References noted within the table.

While the lack of shift towards addressing the sixth wave knowledge and skills is frustrating for many in the field, it is important to acknowledge that the scale and pace of the change required is unprecedented in the education sector. Indeed, while engineering education has undergone periods of curriculum renewal to embed professionalism, ethics and health and safety, the profession has not had to make a significant shift in the knowldege and skills it teaches students across entire programs since the first engineering professionals emerged following the industrial revolution.[30] Internationally a number of insitutions and groups have considered this issue, with leading examples including the 2005 American National Academy of Engineering (NAE) report on educating the engineer of 2020;[31] the 2006 UNESCO workshop on Engineering Education for Sustainable Development;[32] the 2007 UK

Royal Academy of Engineering (RAE) report on educating engineers for the 21st century;[33] the Higher Education Funding Council for England (HEFCE) *Strategic Review of Sustainable Development in Higher Education in England*;[34] and the Chinese Academy of Engineering.[35] In Australia, the 2008 Australian Learning and Teaching Council (formerly the Carrick Institute) report on addressing the supply and quality of engineering graduates for the 21st century[36] also concurred with these international reports, highlighting a lack of progress since the 1997 report, *Changing the Culture*,[37] which had raised concerns with regard to curriculum change and graduate attributes.

The imperative to update engineering education to incorporate sixth wave innovations has also catalysed a number of areas of investigative inquiry, with examples of key research highlighted in Table 2.2.

TABLE 2.2 Examples of discussion about education for sustainability (not including works by the authors of the book)

Discussion	*Example authors in the field*
The term 'sustainability' as it relates to environmental, social science, and higher education	Sauvé,[38] Fien,[39] Leal,[40] Sterling,[41] Corcoran and Wals,[42] Parkin *et al.*,[43] Cortese,[44] Blewitt and Cullingford,[45] and Dawe *et al.*[46]
Education for sustainable development and its priority for the engineering profession	Jansen,[47] Mulder,[48,49] Ferrer-Balas *et al.*,[50] Holmberg *et al.*,[51] Allenby *et al.*[52]
Graduate engineers and knowledge and skills sets	Carroll,[53] Cortese,[54] Crofton,[55] Ashford,[56] Azapagic *et al.*,[57,58] McKeown *et al.*,[59] Pritchard *et al.*,[60] and Allenby *et al.*[61]
Education for sustainable development and pedagogical practices	Timpson *et al*.[62] on integration, Newman and Fernandez[63] on institutionalising such curriculum renewal, Steinemann[64] and Lehmann *et al.*,[65] on problem-based learning, and Crawley *et al*.[66] on sustainability as a meta-narrative.
The larger engineering education for sustainable development agenda	Rowe,[67] who discusses policy direction, Stephens and Graham,[68] who discuss research needs, Steinfeld and Takashi,[69] who discuss the challenge of trans-disciplinarity, and Holdsworth *et al.*,[70] who discuss the need for professional development.

Many of these researchers (including a number of mentors to The Natural Edge Project) have dedicated their careers to understanding and provoking curriculum renewal towards education for sustainability – some for more than three decades – in an environment where it is clearly still 'early days' for embedding sustainability into education. For example, conference themes and journal topics still tend to focus on issues affecting the ability of engineering education to be changed (i.e. organisational, resourcing, funding, timeframe and content issues), rather than discussing and monitoring the extent to which the curriculum has changed.

The majority of academic papers have focused on the experiences of single champions or teams discussing isolated initiatives. Some papers have documented the success of strategically embedding case studies and flagship courses (predominantly in the first year and at postgraduate level), while few papers discuss methods to integrate knowledge and skills across programs and across disciplines. Since 1987 a relatively small number of engineering education institutions have undertaken a process of curriculum renewal in some form, primarily within undergraduate programs.[71] Examples of leading efforts include that of Georgia Institute of Technology,[72] Delft University, Chalmers University and Monterrey University.

There are numerous signals that the time of enforcement to practice in accord with sustainable development is rapidly approaching and that this will have a significant impact on higher education. However, even the proactive departments around the world are doing little more than including one or two 'sustainability' courses within existing programs, leaving isolated individuals or small teams within departments to undertake *ad hoc* curriculum renewal efforts.

> Growing global competition and the subsequent restructuring of industry, the shift from defence to civilian work, the use of new materials and biological processes, and the explosion of information technology . . . have dramatically and irreversibly changed how engineers work. If anything, the pace of this change is accelerating.[73]
>
> Wm Wulf, then President of the American National Academy of Engineering (1998)

> The United States National Academy of Engineering vision of the Engineer of 2020 suggests that engineering curricula be reconstituted 'to prepare today's engineers for the careers of the future, with due recognition of the rapid pace of change in the world and its intrinsic lack of predictability.[74]

In the following discussion we turn to engineering educators to ask what they *want* with regard to curriculum renewal for sustainability. With a growing sense of frustration about the lack of action in transforming engineering curriculum, many (including our research team) have been considering what would be most useful to educators in empowering them to integrate sustainability knowledge and skills within their programs. Our response as young engineers was to work with the world's leading experts to create a series of peer-reviewed textbooks and online lectures to introduce key principles and practices related to sustainable development.[75] Once completed, our team then focused on researching challenges and opportunities within a sub-topic area of sustainability, that of 'energy efficiency', which allowed for specific research to be undertaken as summarised in Table 2.3, that then provided an indication of the larger status of education for sustainabilty.

TABLE 2.3 An example of evolving education for sustainability research in energy efficiency in Australia

Year	*Education research involving the authors*
2007	The National Framework for Energy Efficiency (NFEE) funded TNEP to undertake the first survey of energy efficiency education across all Australian universities teaching engineering, in both undergraduate and postgraduate contexts, and achieved an 82% participation rate.[76]
2007	In response to the findings of the survey, the CSIRO funded TNEP to develop 30 lectures on energy efficiency opportunities in Australia, by major sector and technology, as part of the 'Energy Transformed' flagship program.[77]
2009	In order to identify ways to reduce impediments to the uptake of the materials developed with CSIRO support, NFEE commissioned TNEP to explore barriers and benefits to teaching energy efficiency in the higher education sector, particularly focusing on engineering education. This was based on the

TABLE 2.3 Continued

Year	*Education research involving the authors*
	method of 'community-based social marketing' (CBSM) applied for the first time to a higher education community.[78]
2010	In order to update the 2007 findings, NFEE funded TNEP to undertake a second national survey of industry and higher education sector academics to further explore the knowledge and skills gaps within energy efficiency capacity in Australia.[79]
2010	The federal Department of Resources, Energy and Tourism (RET) formed an energy efficiency advisory group (EEAG) to consider energy efficiency capacity building in the higher education sector, beginning with engineering, with both authors invited to join. NFEE then commissioned a research report to examine 'energy efficiency assessment skills' in Australian industry, called the Long Term Training Strategy for the Development of Energy Efficiency Assessment Skills (LTTS).[80]
2011	RET commissioned two projects through the EEAG led by the authors that (1) updated the 2007 NFEE survey and created a taxonomy for energy efficiency education resource development,[81] and (2) considered graduate attributes and associated learning pathways relating to energy efficiency.[82]
2011	NFEE funded TNEP to undertake an investigation into postgraduate education for energy efficiency, including consideration of connectivity with the vocational education sector.[83]
2012	RET commissioned TNEP to undertake a project to consult with industry and academia on targeted capacity building for energy efficiency, in unprecedented collaboration with Engineers Australia, through nine national engineering colleges and discipline-based groups (biomedical, chemical, civil, electrical, environmental, ITEE, structural, mechanical, mining and metallurgy). This included phone consultation with each, and well-attended workshops in three capital cities.
2013	The Federal Department of Industry, Innovation, Climate Change Science, Research and Tertiary Education (DIISRTE) commissioned TNEP to undertake a project to assess the state of energy efficiency in vocational education programs in Australia under the 'Skills for the Carbon Challenge' initiative. This project involved the research team repeating parts of the previous higher education survey supported by NFEE and RET in the vocational education sector to compare findings.

The findings of these extensive investigations inform an evolving action-research agenda globally spanning many fields of curriculum renewal, behaviour change and sustainable development focusing on energy efficiency.

There are many examples of gaps between what educators think is important and what they actually teach due to time and resourcing constraints. For example, in the 2011 Australian survey of postgraduate educators teaching about 'energy', 'energy generation', 'transmission' and 'distribution losses' was considered fifth most important, but ranked ninth in terms of current in-depth coverage. There are some topic areas that received almost no coverage, despite being considered important by educators, including 'energy recovery' and 'energy rating'. Almost all educators agreed on the importance of a systems focus for curriculum; however, whole system design had a low or medium coverage. Such anomalies would benefit from immediate resource development as they would be readily received by educators.

How can education for sustainability be supported?

The following part summarises the main messages from investigations in Table 2.3 and outlines findings related to 'what educators want' with regard to being empowered to undertake curriculum renewal for sustainability.

1 Curriculum resources relevant to current and anticipated accreditation requirements

Educators have indicated that they want assistance with generating curriculum that is relevent to current and anticipated accreditation requirements of program content. Considering global trends in accreditation, this points to creating resources that show how sustainable development is a 'meta-theme' that grounds learning in existing and emerging contexts as relevant for each profession and sub-discipline. This involves a shift from reductionist thinking towards a whole system approach in addressing problems and creating opportunities. In engineering, for example, this includes collaborative decision-making, interdisciplinary visioning and innovation, and sustainability concepts such as 'whole system design',[84] transformational change towards '*Factor 5*',[85] or higher improvements in resource productivity, and 'net positive development'.[86] With this in mind, assistance would comprise supporting the generation of example learning outcomes, or graduate attributes, through to assessment requirements, and encouraging approaches such as problem- or project-based learning.[87]

2 Curriculum resources easily tailored to existing or new courses and programs

Given the time- and resource-constrained nature of teaching environments, educators often want to reduce their teaching workload by tailoring existing and highly reputable resources rather than creating their own new materials 'from scratch'. Furthermore, educators want these resources to be 'multi-functional' (i.e. adaptable across postgraduate, vocational education and in-house training), providing a number of ways for educators to deliver knowledge and skill areas to varying levels. Given the now growing demand for sustainability literacies, a range of critical knowledge and skill gaps have become apparent, calling for the need to make available education resources that target these gaps – we refer to these as 'take and use resources' as they are packaged in easily accessible file formats and do not require dealing with copyright issues (i.e., for instance, under a common attributes licence or a licence to the world). Such resources could be made available through an online education portal or similar, to provide access to the rapidly growing number of professionals searching for immediate capacity building, while also providing institutions with postgraduate study programs quality resources to embed within their curriculum.

'Take and use resources' should include suggestions for *how* to teach, as well as *what* to teach, providing a comprehensive resource that is pedagogically appropriate. Findings from a comprehensive survey of Australian universities delivering engineering education on energy efficiency suggest a number of preferences for resources that could assist with curriculum renewal, including access to:[88]

- A set of case studies on energy efficiency examples in engineering;
- Worked examples of the application of energy efficiency;

- Lecture notes on energy efficiency opportunities, specifically by sector and technology; and
- A set of mini-lectures (i.e. lecture guides and study materials) on various critical topics.

Additional considerations identified in the survey included ensuring that the resources remain current, assistance with evaluating software for its suitability/application and creating practical resources to address postgraduate needs as many combine work and study. In the development of such resources educators expressed a preference for:

- Clear aims and objectives;
- Succinct learning points;
- Appropriate depth of coverage;
- Easily used/adapted to curricula;
- Incorporates a systems focus; and
- Reputable author or organisation.

3 A strategic approach to curriculum renewal, and to accelerating this process

The challenge of renewing curriculum to include sustainability is complex, particularly as its relevance is wide-reaching across curriculum. Not only are the risks and rewards dependent on many factors external to the university, the risk–reward relationship changes over time as shown in Chapter 1 (Figure 1.7). Where risk increases over time for those slow to respond, the rewards will decrease over this time as they are captured by early movers with successful strategies for curriculum renewal.

Furthermore, it is widely accepted now that legislative and market requirements to reduce environmental pressures such as greenhouse gas emissions will increase and will require both short-term and medium- to long-term strategies, often requiring quite different knowledge and skill sets. For example, in the case of greenhouse gas emissions, as presented in detail in the 2006 Stern review,[89] the short-term strategy is to achieve a 'peaking' in emissions, with renewables, energy efficiency and demand management playing a major role in quickly stopping the annual increase of greenhouse gas emissions globally. Once this has been achieved, the medium- to long-term strategy, referred to herein as 'tailing', involves many other stategies in rethinking activities, products and services to sustain a gradual annual reduction in greenhouse gas emissions globally, over the coming decades.

Hence, given such complexity, educators require assistance to conceive and implement strategies related to curriculum renewal that minimise the risks and assist universities to be placed to capture the rewards. Aspects of this transition have been experienced before in cross-university programs in new topic areas such as ethics, safety and quality assurance; however, the level of complexity and scope is unprecedented for the sixth wave.[90,91,92] Given the timeframes discussed earlier, departments undertaking a process of curriculum renewal should target a timeline of 6–8 years to transition undergraduate programs in order to produce graduates who can assume key decision-making roles in 10–15 years' time, a time when such knoweldge and skills will be critical for society to respond to the sustainability challenge. Furthermore, postgraduate education also needs to be renewed as soon as possible to help equip current practitioners and decision-makers with knowledge and skills to contribute to reducing growing environmental pressures, such as greenhouse gas emissions.

4 Briefing materials on critical topic areas, informed by educator needs

Educators indicate that they could benefit from briefing materials on critical topic areas that are better than conducting self-directed online web searches, which can be time-consuming and frustrating due to the difficulty in identifying rigorous resources. Educators around the world are clearly grappling with sustainability-related terminology and concepts. For example, the findings of the previously mentioned surveys suggest that educators have a low regard for some knowledge and skill areas that are prioritised by governments. The Australian NFEE 2012 report highlighted that this was the case for energy-efficiency auditing and associated activities. Furthermore, fewer than half of the surveyed educators had a high or very high focus on energy efficiency in commercial buildings, despite their signifiant use of electricity and contribution to greenhouse gas emissions. Hence, such areas were identified for awareness raising about the importance of these topic areas in reducing energy consumption and greenhouse gas emissions, and achieving a more sustainable built environment.

Over the past decade, in the workplace the built environment industry has progressed substantially in coming to terms with the complexity of challenges and opportunities in delivering sustainable solutions, and the wide spectrum of knowledge and skill requirements. Going forward, it is clear that 'sustainability'-related initiatives and actions are unlikely to be undertaken by one person in a company. Equally, companies are not seeking training for staff across the full spectrum of sustainability-related knowledge and skills, but are targeting certain immediate needs. For example, in Australia there are a number of principles and concepts that educators are not considering important, or that are not meeting employer expectations, including efficiency, resource efficiency and energy efficiency; climate neutrality or emission mitigation; and performance at part and full load. This is also the case for the topic of energy auditing and energy assessments. This points to an opportunity for resource development to immediately fill this gap.

The literature also suggests that employers already understand the need to identify opportunities, but when it comes to actual projects, the focus is much more towards compliance. This highlights the opportunity for educators to be assisted with generating capacity-building activities that empower graduates as change agents, to focus on substantial improvements rather than just meeting the status quo. There is also an opportunity to develop resources for areas that are critical but not receiving attention from industry or educators, including, for example, in the area of energy, embedded energy of materials and energy mass balances, which would benefit from awareness-raising regarding their importance to life-cycle analysis and identifying opportunities.

5 Connecting curriculum renewal efforts with industry

Educators indicate they would benefit from assistance to connect curriculum renewal efforts with industry through opportunities to interact with employers and also with potential new students. It is common in the curriculum survey literature to find employers complaining about the dated nature of textbooks and traditional university content, wanting more focus on latest technologies and energy options. There is an opportunity here to assist educators with raising awareness about what is already being offered as well as potential future offerings in the curriculum. In the area of energy efficiency, for example, currently there are also cases of principles and concepts being taught that are not yet highly valued by industry; for example, in Australia, these include energy efficiency and low-carbon technologies (renewable energy, fuels) and sustainable energy supply (storage and standby). This could point to the education sector anticipating future needs, not providing adequate notification of the offerings to industry, or could be a mismatch of expectations. Educators are keen to make sure their offerings are future-flexible with regard to tools that students acquire, but not detached from employer needs and expectations.

Factors limiting education for sustainable development

Understanding what educators want and addressing their needs are two very different discussions. The education literature is full of attempts to move forward with one or more of the list above, but very few have been long lasting or mainstreamed. In the following discussion we bring together a synthesis of factors that we have observed to be preventing a sector-wide shift towards education for sustainability. With these in mind, there is almost a sense of relief at understanding why progress in the past has been so fragmented and slow; with this set of conditions there has been little access for educators to mainstream emergent knowledge and skills.

Persistent 'old economy' industry practices

Even after some 100 years of understanding the risk and impacts from environmental pressures there are still organisations who only meet the minimum regulatory requirements to reduce pollution and emissions. Higher education institutions continue to perceive demand for graduates with conventional knowledge and skills as a continuation of the 'norm'. The persistence of old-economy industry practices, even when it is clear they will be short term (i.e. in the face of impending enforcement), is a possible reason for a lack of business- and industry-wide action. The relatively small number of employers calling for graduates who are literate in sustainability are perceived to be foregoing the 'opportunity' to continue with unsustainable practices as long as they can before being required to stop, rather than acting to position themselves for the future.[93] Meanwhile, other companies may still choose to invest in technologies with high environmental pressure while they can, continuing to put a demand on engineering skill sets that do not contribute to sustainable development.

Perceived threat to employability and position

As noted in the UNESCO and World Federation of Engineering Organisations (WFEO) 2012 publication on issues and challenges for the engineering profession in the 21st century,[94] a change in perspective of what constitutes 'engineering practice' to align with sustainability may be perceived as a threat to the employability of graduates and a requirement to invest in capacity building for existing staff with little return. This may fuel both employer and employee resistance to change, and subsequently reduce demand for graduates with these new skills and knowledge. The reality is, however, that for practitioners who do not engage in professional development for sustainable development this may ultimately result in their gradual replacement or transfer to other parts of the workplace that are not as affected; perhaps this may coincide with their retirement! Within higher education institutions, faculty may similarly be threatened by a curriculum transition to incorporate sustainable development knowledge and skills. They may also anticipate that their knowledge may become redundant or superseded within a renewed curriculum, threatening their position in the department. As with practitioners, however, the reality is that those academics who do not engage in professional development in the emerging knowledge and skill areas may also find their courses questioned and potentially being taken over by others with the required expertise.

Short-termism in the higher education sector

It is widely recognised that one of the biggest barriers to corporate sustainability is short-term pressure from the market for ever-increasing quarterly profit results, as evidenced in a 2004 report by the World Business Council on Sustainable Development, which discusses why corporations have been slow to change to sustainable practices.[95]

> According to the 2004 World Business Council on Sustainable Development report on short-termism in Australia, '78 per cent of those surveyed would give up economic value in exchange for reporting smooth earnings growth. Fifty-five per cent of respondents would delay the start-up of profitable investment projects to avoid missing an earnings target while four out of five executives said they would defer maintenance and research spending to meet earnings targets.'

In 2004 the Business Council of Australia (BCA)'s Sustainable Growth Taskforce argued that stock market-driven short-termism is threatening the long-term competitiveness of firms, where increasing demand from shareholders for greater quarterly profits is preventing CEOs from making the investments companies need to position themselves for higher profitability in the medium to longer term.[96] While corporations aim to meet short-term stakeholder expectations for growth and increasing profit margins in quarterly and annual reports, in the higher education sector management is also faced with short-term pressures from annual and semester-based student enrolment numbers, budgets, and typically short fixed-term appointments of two to three years. Heads of department may be faced with the issue that their efforts to prepare for longer-term transitions that will show results past the next program accreditation cycle or student intake may not be supported or recognised. As the timeframe for results is often five to eight years, it may also be difficult to justify major budget allocations for curriculum renewal programs. This may be further complicated by regular restructuring within the academic bureaucracy that may take focus, time and resources away from curriculum renewal.

Growing disconnect between practice and science

While high-profile practitioners across many fields may be advocating the need for change, parts of the economy – from design offices to factory floors and education institutions around the world – are yet to understand the need for a substantial change in their skill set, despite the leading-edge scientific findings that call many current practices into question. Within academia and in practice, some have a pervading belief in established technology or process as the solution, seeing sustainability as a threat which brings into question many current practices. Hence, rather than being challenged by the professional practice implications of the issues raised by the scientific community, sustainable development is considered 'an additional thing to consider' when teaching or doing research – and therefore ignored unless it is clearly required, or if it can be used as a marketing term to win more project work and fund more research.

Lack of convenient access to emerging and rigorously peer-reviewed content

While the emergence of the internet has dramatically improved access to content over the last decade, many countries, particularly developing countries, still have limited and slow access to this information resource. There exist example initiatives such as the Sudanese Virtual Engineering Library (SudVEL) assisted by TNEP working with WFEO and UNESCO,[97] where significant literature resources were made available through a local server on the campus of the University of Khartoum. For those academics who do have fast and reliable access to the internet, renewing curriculum can still be a bewildering experience with literally thousands of websites on topics such as sustainable development technologies, energy efficiency and climate change. Currently there are relatively few such resources recommended by professional bodies or associations that academics can confidently and freely access. The Natural Edge Project has focused on this limitation over the last eight years and has released over 2,000 pages, comprising more than 150 hours, of freely available peer-reviewed lecture materials on sustainability. A sign of the value to the engineering profession of such resources is that this material is provided to members of Engineers Australia though the Environment College webpage, and through the Australasian Association of Engineering Education portal.

In 2005 the content resources developed by The Natural Edge Project won a Banksia Award for leadership in education and training. In 2008 the expanded resources were highlighted in the Australian report on 'Addressing the Supply and Quality of Engineering Graduates for the New Century' as an example of best practice in this area.[98]

Lack of access to information in foreign languages

In addition to the issues associated with internet availability, the lack of content in the first language of the institution is another potential reason discussed in the literature, which could contribute to a lack of curriculum renewal activities in eastern, Asian and sub-Saharan countries. Conversely, where countries that don't speak English might be succeeding in curriculum renewal activities, their strategies and learnings may not be translated for sharing with colleagues in other countries. While organisations such as Japan for Sustainability (JFS) work to translate initiatives between English and other languages, there is still a possible language disconnect that inhibits integration of sustainability content into higher education.

Lack of strong requirements for change

It is clear that swift action to reduce a range of environmental pressures will be required sooner or later, but a lack of certainty on government legislative responses makes it difficult to develop a strategic approach. The signals from industry for graduates who can address these requirements are therefore not strong; however, this demand is growing. This is described in the Stern Review as 'policy-induced uncertainty',[99] an example of which is governments being slow to penalise environmental 'bads' such as carbon dioxide emissions and waste generation. Such uncertainty may inhibit investment in sustainable development, which in turn may inhibit the mainstreaming of demand for professionals with these capabilities. Once there is clear direction on greenhouse gas emission reduction requirements, for example, then this will increase industry demand for the necessary knowledge and skills, sending a signal to higher education institutions to update what their students are learning.

Within academia, there is a parallel limitation in that there is little requirement from the executive of universities for curriculum efforts to keep pace with industry practices or anticipated practices such as those related to sustainable development. According to a comprehensive analysis of engineering education literature by Professor John Heywood, engineering education reform has a relatively long but slow history.[100] A most recent major curriculum change in engineering was the move to what is referred to as 'engineering science', which occurred following the Second World War. Since then, the composition of core knowledge has been seen as largely unchanging.[101] In general, changes to engineering curriculum are made in an internal process, whereby teachers leave out or add material based on their own teaching approaches, preference, experience and level of understanding.

> Like all systems, engineering education has to adjust, albeit slowly, to changes in the socio-economic system in which it functions . . . Nevertheless, outside influences such as changing technology are forcing departments to make changes, and it seems from the engineering literature that research and new practices are having an impact on the curriculum process.[102]
>
> Professor John Heywood

Lack of faculty competencies

Over the last decade there has been an emergence of literature on the issue of lack of capability in sustainable development among academics. Generally speaking, educators teach according to their education and experience, with a relatively low rate of professional development or industry experience in emerging areas.[103] Where sustainability has not formed part of their training, faculty are unlikely to consider it as a skill of value or be prepared to include it in programs. Indeed, some academics argue that students should be given the fundamentals which remain constant over time, and which can be applied to whatever problems arise. In this argument, these 'fundamentals' should not be diluted to include passing 'fads'. Professor Karel Mulder from Delft University reflects that low course enrolments in sustainable development electives are also used by academics in defence of their lack of expertise in this area. He reflects that for these academics,

> Anecdotal evidence of low enrolments in new interdisciplinary programs (such as Industrial Ecology) supports the argument that [education for sustainability] should leave the core of engineering unaffected, instead adding an extra sustainable development course on occasion, as current engineering is supposedly already well equipped to address 21st century challenges.[104]

> The Association for the Advancement of Sustainability in Higher Education (AASHE) is addressing the issue of increasing staff capacity for education for sustainable development, dedicating part of their website to highlighting professional opportunities for faculty.[105]

A shortage of engineering graduates

The current shortage of engineering graduates around the world identified by WFEO[106] is resulting in a push to increase student numbers in engineering programs. Institutions that are already stretched for funding to deliver programs are facing the prospect of more students with little to no increase in institutional support. In this operating environment, integrating new content is a lower priority to ensuring that there are sufficient lecturers, laboratories, tutorial rooms and other such immediate logistical issues. This is complicated by employers being increasingly prepared to train graduates in-house, or outsource professional development through specialised institutions that can fill gaps once the graduates join the organisation. When large organisations recruit graduates and then train them with the required sustainability knowledge and skill sets, they inadvertently send a market signal to engineering education institutions that they are satisfied with the product.

A recent study led by the authors and commissioned by the National Framework for Energy Efficiency in Australia found that six out of ten of the largest engineering companies operating in Australia provide in-house training on energy efficiency to address the current gap in capability.[107] This results in a scenario where employers are paying an additional cost to train their graduates in base competencies that should 'come with the product' and engineering departments show healthy work placements for graduates. This factor will change as a greater number of graduates with sustainable development knowledge and skills are available to employers; however, in the meantime the perverse market signal persists.

Factors driving education for sustainable development

In spite of the constraints affecting curriculum renewal efforts as highlighted above, there is progress being made in a number of leading institutions around the world. Such examples provide valuable lessons in how to achieve such results, as summarised below following reviews of associated reports and interviews with educators involved. Over the years our team has been energised by the growing level of action to renew curriculum for sustainability in engineering and other fields, and we are confident that there is now a critical mass of experience informing development of the model and process that are at the heart of this book.

Shifting requirements by employers

Due to the range of environmental pressures now causing impacts on the world's economies, there is a gradual tightening of government legislation and regulations across most, if not all, sectors especially on greenhouse gas pollution, toxins and air pollution, which are directly affecting employer needs, for example:[108]

- In 2007 the Chinese government's renewable energy law required 16% of primary energy from renewables[109] and solar hot water coverage of 300 million square metres by 2020.[110]
- In India, the government's eleventh five-year plan contained a host of energy targets, including full use of co-generation in the sugar and other biomass-based industry by 2012, and renewable energy accounting for 15% of power by 2032.
- In the European Union regulations such as the 'Restriction of the Use of Certain Hazardous Substances' in Electrical and Electronic Equipment (RoHS) directive[111] and the Waste Electrical and Electronic Equipment (WEEE) directive,[112] are having international impacts on industry pollution controls.

This shift has not been missed by business and many are taking leadership positions to ensure that they are able to continue to compete in the future. For example, in 2007 at the UN Global Compact Leaders Summit in Geneva, chief executives of 153 companies (including 30 from the Fortune Global 500) committed their companies to taking practical actions to increase the efficiency of energy usage and to reduce the carbon burden of products, services, and processes, to set voluntary targets for doing so, and to report publicly on those targets annually.[113]

The United Nations Global Compact is a strategic policy initiative for businesses that are committed to aligning their operations and strategies with ten universally accepted principles in the areas of human rights, labour, environment and anti-corruption.[114]

Corporate leaders are also creating long-term visions for action. The World Business Council for Sustainable Development (WBCSD) was formed in 1992 and now consists of more than 200 companies from more than 20 major industrial sectors including many engineering disciplines.[115] In 2010 the Business Council released *Vision 2050* at the World CEO Forum. The report was compiled by 29 leading companies from around the world, representing 14 different industries, answering the question of what the world will look like in 2050. This includes new opportunities for business as well as critical actions that must happen over the coming decade to make a sustainable society possible.[116]

The world already has the knowledge, science, technologies, skills and financial resources needed to achieve Vision 2050. However, concerted global action in the next decade will be required to bring these capabilities and resources together, putting the world on the path to sustainability.

WBCSD President Bjorn Stigson, in *Vision 2050*[117]

Meanwhile, large engineering and design consulting firms such as ARUP, HATCH, KBR, Parsons Brinkerhoff, HASSELL and GHD are undertaking internal staff development while universities catch up with producing graduates who can meet their needs. Jonathon Porritt, chairman of the UK Sustainable Development Commission and Founding Director of the UK's Forum for the Future, told a 2007 Global Sustainability Forum on the Future for Engineering Education, 'Big companies now are saying, "if you can't give us engineers who understand the centrality of sustainable issues . . . then you're not giving me the type of engineer that I need"'.[118]

Jeroen van der Veer, Chief Executive, Royal Dutch Shell also reflects that in this rapidly changing environment, 're-engineering engineers' to meet such needs will be a significant challenge for institutions:[119]

> Universities and colleges of technology should supply tomorrow's technical professionals with a broad mindset about their role in society. Like all education it's about the integration of knowledge, skills, attitudes, values and diversity. Plus essentials for engineers like creative and critical thinking, and being able to work in multidisciplinary teams in globalised organisations. Those re-engineered engineers can be the change agents for sustainable global development. Can there be anything more important than that?

Increasing student demand

A number of studies and surveys are showing that there are emerging signs that the consumers of education – the students – are seeking sustainability content within their programs.[120] A 2008 survey in the UK found that almost two-thirds of the just under 500 graduates surveyed felt sustainability was either important or very important to their job role today, with more than 90% of the respondents wanting to understand the impacts of their decisions and understand practical methods to incorporate sustainability into their work.[121]

> Whilst the demand is there, it was found that almost a quarter of the engineers had learnt nothing about sustainability at university . . . The survey targeted young engineers who had graduated two to ten years ago as they have experience of working in industry.
>
> Engineers of the 21st Century Engineering Education Project[122]

In 2009 the US Kaplan College Guide focused for the first time on sustainability as a defining consideration for prospective students and named 25 of the top 'Green Colleges' and identified 10 'Hot Green Careers', namely, environmental conservation; design, engineering and science; geothermal development; green interior design; hydrology; organic agriculture; solar energy engineering; and transportation systems planning. According to Jason Palmer, contributing editor to the guide, 'Our guide is designed to help students get into schools and prepare for careers that will enable them to make a meaningful and positive impact on our planet'.[123]

Sustainability vocabulary is increasingly being used as a marketing device on department websites and through program descriptions in recruitment handbooks. For example, the Faculty of Engineering at Imperial College London initiated the EnVision program to promote excellence in teaching, learning and academic leadership, focusing on students making links between careers (in engineering) and aspirations of 'making a difference to the world'.[124] EnVision activities included the set-up and support of projects which enabled students to experience aspects of sustainable engineering, interdisciplinary work and real-world issues. Senior lecturer and coordinator of the program, Esat Alpay, reflects that, 'Sustainable engineering is promoted through initiatives such as the "Engineering Impact" series of lectures by eminent speakers on issues such as climate change, health, energy and design, with the aim of inspiring students towards the engineering profession and their potential impact on society as future engineers'.[125]

At Zurich's Swiss Federal Institute of Technology's (ETH) Center for Sustainability (ETHsustainability), the Youth Encounter on Sustainability (YES) program collaborates with academics from a number of institutions internationally to offer a course catering for students demanding such capacity. With over 900 alumni from 100 different countries, the course has played an important role in sensitising upper-level university students to the complex issues of sustainable development in a multicultural and multidisciplinary setting. Program Director Michelle Grant reflects that,[126]

> The fact that students actively seek out such programs highlights the deficiencies in the prevailing approach of many institutions of higher learning. There is a growing personal awareness among university level students of the important challenges we are facing in the world, due in part to increased media exposure, increasing globalisation and study exchange programs and youth driven initiatives. This is leading to increasing interest in stand-alone programs such as the YES course, which receives a growing number of applications each year from upper level university students from many disciplines from all around the world. This leads to even greater pressure on institutions to evolve curriculum and teaching and learning methods to meet the needs of students who are hungry for an education that will build their capacity to address global challenges.

The then Head of Engineering Sciences at UNESCO, Dr Tony Marjoram, acknowledged in 2010 that the high level of interest from students in sustainability challenges such as the Mondialogo Engineering Award, and student-led organisations such as Students for Sustainability, Engineers for a Sustainable World, and Engineers Without Borders, was another indicator of growing student demand for such curriculum. He concludes, 'To promote engineering and attract young people we need to emphasise these issues in teaching curricula and practice'.[127] This has begun to be reflected within formal gatherings of professional associations and academia. For example, at the second International Federation of Engineering Education (IFEES) conference in 2008, held in collaboration with the 7th Global Colloquium on Engineering Education, the Global Student Forum Working Session focused on Sustainability and Engineering Education where participants joined 90 global student leaders participating in the Global Student Forum to 'discuss solutions related to Sustainability of Engineering Education, Sustainability in Engineering Education, and Sustainability through Engineering Education'.[128]

Increasing level of availability of research funding related to sustainable development

As the imperative for society to address the growing levels of environmental pressures grows, a number of governments and organisations are now offering sizable research and teaching grants in the area, as shown in Table 2.4. This trend is set to continue with sustainable development becoming a core consideration for research and teaching funding in the future.

TABLE 2.4 Examples of research grant funding related to education for sustainability

Country	*Description*
United Kingdom	• 'Leading sustainable development in higher education' initiative (2009–2010), through the Higher Education Funding Council for England (HEFCE) offers higher-value projects (up to £250,000) to small projects (up to £75,000).[129] • 'Learning for Change: Scotland's Action Plan for the Second Half of the UN Decade of Education for Sustainable Development' sets out a range of initiatives for universities to incorporate sustainable development within their institutions through estate management and curriculum integration. From 2005 to 2015, The Scottish Funding Council has committed over £400M to be invested in the program, with £192M for research. Funding is also provided to the Scottish Further Education Unit (SFEU), which supports universities integrating sustainable development into their courses.[130] • Education for Sustainable Development and Global Citizenship (ESDGC) is a framework developed for universities by the Higher Education Funding Council for Wales (HEFCW). The framework identifies actions which cover commitment and leadership, teaching and learning, institutional management and audits on curriculum comparing with international best practices. In 2008 £359,000 was allocated to support the delivery of the program.[131]
Australia	• 'Special Sustainability Round, 2010' as part of the federal government's Education Investment Fund (EIF). This funding (AU$650 million) is aimed at assisting with building infrastructure that supports teaching and research for sustainability.[132] • 'Education for Sustainability Grants Program' supports national strategic projects that facilitate change in community attitudes and behaviours in support of sustainable development in Australia. Beginning in 1997, this initiative has allocated more than AU$3 million to over 96 projects to community groups, industry associations, rural organisations, local governments and universities.

TABLE 2.4 Continued

Country	*Description*
	• The Sustainable Research Excellence (SRE) initiative was announced by the Australian Government in the 2009/2010 budget. Over four years (09/10–12/13), Australian competitive grants will provide AU$510 million to support universities engaged in research activities related to sustainability and including capacity-building components. • National Climate Change Adaption Research Facility (NCCARF) aims to provide strategic information to policy makers on climate change. Up to AU$27 million in seed funding is provided by the Department of Climate Change and Energy Efficiency (DCCEE) and is administered through the Climate Change Adaptation research grants program. Research fields cover health, biodiversity, infrastructure and agriculture.[133]
USA	• Sustainable Agriculture Research and Education (SARE), professional development program. Beginning in 1994, SARE provides sustainable agriculture education and professional development. SARE receives US$4.5 million per year from congressional allocations.[134] • Higher Education Sustainability Act (HESA) supports universities to develop, implement and evaluate academic sustainability curricula, programs and practices. A university pool of funding worth US$28.8M is available for (amongst eight priorities) sustainability initiatives as described in section 881 of HEA.[135] • RE-ENERGYSE (Regain our Energy Science and Engineering Edge) is a federal initiative that includes funding of US$40M in clean energy-related courses and scholarships for higher education.[136] • Climate change education grants of US$10M (2009 and 2010) from the National Science Foundation (NSF) are aimed at educating and funding the creation of learning resources to develop a new workforce generation by targeting higher education.[137] • Science, Engineering and Education for Sustainability (SEES) is an NSF initiative funded with US$765.5M to integrate work in climate and energy science and engineering. This is delivered through systems approaches to research programs, education and workforce development.[138]

In Australia, a 2008 national report on 'Addressing the Supply and Quality of Engineering Graduates for the New Century' made it clear that governments are expecting university-based engineering research to be contributing towards innovation, ultimately providing economic value and contributing to solving environmental, security, healthcare and other significant problems.[139]

The number of research institutes and centres focused on sustainable development and on integrating sustainability in undergraduate courses has rapidly grown over recent years, along with the growth in publications in the field (as outlined in Table 2.5). These include, for example:

- University of Plymouth's Centre for Sustainable Futures, UK
- Cambridge University Engineering Department's Centre for Sustainable Development, UK
- University of Gloucestershire's International Research Institute in Sustainability (IRIS), UK
- Curtin University Sustainability Policy Institute (CUSP), Perth, Australia

- Australian Research Institute for Environment and Sustainability (ARIES), Melbourne, Australia
- Monash Sustainability Institute, Melbourne, Australia
- Sustainable Built Environment National Research Centre (SBEnrc), Australia

Further, a growing number of universities are forming alliances such as the Alliance for Global Sustainability (AGS), including Massachusetts Institute of Technology (MIT), the Swiss Federal Institute of Technology-Zurich, the University of Tokyo and Chalmers University of Technology in Sweden. In Japan, the University of Tokyo initiated the Integrated Research System for Sustainability Science collaboration or IR3S, which involved lucrative research collaborations amongst a number of the country's top universities.

TABLE 2.5 Examples of interesting education for sustainability initiatives

Country/region	*Example author and institution details*
Europe	Kamp[140] and Mulder[141] in Netherland's Delft University; Lundqvist *et al.*[142] in Sweden's Chalmers University; Fenner *et al.*[143] in the UK's Cambridge University; Humphries-Smith[144] in the UK's Bournemouth University; Lozano[145] in Wales' Cardiff University; Fletcher *et al.*[146] in England's Aston University; Ferrer-Balas *et al.*[147] in Spain's Technical University of Catalonia.
America	Epstein *et al.*[148] in the Massachusetts Institute of Technology; Mihelcic *et al.*[149] in Michigan Technical University.
South America	Lozano-Garcia *et al.*[150] in ITESM Monterrey; Wright *et al.*[151] writing about the collaboration between Michigan University and Chile's University of Concepción.
Asia	Onuki and Takashi[152] in Japan's University of Tokyo; Uwasu *et al.*[153] in Japan's Osaka University; Kuangdi[154] in a Chinese national overview.
Africa	Olorunfemi and Dahunsi[155] in Lagos State Polytechnic and the University of Ibadan, Nigeria; Ramjeawon[156] in the University of Mauritius.
Australia	Davis and Savage[157] in Queensland University of Technology; Goh[158] in the University of Southern Queensland; Bryce *et al.*[159] in the University of Technology Sydney; Mitchell[160] in the University of Sydney; Carew and Therese[161] in the University of Wollongong; Koth and Woodward[162] in the University of South Australia; Daniell and Maier[163] in the University of Adelaide; Carew and Lindsay[164] in the University of Tasmania and Curtin University;[165] Rose[166] about Monash University.

Source: References noted within the table.

Supporting such activities, there are a number of institutes, centres and collaborations addressing sustainable design and practice that higher education institutions can draw on and collaborate with. These include the Rocky Mountain Institute and the Pew Climate Centre on Global Climate Change in the United States; the Climate Group in the UK; the Wuppertal Institute for Climate, Environment and Energy in Germany; The Natural Edge Project (TNEP) in Australia; and the Energy and Resources Institute (TERI) in India.

Increasing focus in declarations and conference topics

There are now a plethora of declarations and action plans related to education for sustainable development. However, such documents are not necessarily significant drivers by themselves, nor a measure of actual progress. As many declarations and action plans are non-binding agreements, there is potentially little impetus for the institution to make substantive changes, especially as there is typically little current pressure from accrediting bodies (despite a growing number of leading accrediting bodies incorporating sustainability graduate attributes such as the Institution of Engineers Australia). In addition, signatories can be lulled into a false sense of achievement having signed the agreement (which can be a lengthy process) leading to reduced action on the ground. Professor Don Huisingh, Editor-in-Chief of the *Journal of Cleaner Production* reflects:[167]

> When one looks into the changes that are or are not occurring within universities, it becomes evident that many signatories to university leader's declarations are not doing much – or anything – within their institutions that can be traced directly to their signatures to such documents. Fortunately, there are some exceptions but they are far from the norm!

However, these types of commitment statements are a useful indicator of increasing awareness and willingness to take action by higher education institutions. In Table 2.6, a number of often-referenced declarations and action plans targeting engineering education from technical institutions, professional organisations and university alliances are highlighted. Most include a preliminary discussion on the importance of professional capacity building to address 21st century challenges and a call for the higher education sector to re-orient its teaching as a matter of priority. For instance, the Barcelona Declaration on Engineering Education for Sustainable Development calls for institutions and universities to redefine their missions 'so that they are adapted to new requirements in which sustainability is a leading concern'. The declaration also calls for universities to 'redirect the teaching–learning process in order to become real change agents who are capable of making significant contributions by creating a new model for society'.[168]

From a qualitative review of mainstream international and regional engineering conference programs spanning the last decade (including the Australasian Association of Engineering Education annual conferences, the Global Colloquia on Engineering Education, and the International Conference on Engineering Education) it is clear that major refereed forums are now featuring engineering education for sustainable development as both a content theme and a pedagogy theme. Topics covered include issues affecting the ability of engineering education to be changed, including, for example, organisational issues, resourcing issues, personality issues, funding issues, timeframe issues and content issues. Papers discussing overstretched resources and declining student intake into environmental disciplines are common features within the programs. Some of the papers appearing in such conferences document success, including case studies and flagship courses (first year and Master's level), with such efforts rarely documented as part of a longer-term strategic plan for curriculum renewal.

Over the past several years there has also been a shift in some global 'mainstream' engineering education conferences, with regard to the themes and requests for papers. For example, the 2008 7th Global Colloquium on Engineering Education (GCEE) theme was 'Excellence and Growth in Engineering Education in Resource Constrained Environments', with a research track focused on 'Inferring and Designing Engineering Education Practice from Research and Societal Context: To what extent should engineering educators collaborate globally to re-engineer their programs?'.[169] However, although the 2008 International Conference on Engineering Education (ICEER) theme was 'New Challenges in Engineering Education and Research in the 21st Century', including invited topics on environmental challenges and the role of engineering education in sustainable development, only two out of more than 235 presentations, and three out of more than 65 posters explicitly addressed either of these topics, and these were presented as case studies.[170] This is indicative of conferences around the world today, highlighting that we still have a way to go in addressing sustainability within curriculum agendas.

TABLE 2.6 Examples of declarations and action plans for engineering education and sustainability

Year	*Declaration/action*	*Brief Description*
1991	Arusha Declaration,[171] World Federation of Engineering Organisations (WFEO)	This declaration on the future role of engineering was based on the 1987 World Commission on Environment and Development report, *Our Common Future*,[172] and other documents. The declaration called for specific actions by government, industry and individual professional engineers in their projects, stating 'that education on the issues involved in sustainability be given the highest priority'.
1994	Engineering Education Workshop (Asia/Pacific) New Zealand[173]	An international workshop of educators from the Asia Pacific region examined 'fundamentals of environmental education in engineering education', finding that *all* engineers need to be environmentally educated to understand the issues involved in sustainable development and cleaner production. The report noted the need for a holistic approach which includes developing skills and knowledge, attitudes, systems skills, interaction skills, broad knowledge in specific areas, and exposes students to significant issues.
1997	UNEP Industry and Environment Centre, WFEO, the World Business Council for Sustainable Development (WBCSD), and the French Ecole des Ponts[174]	Attendees at a joint conference on the topic of 'Engineering Education and Training for Sustainable Development' concluded that many practising engineers currently have little or no education in sustainable development, and that in future it should be included in both undergraduate and postgraduate courses. They also concluded that the transition to sustainable development will require major changes in ongoing education, and furthermore, practising professionals will need retraining.
1997	American Society of Engineering Educators – International Conference[175]	Attendees at the ASEE conference identified technology in engineering education (the virtual university), sustainable development, and the impact of globalisation on engineering as the primary influences on future engineering education. They also recommended targeted action to guide curricula, provide teaching materials, and develop networks.
2002	National Academy of Sciences – Declaration by US Engineering Community for Sustainability [176]	Representatives of the American National Academy of Engineering met with representatives of the major American engineering organisations to consider how to unify American engineers in support of the goals of the 2002 World Summit on Sustainable Development in Johannesburg, South Africa and to work together after this conference. The group committed to moving forward in supporting the US engineering community to meet societal needs through capacity building, improved education, training, information development dissemination, and engaging the engineering profession in all stages of the decision process.
2004	Barcelona Declaration on Engineering Education for Sustainable Development[177]	In this declaration, delegates concluded, 'Engineering has responded to the needs of society and without a doubt, today's society requires a new kind of engineer . . . There is evidence that sustainable development has already been incorporated in engineering education in a number of institutions around the world.'

TABLE 2.6 Continued

Year	*Declaration/action*	*Brief Description*
2004	Shanghai Declaration on Engineering and a Sustainable Future[178]	The World Engineers' Convention in Shanghai had the theme 'Engineers Shape the Sustainable Future'. The Declaration called upon the engineering community, governments and international organisations to promote engineering for our sustainable future, proclaiming, 'Engineering and technology are vitally important in addressing poverty reduction, sustainable development and the other UN Millennium Development Goals, and need to be recognized as such'.
2006	UNESCO Workshop on Engineering Education for Sustainable Development[179]	This workshop identified and emphasised the need for learning and teaching materials, methods and capacity in sustainable development, and for better incentives for engineers to work, research and publish in the field of engineering and SD (ESD, e.g. accreditation, work opportunities, research grants, peer-reviewed publications).

Source: Compiled by The Natural Edge Project, references noted within the table.

Increasing professional advocacy

Major calls for changes in the curriculum began approximately 20 years ago with the Brundtland Commission's report *Our Common Future*.[180] In the last decade, there have been a number of significant additional calls for action from the scientific, economic, political and media communities, which have brought sustainability issues to the forefront of the global consciousness. A growing number of professional organisations around the world have been declaring an urgent need to keep up with the pace of change and have been forming collaborations to make progress, in particular since the early 1990s. For example, in 1992, together with the International Union of Technical Associations and Organizations (UATI) and the International Federation of Consultant Engineers (FIDIC), WFEO created the World Engineering Partnership for Sustainable Development (WEPSD),[181] which has since been active in promoting a new vision for 21st-century engineering.[182] The 1997 report of the Joint Conference on Engineering Education and Training for Sustainable Development in Paris called for sustainability to be 'integrated into engineering education, at all levels from foundation courses to ongoing projects and research' and for engineering organisations to 'adopt accreditation policies that require the integration of sustainability in engineering teaching'.[183]

Further, in 2003, members of the Sustainability Alliance stated in a submission to a UK national environmental audit committee inquiry into education for sustainable development:[184]

> Among the 5.5 million people in the UK who call themselves professionals, there is a growing realization that they need help in understanding how to put the principles of sustainability into practice. The reason they are beginning to call for more help and guidance from their professional associations is because they are being required to demonstrate their competency throughout their professional life, in complying with a growing and complex set of environmental, social and ethical issues. In short, for doctors, engineers, accountants and many other professions, there is a growing emphasis on occupational standards, competency, ethics and codes of conduct.

In 2004 the Barcelona Declaration outlined how universities and engineering educators need to change in order to[185]

> prepare future professionals who should be able to use their expertise not only in a scientific or technological context, but equally for broader social, political and environmental needs. This is not simply a matter of adding another layer to the technical aspects of education, but rather addressing the whole educational process in a more holistic way, by considering how the student will interact with others in his or her professional life, directly or indirectly. Engineering has responded to the needs of society and without a doubt, today's society requires a new kind of engineers.

> Engineering and technology are vitally important in addressing poverty reduction, sustainable development and the other UN Millennium Development Goals, and need to be recognized as such . . . Curricular and pedagogical reform in engineering education and continuous professional development to encompass wider social and ethical concerns are needed.
>
> The Shanghai Declaration on Engineering and the Sustainable Future[186]

Senior government and industry leaders around the world have been discussing the prominent role that engineering will need to take in assisting the world's communities to address 21st-century challenges for at least the last decade, highlighted in Table 2.7. There seems to be a clear consensus that action needs to be taken and that a strong basis for such action is forming internationally.

TABLE 2.7 Examples of professional advocacy for engineering education for sustainability

Year	*Organisation*	*Statement*
1997	World Federation of Engineering Organisations (WFEO)	The Engineer's Response to Sustainable Development: 'Engineers around the world understand that they have a tremendous responsibility in the implementation of sustainable development. Many forecasts indicate there will be an additional five billion people in the world by the middle of the 21st century. This future "built environment" must be developed while sustaining the natural resources of the world and enhancing the quality of life for all people.'[187]
2007	UK Higher Education Academy (Professor Julia King, Vice Chancellor, Aston University)	'Amongst the greatest challenges we face in the world today are those of delivering growing, secure and affordable supplies of clean water and of energy, to meet the needs and expectations of an expanding population, whilst reducing our CO_2 emissions and the human contribution to climate change. The implementation of innovative engineering solutions is fundamental to addressing these challenges, whilst also offering exceptional opportunities for economic growth to the nations which are able to deliver them.'[188]
2008	Institution of Engineers Australia (Peter Taylor, CEO)	'The concept of sustainability will influence almost all engineering developments and the potential effects on the environment, long term and short term, proximate and remote, will be integrated routinely into engineering design and planning.'[189]

TABLE 2.7 Continued

Year	*Organisation*	*Statement*
2008	Chinese Academy of Engineering (Professor Kuangdi Xu, President of the Chinese Academy of Engineering, 2008)	'If China continues with a traditional development pattern, its resources will not be sufficient to support its growth, and the environment will be unable to bear the added burden, which will make it extremely difficult for China to realize its goal of becoming a prosperous society . . . Faced with the dilemma that it cannot undertake a traditional industrialization program like those adopted earlier by developed countries at a cost of huge energy consumption and severe environmental pollution, nor can it immediately become a post-industrialization society, China must develop new methodologies and approaches to industrialization that are characterized by high technological content, desirable economic effects, low resource consumption, little environmental pollution, and effective human resources.'[190]
2009	WFEO (Barry Grear, President, 2009–2010)	'In light of the wealth of information available to the engineering profession, there is significant impetus to review what we do and how we do it. However, our references to Sustainable Development are for the most part still at too high a level. There must be a greater degree of detail provided by educators so that students have to think very carefully about the issues at hand. It is sobering for our profession to realise that this is not yet the norm for most of our engineers in training.'[191]
2010	UNESCO, WFEO, CAETS, FIDIC (Gerard van Oortmerssen, President CAETS, 2008)	'Talented engineers are needed to provide solutions for these problems through greater efficiency in production processes and transportation systems, new sustainable energy sources, more efficient use of materials, the recovery of materials from waste . . . the list is long . . . There is growing demand for engineering talent . . . and the nature of engineering is changing. Engineers, more and more, have to be aware of the social and environmental impacts of technology, and have to work in complex teams, interacting and cooperating with society.'[192]

Source: Compiled by The Natural Edge Project, references noted within.

Shifting codes of practice for engineers

Responding to a growing awareness of the issues surrounding sustainability, a number of high-profile professional engineering institutions (PEIs) internationally have placed increasing emphasis on policies and code of ethics statements relating to the role of engineering in addressing 21st-century challenges, as can be seen in Table 2.8. Many engineering organisations are members of at least one of these 'umbrella organisations', which have published documents outlining the changing role of the engineering profession. WFEO has also published a model code of ethics that includes reference to sustainable development,[193] and most national and international PEIs follow a code of ethics which reads along similar lines, including, for example, Engineers Australia in 2000,[194] IPENZ in 2005,[195] and Engineers Ireland in 2010.[196]

TABLE 2.8 Examples of strengthening professional requirements for engineering education and sustainability

Date	*Key documents outlining professional requirements*
1990	The International Federation of Consulting Engineers (FIDIC) introduced environmental policies including guidelines on the obligations of the consulting engineer with respect to their projects and clients, stating, 'engineers should provide leadership in achieving sustainable development'.[197] FIDIC, the United Nations Environment Program (UNEP), and the International Chamber of Commerce (ICC) subsequently developed training programs to provide guidance on how to describe and analyse environmental issues as well as setting up environmental management systems.[198]
1994	The Institution of Engineers Australia developed a policy on sustainability requiring that 'members, in their practice of engineering, shall act in a manner that accelerates achievement of sustainability'.[199]
1992–1996	World Engineering Partnership for Sustainable Development (WFEO, FIDIC and the International Union of Technical Associations – UATI) formed a collaboration to lay the groundwork for the many programs in support of sustainable development, which have subsequently been pursued by WFEO, FIDIC and other international organisations through their members and committees.
1997	Eighteen national and international institutions representing the chemical engineering profession globally signed the London Communiqué which pledged 'to make the world a better place for future generations'.[200]
1997	Joint paper entitled 'Role and Contributions of the Scientific and Technological Community to Sustainable Development', produced by the International Council for Science (ICSU), WFEO, Third World Academy of Sciences (TWAS), the InterAcademy Panel (IAP), and the International Social Science Council (ISSC),[201] following the 1996 World Congress of Engineering Educators and Industry Leaders, organised by UNESCO, UNIDO, WFEO and UATI. WFEO also produced 'The Engineer's Response to Sustainable Development'.[202]
1998	The Royal Academy of Engineering (London) (RAE) sponsored a visiting professors scheme in the UK 'to embed the topic of engineering for sustainable development into engineering courses and not to create a separate subject'.[203]
2001	WFEO Model Code of Ethics published, which states that, 'Engineers whose recommendations are overruled or ignored on issues of safety, health, welfare, or sustainable development shall inform their contractor or employer of the possible consequences'.[204]
2001	Twenty chemical engineering institutions signed the Melbourne Communiqué, a one-page document committing each of them to work towards a shared global vision based on sustainable development at the World Congress on Chemical Engineering.
2004	The United States National Academy of Engineering formulated its vision of the Engineer of 2020. The report suggests that engineering curricula be reconstituted 'to prepare today's engineers for the careers of the future, with due recognition of the rapid pace of change in the world and its intrinsic lack of predictability'.[205]
2005	The UK Royal Academy of Engineers published a set of twelve 'Guiding Principles' for engineering for sustainable development, in a document which also provided examples and applications for curriculum implementation.[206]
2006	International Federation of Engineering Education Societies (IFEES) – a network of 35 engineering organisations including WFEO and FIDIC – formed to establish effective engineering education processes of high quality around the world, to assure a global supply of well-prepared graduates with a focus on sustainable development.
2006	The Canadian Council of Professional Engineers published a 'National Guideline on Environment and Sustainability',[207] which outlined nine tenets that professional engineers should adhere to.

TABLE 2.8 Continued

Date	*Key documents outlining professional requirements*
2007	The Institution of Chemical Engineers created 'A Roadmap for 21st Century Chemical Engineering'.[208] Each of its six themes, which include 'sustainability and sustainable chemical technology' and 'health, safety, environment and public perception of risk', incorporates strong sustainability threads.
2007	The Institution of Engineers Australia launched a formal sustainability charter.[209] This takes a broad view, placing a particular emphasis on the social sphere, an area where engineering has traditionally been weakest. The charter states that 'sustainable development should be at the heart of mainstream policy and administration in all areas of human endeavour'. It also notes that achieving this will not be easy and 'requires a fundamental change in the way that resources are used and in the way that social decisions are made'.
2008	The National Academy of Engineering convened a committee of leading technical thinkers to create a list of the grand challenges and opportunities for engineering at the commencement of the 21st century. This included making solar energy economical as the number one challenge.[210]
2009	Engineering Council UK set out six guidance principles on sustainability for the engineering profession which it suggests respective professional engineering institutions may wish to use in developing guidance for their members.[211]

Source: Adapted from Byrne *et al.* (2010). References noted within the table.

Among these published codes it is clear that sustainable development/sustainability is envisaged as an area of ethical responsibility for practising professional engineers. Australia is often discussed internationally as a leader in professional sustainability-related policies, having begun in the mid 1990s and having instituted a sustainability charter in 2007. The Australian Engineering Code of Ethics statement is also quite direct, stating in Tenet 6 that

> Members shall, where relevant, take reasonable steps to inform themselves, their clients and employers, of the social, environmental, economic and other possible consequences which may arise from their actions.[212]

However, for the most part, rather than the codes of ethics setting sustainability/sustainable development as the very *context* of engineering practice, most policy statements and codes include sustainability more by way of add-on statements that may accompany terms such as 'social', 'environmental', 'health' and 'safety'. This is perhaps suggestive of a larger issue with incremental rather than holistic changes to code of ethics documentation as issues emerge in the profession, though it may also be a function of the relatively recent emphasis on this issue and one which will be addressed among future versions of codes of ethics, as they naturally evolve to reflect evolving PEI policies. At any rate, PEIs such as the UK's Engineering Council appear to acknowledge this issue when they issue among their six guidance principles on sustainability for the engineering profession that engineers should 'do more than just comply with legislation and codes'.[213]

According to the Engineering Council of the United Kingdom, engineering education should incorporate a number of sustainable development-related principles:[214]

- Contribute to building a sustainable society, present and future
- Apply professional and responsible judgement and take a leadership role
- Do more than just comply with legislation and codes
- Use resources efficiently and effectively
- Seek multiple views to solve sustainability challenges
- Manage risk to minimise adverse impact to people or the environment

Shifting accreditation requirements for graduate engineers

In 2010, we had the pleasure of working with Dr Edmond Byrne and Dr John Fitzpatrick from Ireland's University College Cork, to prepare a white paper which reviewed international progress in accreditation requirements for engineering education for sustainable development, for the 2010 International Symposium on Engineering Education (ISEE 2010).[215] The paper informed our workshop on the role of accreditation and the paper includes an overview of current accreditation guidelines with respect to sustainability for a number of professional engineering institutions globally. We summarise part of the paper here and invite readers to refer to the paper for further details.

Alongside shifting professional expectations, the accreditation process is a powerful instrument in directing the education of engineers and, over the longer term, the capacity of the engineering profession. Many engineering job descriptions require that engineering employees have graduated from accredited undergraduate degree programs, and requirements for registration to practise engineering, professional indemnity and personal liability certification for practising favour successful completion of an accredited engineering degree program. Internationally, engineering qualifications may be recognised in different countries, based on international agreements and accords. In addition, the professional engineering career path includes the opportunity for graduates to gain chartered status certification, where certification provides a measure of quality assurance that graduates will have attained base 'competencies' or 'attributes'.

The UK's Royal Academy of Engineering highlights the importance of accreditation as an agent for evolution and change, observing that 'the accreditation process for university engineering courses should be proactive in driving the development and updating of course content, rather than being a passive auditing exercise'.

Over the years, accreditation guidelines have evolved to reflect national legislative requirements and strategic policy direction in addition to the ethos of the accrediting PEIs.[216] Examples of major accreditation triggers in the past include the American Accreditation Board for Engineering and Technology (ABET) SUCCEED (Southeastern University and College Coalition for Engineering EDucation) initiative during the 1990s,[217] SARTOR 97 (Standards and Routes Towards Registration) in the United Kingdom,[218] and the European EUR-ACE (EURopean Accredited Engineer) initiative (2004–2006).[219]

Private industry is also becoming increasingly involved in directing accreditation through their membership of professional organisations. For example, in the UK the Sustainability Alliance represents 12 professional bodies which play a role in accrediting higher education courses. With over 200,000 members, this includes the Chartered Institution of Wastes Management, Chartered Institution of Water and Environmental Management, Institution of Civil Engineers, Institution of Electrical Engineers, Institution of Environmental Sciences, Institution of Incorporated Engineers, Royal Institute of British Architects, and Royal Town Planning Institute.[220] However, the pace at which accreditation guidelines incorporate various declarations, initiatives, communiqués, charters and policies appears to be often quite slow. Furthermore, on top of any delays in updating policy and codes, time must be given to institutions to address the new criteria, over perhaps one or more accreditation rounds (each typically of four years or more). Hence, there may be a considerable lag between the institutional accreditation requirements and actual curriculum renewal. Within this context and given the recent emergence of sustainability-related imperatives for engineering education, a significant challenge exists to incorporate such aspects into accreditation requirements in a timely manner.

Despite these challenges, over the last 20 years, many institutions responsible for accrediting engineering degrees and co-ordinating agreements have strengthened their position through more rigorous accreditation requirements that deal with the environment and sustainable development, as highlighted in Table 2.9.

TABLE 2.9 Example accreditation trends in sustainability competencies

Country	*Graduate numbers and shifting accreditation trends*
China	China alone produces nearly half of the world's engineering graduates, with some 600,000 college and university graduates in 2005.[221] While China's engineering accreditation system is still being formalised, some leading HEIs in China have been proactive in sustainability initiatives,[222] including hosting forums supported by the China Association for Science and Technology, the Chinese Academy of Engineering and the World Federation of Engineering Organizations, and engaging with universities around the world particularly in Europe to foster inter-university student and staff exchanges.
India	India graduates 350,000–500,000 engineers each year, depending on what is used to define graduation requirements.[223] In 2007, India's National Board of Accreditation was inducted into the Washington Accord, which requires the country's accreditation system to quickly align with the Accord's requirements in order to acknowledge professional engineering qualifications, which includes sustainability competencies.
USA	America produces approximately 70,000 engineering graduates each year. The American Accreditation Board for Engineering and Technology (ABET) state in their engineering accreditation criteria that engineering programs must demonstrate that their students attain, among other things, 'an ability to design a system, component, or process to meet desired needs within realistic constraints such as economic, environmental, social, political, ethical, health and safety, manufacturability, and sustainability'.[224]
United Kingdom	The United Kingdom graduates approximately 12,000 engineers each year, from the pool of approximately 100,000 European engineering graduates. The Engineering Council is responsible for the UK Register of Chartered Engineers, Incorporated Engineers and Engineering Technicians. The importance of sustainable development is clearly identified in the Council's Standard for Professional Engineering Competence (UK-SPEC) which came into force in 2004. It includes a statement that chartered engineers must, 'undertake engineering activities in a way which contributes to sustainable development'.[225]

TABLE 2.9 Continued

Australia[226]	Australia produces approximately 6,000 domestic bachelor of engineering graduates each year.[227] In 2006, the Institution of Engineers Australia (EA), the industry body which accredits engineering education, revised the accreditation criteria, system and processes for professional engineering qualifications, focusing on industry liaison and broad graduate attributes, encouraging engineering schools to devise innovative curriculum and pedagogy to meet 'alternative missions'.[228] EA has incorporated specific competencies related to sustainable development into the associated Australian Engineering Competency Standards – Stage 1 Competency Standards for Professional Engineers,[229] including statements such as, 'Professional Engineers are responsible for bringing knowledge to bear from multiple sources to develop solutions to complex problems and issues, for ensuring that technical and non-technical considerations are properly integrated, and for managing risk as well as sustainability issues . . . One hallmark of a professional is the capacity to break new ground in an informed, responsible and sustainable fashion.'

Source: References noted within the table.

As can be seen from Table 2.9 there is a broad range of outcome requirements among PEIs internationally, using terms such as 'sustainability/sustainable development', 'environmental or social issues', 'ethical issues', 'multi-disciplinarity', 'complexity' or 'complex systems', and 'open ended' and 'wicked' problems. Requirements range from the most detailed and explicit descriptors based on a learning outcomes approach in countries such as Australia, the UK, Canada, Germany and Ireland, to a more generalised learning outcomes approach which simply provides a number of headings (without further explanatory detail) in countries such as the USA, Japan and Taiwan. There are also accreditation processes which appear not to be based on program learning outcomes (e.g. India), to places which appear to have had no accreditation procedure in place (e.g. China).

Considering the graduation figures in Table 2.9 it is concerning that areas with the least stringent accreditation requirement produce the highest numbers of engineers, with the most explicit sustainability accreditation requirements, such as those in the United Kingdom and Australia, affecting only 18,000 of more than 1.25 million graduates each year. However, there are international efforts being undertaken by groups such as the International Engineering Alliance (IEA). The IEA currently has three key agreements covering mutual recognition related to tertiary-level qualifications in engineering, known as the 1989 Washington Accord for professional engineering, the 2001 Sydney Accord for engineering technology, and the 2002 Dublin Accord for technician engineering.[230] These three accords now have one set of rules and procedures, and 'exemplar' statements of the 13 graduate attributes for each of the three occupations.[231] Signatories are required to operate accredition systems that deliver graduates that substantially meet the relevant exemplar. The graduate attribute exemplars contain many references to aspects of sustainability such as systems design, 'the engineer and society', ethics and 'environment and sustainability'. Specifically the exemplars state that graduates should 'understand the impact of engineering solutions in a societal context and demonstrate knowledge of, and need for, sustainable development'.[232]

Signatories of the Washington Accord include Australia, Canada, Chinese Taipei, Hong Kong, Ireland, Japan, Korea, Malaysia, New Zealand, Russia, Singapore, South Africa, Turkey, the UK and the US. The Washington Accord acknowledges provisional members, who are given two years to ensure their academic systems are at an international level, to then be considered for full membership.[233] These countries currently include Bangladesh, Germany, India and Sri Lanka. A notable absence from the Washington Accord is China, which is currently working towards provisional membership of the accord requirements.

3

DELIBERATIVE AND DYNAMIC CURRICULUM RENEWAL

Despite decades of attempts to embed sustainability within higher education, literature clearly suggests that, on the whole, the education sector has been relatively slow to incorporate sustainability knowledge and skill areas, and is generally poorly prepared to do so. Over the last decade our research team at The Natural Edge Project have developed many educational resources to accelerate the transition to education for sustainability. This chapter outlines a model that has been informed by these efforts working with our mentors around the world, who are leading the way.

With current approaches, education for sustainability may never eventuate across the entire higher education sector – clearly a continuation of 'business as usual' is not a palatable option for the sector or society at large. In light of this reality, we have sought to develop a whole-of-system approach to implement systematic, intentional and timely curriculum renewal that is responsive to emerging challenges and opportunities, encompassing curriculum renewal and organisational change. This includes action at the level of the staff and at the level of the institution.

This chapter takes a step away from the topic of sustainability to consider the processes for embedding any emergent knowledge and skill set into the curriculum. We present a model for deliberative and dynamic curriculum renewal that has evolved from our exploration of literature, case studies, pilot trials and a series of workshops with sustainability educators from around the world over the last decade. We also pay homage to the curriculum theorists who have inspired this model, considering the evolution of curriculum renewal processes since the early 1900s, including Tyler, Taba, Wheeler, Kerr, Walker, Stenhouse and Egan.

It is important to ensure the context for including sustainability is clear, to avoid renewed curriculum being susceptible to random change. Critical sustainability knowledge and skill areas could subsequently be deleted or replaced without consideration of the overall impacts on the curriculum. Language associated with this type of curriculum includes '*ad hoc*', 'champion-based', 'vulnerable', 'isolated', 'duplicated' and 'expensive'. In contrast, where knowledge and skill sets are considered within the overall context for the program of study (i.e. 'deliberative') and in the systematic development of particular graduate attributes (i.e. 'dynamic'), the resultant curriculum is resilient to incidental change that is not aligned with program aspirations. Language associated with this type of curriculum includes 'systemic', 'team-based', 'robust', 'connected', 'complementary', and 'cost-effective'.

We conclude by discussing methods of managing timeframes for curriculum renewal. Beyond the deliberative and dynamic model described in this chapter, timing depends on numerous other factors. There is a clear need for institutions to consider 'by when' the renewal process should be complete, and who should be involved. This may then highlight the need for considering 'rapid' curriculum renewal – for example, if there is a significant backlog of content that needs to be embedded in a contracted period of time. Chapter 4 then presents an organisational change based model to manage such a procsess.

A journey to curriculum renewal

The growing imperative to act

In his work on curriculum development Professor John Heywood, Trinity College, Dublin, concludes that, irrespective of the model used, the combination of elements in curriculum design, assessment and evaluation will require a substantial change in the culture of the organisational unit responsible for the delivery of the curriculum.[1]

Some 60 years on from Rachel Carson's description of a 'silent spring'[2] and 20 years on from the first Rio environment summit,[3] issues such as climate change, deforestation, ocean acidification and biodiversity loss provide daily reminders of the urgency for action to restore a number of significant planetary systems. As discussed in Chapter 1, it is imperative that society transform its relationship with the use of resources, to do much more with much less and at the same time play a role in restoring degraded systems. Economy-wide efforts will be required to reduce environmental pressures on a meaningful scale. Such efforts need to ensure that economic performance is maintained, referred to as the challenge to decouple economic growth from environmental pressure. Furthermore, positive environmental impacts (for example reforestation, aquifer recharge, etc.) need to be 'recoupled' to economic performance, towards sustainable development that ensures the needs of future generations can also be met.

It is highly likely that there will be a swift increase in enforcement efforts related to reducing environmental pressures, with such efforts having a significant effect on the higher education system. In light of this, many are now asking, 'How can sustainability knowledge and skills be effectively embedded into curriculum?', and further, 'What can be done to accelerate the process?'. These questions stand to join a number of other key areas that will dominate much of the agenda in the higher education sector internationally over the coming two decades, with the quality of the response significantly affecting the viability of programs. Some institutions will make strong commitments to education for sustainability that will position them well for avoiding future risk and capture the associated rewards, as outlined in Chapter 1. Others will decide to continue to meet minimum compliance, placing them in a position of growing risk of not being able to keep up with certain increases in compliance soon to come.

A response by young engineers and scientists

Understanding that the challenge to significantly reduce environmental pressures would dominate our future careers, in 2002, as a team of young engineers and scientists, we formed The Natural Edge Project, incubated within Engineers Australia and now co-hosted by Queensland University of Technology, Curtin University and the University of Adelaide. The project is a collaborative partnership for research, education, policy development and strategy for sustainable development. The team comprises researchers in a number of Australian universities and its mission is to 'contribute to, and succinctly communicate, leading research, case studies, tools and strategies for achieving sustainable development across government, business and civil society'.

Our initial intention was to invest three years into learning from the experience of the world's leading sustainable development practitioners, researchers and policy makers, to investigate the possibility of merging the agendas of industrial development and environmental management in such a way that both society and the environment would prosper. This involved a detailed investigation of both mainstream industrial development theory and practice, notably Professor Michael Porter's work on *The Competitive Advantage of Nations*,[4] and the emerging 'green' business theory and practice, notably works by Professor Ernst von Weizsäcker, Amory Lovins, and Hunter Lovins, on *Factor 4: Doubling Wealth, Halving Resource Use*,[5] and *Natural Capitalism*.[6] The result of the investigation was the release of the book *The Natural Advantage of Nations: Business opportunities, innovation and governance in the 21st century*.[7]

Books by The Natural Edge Project intended to inform education for sustainability:

- *The Natural Advantage of Nations: Business opportunities, innovation and governance in the 21st century* (Earthscan 2005)[8]
- *Whole System Design: An integrated approach to sustainable engineering* (Earthscan 2008)[9]
- *Factor 5: Transforming the global economy through 80% increase in resource productivity* (Earthscan 2009) with Professor Ernst von Weizsäcker[10]
- *Cents and Sustainability: Securing our common future by decoupling economic growth from environmental pressures* (Earthscan 2010), with a foreword from Gro Brundtland[11]

Following the release of our first book in 2005 we were then asked if we had materials that might assist engineering and other built environment education programs to incorporate the materials in the book. This then led to a dual focus for our team, first on content development through the publication of further collaborative books[12] and the development of peer-reviewed open-source lecture materials,[13] and second on working with higher education institutions to prepare, explore, trial and integrate such content into programs.

Open-source online education materials by The Natural Edge Project intended to inform education for sustainability:

- Sustainability Education for High Schools: Year 10–12 Subject Supplements (12 Lessons and a Teachers Guide)

- Introduction to Sustainable Development for Engineering and Built Environment Professionals (12 Lectures)
- Principles and Practices in Sustainable Development for the Engineering and Built Environment Professions (12 Lectures)
- Energy Transformed: Sustainable Energy Solutions for Climate Change Mitigation (30 Lectures)
- Water Transformed: Sustainable Water Solutions for Climate Change Adaptation (24 Lectures)

Along the way, as we developed educational resources, our team was continually challenged by colleagues around the world lamenting that while the content was great, it was difficult to get traction in integrating it within their current programs of study.[14] This led us to search for models of curriculum renewal that could assist our colleagues in renewing their programs, along with those at our host universities, and we were surprised by the lack of guidance in such processes. It came as a shock to us when we realised that the systems of curriculum renewal themselves were not well documented, and realised that this was a key gap that needed to be filled, creating the impetus for this book.

In this book, we define curriculum renewal to mean the redevelopment of curriculum, which may involve for one or more existing or new courses in a program, the review of past syllabi (i.e. one or more documents that include statements of the aims and objectives of course and its content) and pedagogy (i.e. the way in which the course is taught).[15]

Responding to our need for clarity on how sustainability resources could be integrated into existing programs, this book presents two new models to describe the processes involved in curriculum renewal at the level of both the program and the institution, as shown diagrammatically in Figure 3.1. We discuss the curriculum model in this chapter (the inner circle) and the organisational process in Chapter 4 (the outer circle).

The following sections describe the curriculum model and its component elements, highlighting the importance of a holistic approach that builds upon both curriculum theory and organisational theory. Without either of these foundations, timely curriculum renewal becomes an *ad hoc* process reliant on individual champions, an energy intensive and frustrating experience that many leaders in the area face. The intention is to inform efforts as to make the process understandable, achievable and valuable to the university. The models are based on process considerations and are not prescriptive, lending themselves to be adapted within any curriculum context. We look forward to hearing from readers about insights into what becomes possible when considering curriculum renewal through these structures.

The development of a new approach to curriculum renewal practices

Building on from business-as-usual curriculum renewal practices

The most common form of curriculum renewal in universities around the world could perhaps be summarised in Figure 3.2, with individual academics undertaking their own development and updating

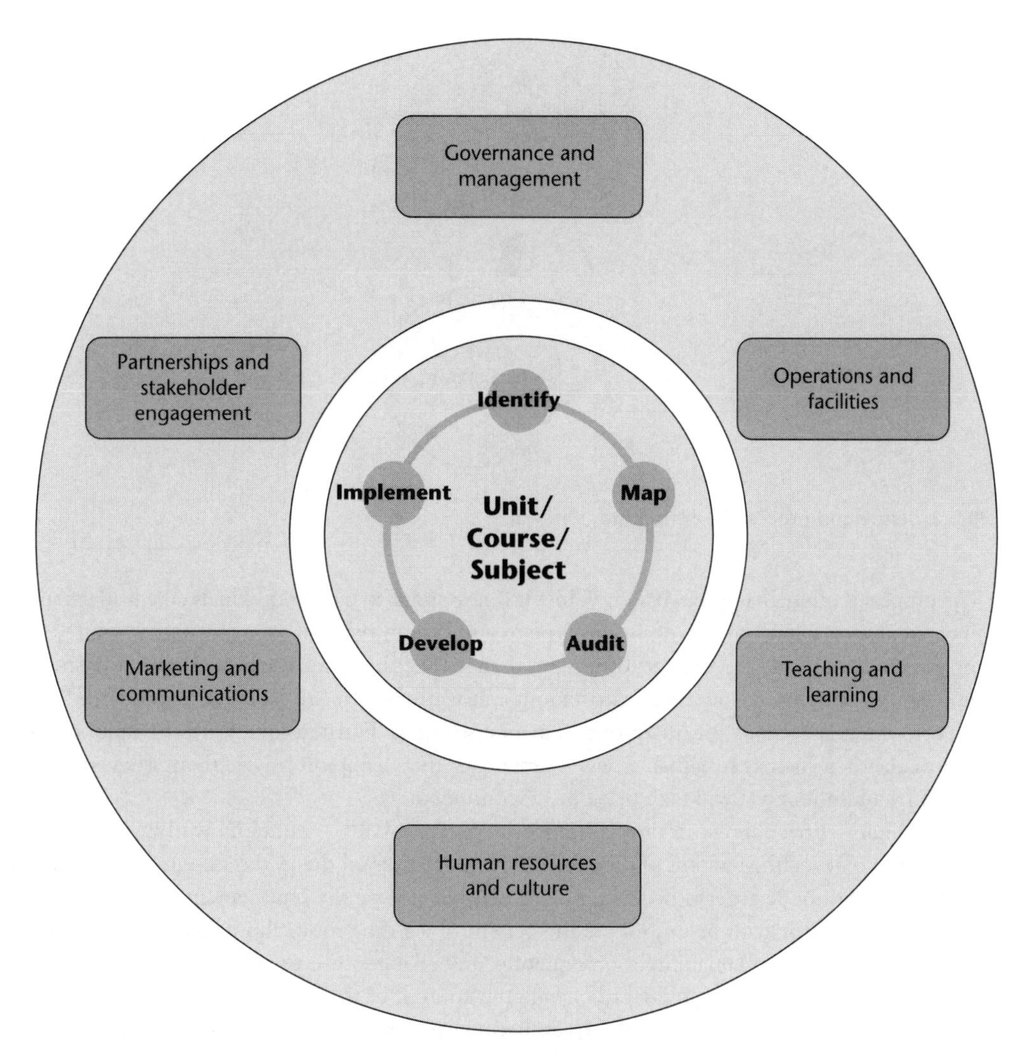

FIGURE 3.1 A stylistic representation of a systems approach to curriculum renewal

of courses in the absence of interaction with other colleagues, industry or the wider university. The course is then implemented and the academic may or may not then review and revise the offering over time.

There are several issues that we have observed with this process of curriculum renewal that have led us to develop and trial a broader approach:

- The process is *ad hoc*: This type of curriculum renewal is based on the interests of the academic who is given primary responsibility for the subject, usually with a high level of autonomy. Depending on the institutional requirements, the subject may or may not undergo review at a program level, making it difficult to track the details of what 'curriculum renewal' took place. This is particularly challenging when key performance indicators focus on *whether* curriculum renewal has occurred (i.e. box-ticking), rather than *what* occurred and its impact on the content of the course.

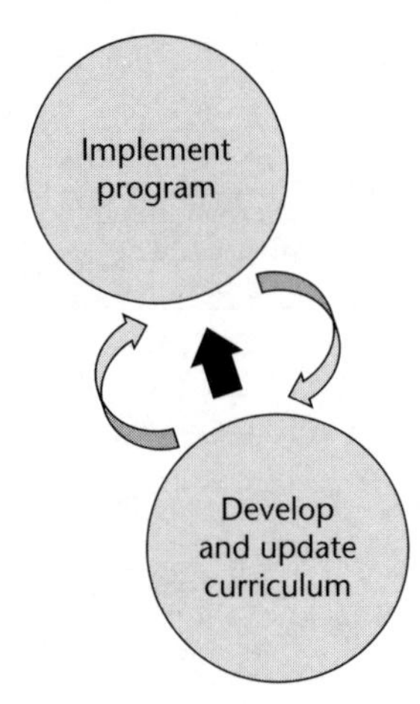

FIGURE 3.2 Status-quo process for curriculum renewal

- The process is often champion-based: While it is important to have individuals take leadership roles in curriculum renewal initiatives, in this type of curriculum renewal process, there is limited access for these 'champions' to engage others in their curriculum innovation. This is evidenced in numerous examples of successful curriculum innovations that are isolated, experienced in one subject, one year of study, or one program in an institution. Furthermore, if no champions exist for a particular area such as sustainable development, or the champion leaves, then it is very unlikely that curriculum renewal will take place or be continued.
- The resultant curriculum structure is 'fragile' and expensive to maintain: Within this pattern of curriculum renewal the content of the subjects often depends on the lecturer continuing in this role. In the case that the lecturer is no longer delivering the course the replacement may drop the areas they are not familiar with or cannot see the benefit of learning more about, typically topics related to environment and sustainability. Subsequently, any changes to program management, budget or human resourcing has the potential to change the content of the subject being offered.
- Potential for duplication and mismatched content: In the case where individual courses are updated there is often little communication with other courses in the program. This can result in emerging topics being duplicated as their value is recognised by more than one lecturer in the program and added to their courses. Conversely there is also the risk that as a program has a 'sustainability champion', other lecturers will feel that they do not need to add such material to their courses, resulting in a lack of content related to sustainability.
- Curriculum renewal is confrontational and fatiguing: This process is particularly challenging for curriculum renewal advocates hoping to influence the content of programs, such as research staff, industry advisors and those at a faculty or university level that see the value of such efforts. A key challenge is identifying academics who are interested in incorporating sustainability into programs and providing support to do so, in a manner that enhances the overall program offering. Without a strategic approach that appropriately involves key players internally and externally to support the process to integrate sustainability into programs course by course, champion by champion, the process can be frustrating and fatiguing, leading to withdrawal of efforts and exacerbation of the other issues noted above.

New areas of knowledge and skills that are considered in isolation of the overall program curricula tend to be vulnerable. Critical emerging areas can be inadvertently deleted or replaced with subsequent gaps in program outcomes.

Considering 'learning pathways' for graduate attributes

Understanding these shortcomings of typical curriculum renewal processes, there is a clear need for a more strategic approach. Over the last two decades in particular, the literature shows an emergence of conversation around the need to identify what students should be able to do by the time they graduate (i.e. graduate competencies). This is being influenced by a number of professional bodies and accreditation agencies describing desired competencies that will be considered as part of professional status reviews and university accreditation processes. These are subsequently being translated into desired 'graduate attributes' for programs of study.

For example, a competency statement of 'In-depth understanding of specialist bodies of knowledge within the engineering discipline'[16] could translate to a civil engineering graduate attribute of 'In-depth understanding of construction materials, their properties and performance'.

A problem in the past has been that while these competency statements and graduate attribute statements are documented, it can be difficult to assess if they are being developed in programs. For the above example, in which year of the program and in which subjects do students learn about construction materials with regard to their properties and performance? Where are the topics and key concepts introduced, then developed to an 'in-depth' level of understanding?

Research suggests that a valuable response to the introduction of such formal 'competency' requirements by accreditation bodies is to attempt to document – or map – the evidence that graduate attributes in the area are being developed across the program. Such a process is shown simplistically in Figure 3.3. In the figure the term 'learning pathway' describes the process to develop the graduate attribute. This could be as simple as a paragraph of text outlining the progression through the program of the graduate attribute, through to complex spreadsheets and flow diagrams for each year level showing how subjects are intended to contribute to the overall program (as shown in Table 8.5 as part of our work with James Cook University in Townsville, Australia).

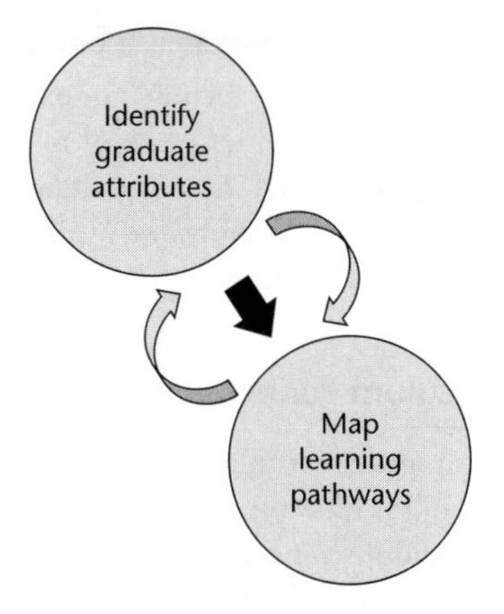

FIGURE 3.3 Emergent processes for curriculum renewal

However, even with a learning pathway developed, without having some sort of 'check-in' process, it is difficult to know whether the intended curriculum matches what is delivered. Hence, it is important to include an internal checking/auditing process for desired learning outcomes, a process that asks two important questions: (a) have the learning pathways been addressed in the intended curriculum for a given course (e.g. through the content and assessment); and (b) have the students attained the intended learning outcomes of the subject. This process is shown simplistically in Figure 3.4, highlighting the need for auditing to connect the mapped learning pathways with the individual course development.

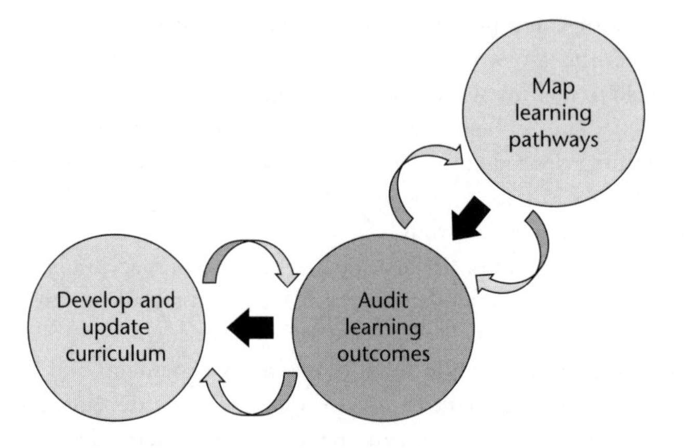

FIGURE 3.4 Auditing learning outcomes as part of curriculum renewal processes

Understanding the need for a curriculum renewal strategy

Central to any curriculum renewal process should be a curriculum renewal strategy that is developed through several elements as shown in Figure 3.5. This includes internal and external collaboration, for example with stakeholders such as potential employers, students, professional bodies and internal program managers. It is critical to have an evolving capacity for teaching the renewed curriculum through raising awareness and building capacity among educators, and to continually monitor the curriculum renewal process for improvement opportunities.

Such context provides clarity in answering the question, 'why take action at all?'. The multiple feedback loops provide an alert as to when curriculum renewal would be appropriate, in responding to or anticipating changing circumstances. This could include, for example, if there is a shift in graduate attributes, which could come about through the emergence of a new capability (e.g. being able to embed carbon accounting within computer-aided drawing design). Alternatively, it could be if monitoring and evaluation shows that subject not performing against learning outcomes, or if there is a change of responsibility for convening the course.

Deliberative and dynamic curriculum renewal

With these issues for current curriculum processes in mind, our team sought to identify a process that created conditions conducive to curriculum renewal. Specifically, we enquired into opportunities for team-based, systemic and robust curriculum renewal that is connected from subject to program, complementary to existing curriculum and cost effective to implement. Following extensive development and trialling of a number of elements of curriculum renewal, including those outlined above, a

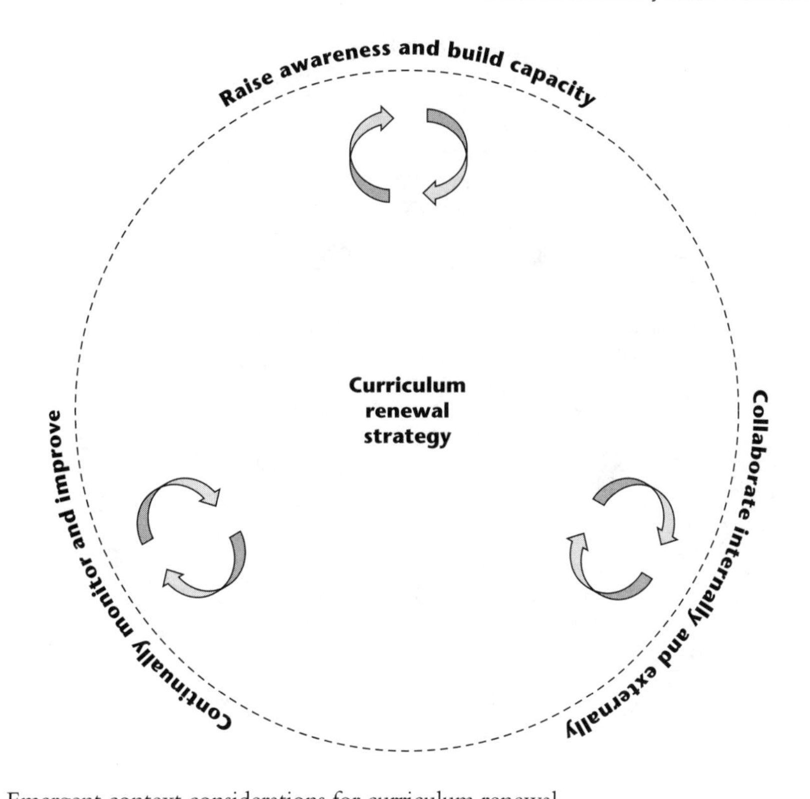

FIGURE 3.5 Emergent context considerations for curriculum renewal

model was created to provide a strategic approach to curriculum renewal as shown in Figure 3.6. The model builds on our experiences and that of our partners and mentors, and is informed by the leading efforts in the field over the last half century, including models developed by Professor Ralph Tyler (1949),[17] Dr. Hilda Taba (1962),[18] Daryl Wheeler (1967),[19] Dr. John Kerr (1968),[20] Dr. Decker Walker (1971),[21] Professor Lawrence Stenhouse (1975),[22] and Professor Kieran Egan (1978).[23]

In this model, deliberative and dynamic curriculum renewal begins with a central curriculum renewal strategy, which then informs cycles of identifying graduate attributes, mapping learning pathways, auditing for learning outcomes, developing and updating courses, and implementing them. These steps are undertaken in parallel with efforts to raise awareness and build capacity among faculty, collaborate internally and externally with staff, students and the wider community, and to continually monitor and improve the process and curriculum itself.

In the following sections we discuss the model with regard to the curriculum context for creating a 'deliberative' and 'dynamic' process of renewal. We then move to considering what is required at the institution level to support such curriculum renewal efforts.

How can curriculum renewal be 'deliberative'?

There is a growing need to improve the process of curriculum renewal to ensure that a systemic approach is undertaken to considering emerging waves of innovation, and the associated knowledge and skill areas. Such a strategic approach would include internal staff capacity building, an integrated approach across the program, and providing opportunities to engage with real projects on campus, in

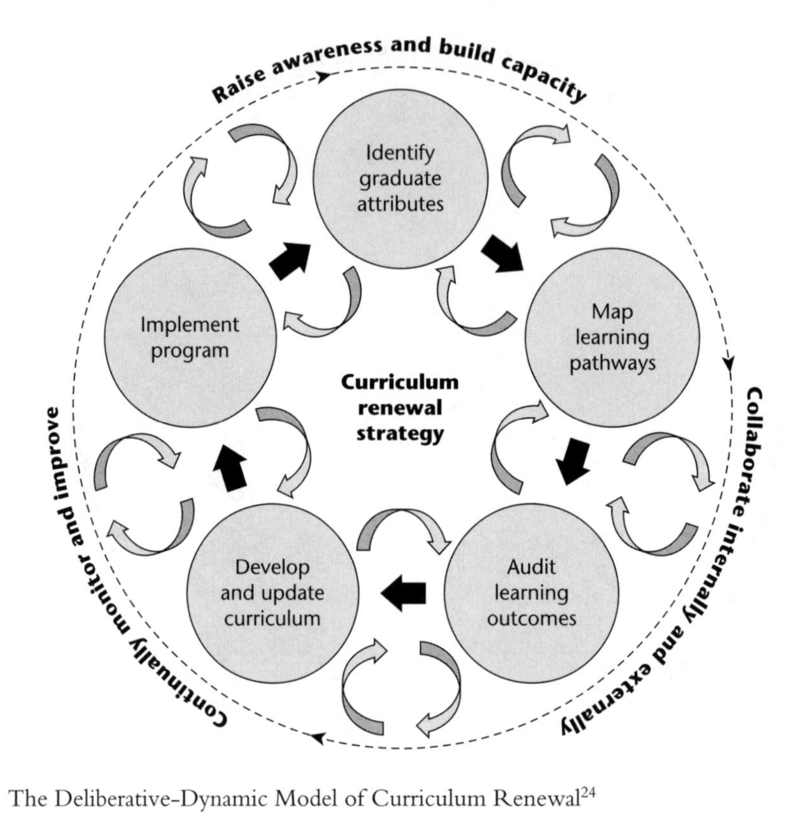

FIGURE 3.6 The Deliberative-Dynamic Model of Curriculum Renewal[24]

industry and in the community. Further, such a strategy would carefully consider current and anticipated legislative changes and market trends in the industries and sectors that the graduates are intended for. With this in mind, the model shows that deliberation takes place at each of the five main elements of the curriculum renewal process, both internally and externally. In effect, the elements by themselves are unlikely to result in effective curriculum renewal without a deliberative approach that allows ongoing internal and external interactions, monitoring and capacity building.

The consideration of context (i.e. temporal issues including environmental and social trends) in curriculum renewal is increasingly commonplace. Curriculum design processes are beginning to acknowledge the importance of a situational analysis of the field being taught, with regard to new knowledge and skills, and its impact on employer needs regarding graduate attributes. However, this is a relatively recent phenomenon. In the 1970s Reynolds and Skilbeck were among the first curriculum writers to recognise the importance of keeping track of the external context in the curriculum renewal process.[25] Their work emphasises a thorough consideration of relevant external and internal factors, although the pace at which curriculum renewal should take place does not feature. In the 1990s, Cornbleth wrote of curriculum as an ongoing social process, comprised of the interactions of students, teachers, knowledge and their situation (i.e. 'milieu').[26]

However, despite such conversations, there is still an absence of guidance on how to undertake curriculum renewal in a manner that is informed by the surrounding context. Back in the 1970s, education researcher Walker developed a starting point for 'how', in the form of a dynamic model for curriculum development, shown in Figure 3.7, that includes 'deliberations' about context including beliefs, theories, conceptions, points of view and aims and objectives for the intended course. In this

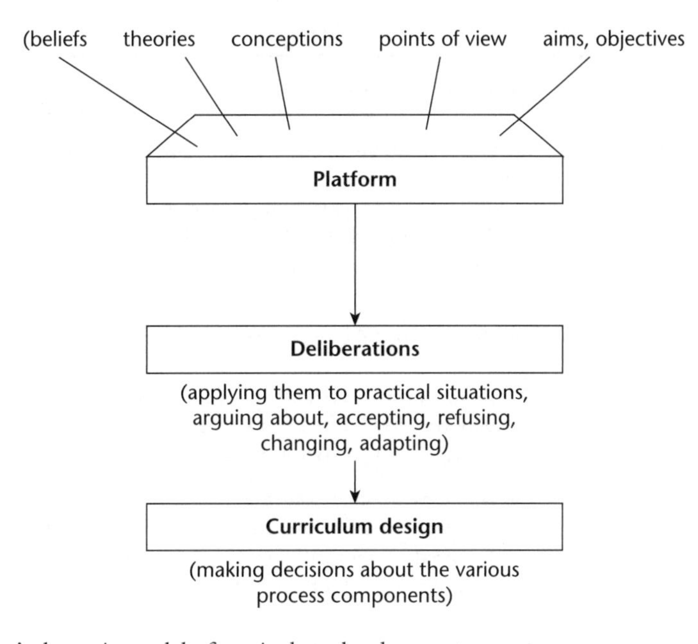

FIGURE 3.7 Walker's dynamic model of curriculum development

Source: Walker, D. (1971)

model, the curriculum design process would be informed by considering how these factors might affect the intended curriculum.

This important model by Walker effectively highlights the need for the educator to recognise inputs into the curriculum design process, such as beliefs and points of view, before entering a process of curriculum renewal, and ensuring interaction with others during the deliberations. However, in the funnel-like approach, there is little opportunity for the resultant curriculum design to be evaluated with regard to how it addresses the issues identified in the 'platform' and 'deliberations', i.e. in an iterative capacity. Furthermore, the model doesn't make it clear that different 'deliberations' may be needed at different points in the curriculum design process, as highlighted in Table 3.1.

Identifying graduate attributes for students requires meaningful interaction by program leaders and educators with external stakeholders such as professional bodies, potential employers, government agencies and the students themselves. In contrast, the *mapping of learning pathways* to deliver these graduate attributes requires consideration of both content and pedagogical issues by educators and their colleagues, with occasional involvement of other stakeholders. *Developing and updating course* content might include focused interactions with students, along with teaching and learning experts. Despite the additional 'effort' required to undertake such deliberations, many of these interactions will raise awareness and build capacity amongst the deliberators themselves.

With the level of complexity of each of the elements of the curriculum renewal process proposed here, the need for a 'deliberative approach' that considers a variety of internal and external stakeholders is clear. Building on investigations by early pioneers such as Reynolds, Skilbeck and Walker, and other

TABLE 3.1 Examples of 'deliberation' through the curriculum renewal process

Element	*Example focus of deliberations*
Identify graduate attributes	Interaction with professional bodies, potential employers and government agencies to consider: • Current and anticipated legislative and market context • Current and anticipated accreditation requirements • Current and anticipated professional status requirements (such as Chartered) Interaction across the university, including alumni and future students, to consider: • Alignment with the identity of the university • Opportunities for niche or specialist offerings • Available strengths across various content areas
Map learning pathways	Interaction with other academic staff in program, teaching and learning experts, professional bodies and potential employers to consider: • Required component knowledge and skill sets (technical and enabling) • Appropriate levels of coverage at each year level • Opportunities for synergistic efforts between courses and year levels • Pedagogical challenges and opportunities for developing graduate attributes
Audit learning outcomes	Interaction with institutional management and other educators involved with the curriculum to consider: • Knowledge and skill development gaps in current curriculum • Resourcing implications and priorities for moving forward in curriculum renewal • Appropriate ways to respond to identified gaps in courses and programs
Develop and update curriculum	Interaction with teaching and learning experts, past and current students to consider: • Teaching and learning opportunities to develop the required knowledge and skills • Content and assessment implications for the course • Opportunities for new platforms and media for curriculum
Implement program	Interaction with current students and peer educators to consider: • How well the renewed curriculum addresses the intended learning outcomes • Implications for future iterations of the course and overall curriculum renewal effort

researchers such as Cornbleth, a deliberative approach can enhance the curriculum renewal process to deliver a current, relevant and future-flexible product that meets the needs of the education provider, potential employers and current and future students.

How can curriculum renewal be 'dynamic'?

The way curriculum renewal is understood and theorised has progressed significantly over the last century; however, there remains considerable debate as to its meaning and practice. Indeed, the very concept of curriculum renewal is continually changing, which offers both challenges and opportunities for higher education research in this field.[27] Over the years the term has been used broadly to mean such things as curriculum change and curriculum improvement,[28] curriculum redesign, revision, redevelopment, reform,[29] transformation[30] and restructuring of curriculum.

In this book we use the underlying explanation of 'curriculum' by Billet and Stevens, which captures three core components – that of what is *intended* (i.e. the 'intended curriculum'), what *happens* (i.e. the 'enacted curriculum') and what is *experienced* (i.e. the 'experienced curriculum').[31] These three components can then provide a means to understand the different sets of concerns that may give impetus for curriculum renewal, such as government and potential employers who may focus on intents or outcomes (e.g. competency standards); teachers and trainers who have to implement or enact the curriculum and who may be concerned with how the students are taught; and students themselves, who may be most interested in the quality of experiences that lead to desired outcomes.

Considering this perspective, it is clear that the process needs to be strategic, continually attempting to capture the multiple perspectives to deliver curriculum that is responsive, targeted and regularly reviewed. In the following paragraphs, we pay homage to the early innovators of last century, highlighting how the field of the curriculum renewal theory has evolved and how this has informed development of the new model presented in this book.

Curriculum renewal has essentially evolved from a linear approach through to a linear-circular approach, then to an iterative-circular approach. The latter approach allows for opportunities to address multiple perspectives throughout the curriculum renewal process. In essence, the central elements forming the backbone of a traditional linear approach are made 'dynamic' in the new model, through iterative review to incorporate internal and external considerations in addition to monitoring and evaluation between each key step.

Taylor and Bobbitt (Early 1900s): Thinking about curriculum theory and practice as 'product' was heavily influenced by the development of management thinking and practice by Frederick Taylor in the early 20th century, a mechanical engineer by training whose career was focused on making American industry more efficient. Taylor reasoned that the emerging profession of efficiency experts should be able to identify precise steps that each worker should undertake as part of a larger system, then optimise their performance.[32] John Franklin Bobbitt's curriculum planning procedures (referred to as 'job analysis') developed in the early 1900s adapted Taylor's work, where the curriculum was comprised of the school experiences needed to enable children to do the activities that adults undertook to fulfil their various roles in society.[33,34] This work set the basis for considering an ongoing process to make education programs more 'efficient' to support industry, which is at the heart of this book's intention.

Tyler (1940s): In the 1940s, Professor Ralph Tyler documented a linear approach to curriculum renewal, involving four steps as shown in Figure 3.8. In Tyler's model the steps outline the core of the approach and provide a clear sense of progression through the curriculum renewal process. This influential model by Tyler subsequently became the accepted approach to curriculum development for almost three decades, and still guides the essential questions of curriculum development.[35]

Despite providing a valuable framework at the time, a number of issues have since been identified and discussed with regard to this model, including a lack of social vision or program to guide the process of curriculum construction, a linear 'ends-means' view of education that is heavily focused on employer needs, and reliance on measurement despite the uncertainty about what is being measured and how

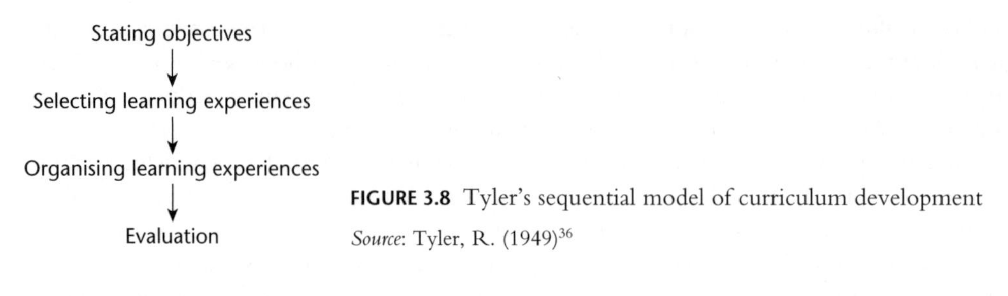

FIGURE 3.8 Tyler's sequential model of curriculum development

Source: Tyler, R. (1949)[36]

the impact of particular experiences can be accurately measured.[37,38] The value of the model is to demonstrate that curriculum renewal can be undertaken as a series of steps that allow for a strategic approach to be undertaken and replicated across different programs and institutions.

Kerr (1960s): Responding to the issues outlined by Tyler, in his seminal 1968 book, *The Changing Curriculum*, Dr John Kerr, a schoolteacher and headmaster, called for urgent research into the dynamics of change in the curriculum development process, a call that has largely been unanswered. He reflects in the introduction to the book that reform in education often involves persuasive and enthusiastic innovators unintentionally hampering systemic progress because there is no rational, coherent theory, set of concepts, or 'elements' on which to base curriculum modifications.[39]

> Curriculum making has developed rather as a craft develops from long practical experience, [where] outmoded techniques and materials tend to be retained [and] there is no planned strategy for making curricular decisions or evaluating changes introduced . . . There is need to analyse the process of curriculum planning and to identify the elements that should be the determinants of curriculum design.[40]
>
> J. Kerr

Numerous curriculum theorists over the last half-century have produced models that include a number of the 'elements'[41] that Kerr calls for. Much of the work that has gone into the development of this book has been inspired and informed by Tyler's work, as part of a growing response to his call for a better understanding of the dynamics of curriculum renewal.

Taba (1960s): Dr Hilda Taba (a former student of Tyler) built on Tyler's work to highlight the interconnectivity of elements within the curriculum renewal process. The work explores how each of the core elements in the process interact with each other, with a monitoring and evaluation component, as shown in her Rational Planning Model in Figure 3.9.

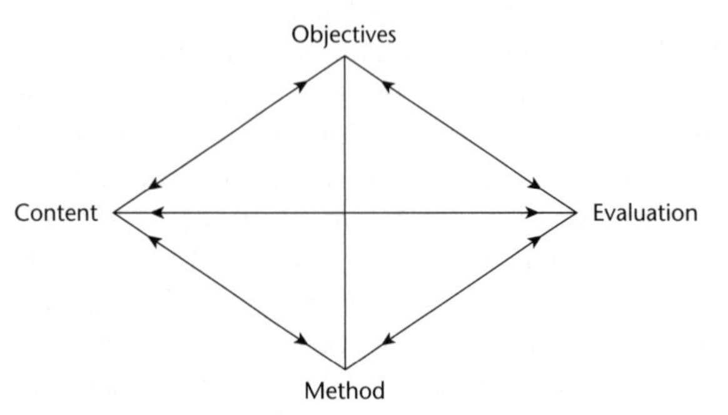

FIGURE 3.9 Taba's rational planning model of curriculum development

Source: Taba, H. (1962)[42]

This model allows the curriculum developer the freedom to change the order of the process and creates the basis for an iterative approach, and allows the developer to react to situations in determining what sequence of elements is most appropriate.[43] However, it is a simplistic view of the process that is relatively rigid with only four main considerations and little guidance as to the application to programs.[44] There is also little focus on implementation and the lessons learned from this process. The value of the model is to show that elements of curriculum renewal can be iterative and inform each other in a dynamic way.

Wheeler (1960s): Inspired by Taba and based on his experience as Professor of Curriculum Development in Beirut, Dr Decker Wheeler developed a cyclical approach to curriculum renewal, which he described in his preface as 'an attempt to lay out guidelines for a first approximation to a rationale whereby curriculum problems may be systematically considered'.[45] In his model shown in Figure 3.10, the core linear process is designed to be cyclical to enable it to be repeated for subsequent updates.

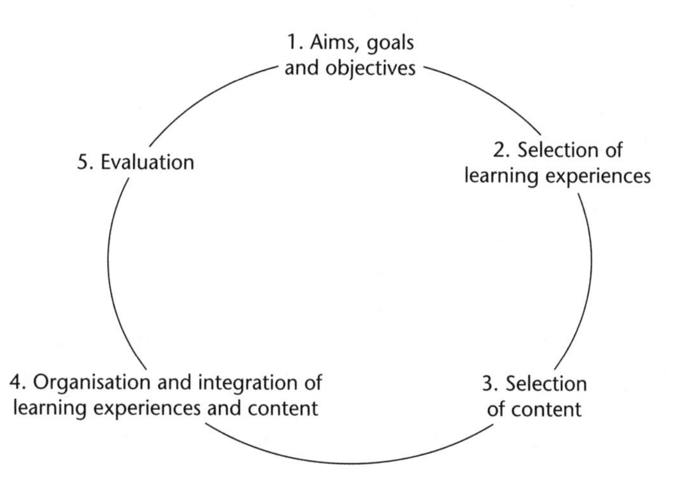

FIGURE 3.10 Wheeler's interactive model of curriculum development

Source: Wheeler, D. (1967)[46]

This model is similar in structure and intent to the deliberative and dynamic model presented herein and our work has focused on adding rigour and depth to each of the main steps.[47] Given the model was developed in the late 1960s in the absence of any impetus to accelerate such processes, this aspect has been a major focus and contribution of the new model.

Stenhouse (1970s): In the 1970s Professor Lawrence Stenhouse documented principles that emphasised empiricism and process for selecting content, developing teaching strategies, sequencing learning experiences and assessing student strengths and weaknesses, with an emphasis on what actually happens in the classroom and what teachers do to prepare and evaluate rather than on the behavioural objectives and tight hierarchical learning tasks associated with the objective approach.[48,49] This work has been very informative to our investigation and provides a key basis for curriculum renewal. In 1975 Stenhouse presented a number of principles in a seminal text on curriculum development that aligned with Tyler, namely:[50]

- Formulating the intention: A formulation of the intention or aim of the curriculum which is accessible to critical scrutiny,

- Selecting learning experiences: A principle for the selection of content – what is to be learned and taught,
- Organising learning experiences: Principles for the development of a teaching strategy – how it is to be learned and taught,
- Evaluating: Principles on which to study and evaluate the progress of students and teachers.

Stenhouse, on curriculum renewal as a recipe: 'A curriculum, like the recipe for a dish, is first imagined as a possibility, then the subject of experiment. The recipe offered publicly is in a sense a report on the experiment. Similarly, a curriculum should be grounded in practice. It is an attempt to describe the work observed in classrooms that it is adequately communicated to teachers and others. Finally, within limits, a recipe can be varied according to taste. So can a curriculum.'[51]

As with Wheeler's model, the Stenhouse principles provided the starting point for development of the deliberative and dynamic model that allowed our team to then trial each element and its dynamic components with partners around the world to provide the level of depth and rigour needed to support universities to undertake such a process.

Reynolds and Skilbeck (1970s): A process- and practice-focused approach can lead to very different methods for curriculum renewal depending on the quality and enthusiasm of the educators, and can result in a high degree of variety in content. A popular criticism of the approach was that it did not place a strong enough emphasis upon context – that is, the environment ('milieu') within which the curriculum takes place.[52] In the 1970s, researchers John Reynolds and Malcolm Skilbeck were among the first curriculum writers to recognise the importance of context in the curriculum renewal process.[53]

Their 'Situational Model' emphasises considering external and internal factors, summarised in Table 3.2 and provides key aspects of this process, namely:

1. Situational analysis;
2. Goal formulation;
3. Program building;
4. Interpretation and implementation; and
5. Monitoring, assessment, feedback, reconstruction.

TABLE 3.2 Reynolds and Skilbeck's Situational Analysis Model[54]

External factors	*Internal factors*
Societal expectations and changes	Students
Expectations of employers	Teachers
Community assumptions and values	Institutional ethos and structures
Nature of subject disciplines	Existing resources
Nature of support systems	Problems and shortcomings in the existing curriculum
Expected flow of resources	

Since this leading work the curriculum literature has expanded with discussion about how context can inform curriculum renewal activities. Tertiary educators have begun to pay attention to address what is taught, how it is taught and how learning is assessed.[55,56] It is on the building blocks of these curriculum theorists and practitioners that we have sought to document the iterative and vibrant nature of curriculum renewal in the 21st century, and the central role of an overarching strategy to drive the curriculum renewal process.

What are the core elements of 'deliberative and dynamic' curriculum renewal?

The model presented in Figure 3.6 is arranged on a scaffold of core 'elements' of curriculum renewal that have been distilled, developed and trialled through collaborations with a number of universities over the last decade, summarised in Table 3.3. As discussed above, there are six essentially 'linear' deliberative elements, accompanied by four 'cyclic' dynamic elements that incorporate internal and external context throughout the process, in a manner that both informs the curriculum and builds a community of learning around the process.

Each of the elements is to some extent intuitive and may already be part of current practices in many universities. The value of the model is the holistic approach to blending a *deliberative process* with a *dynamic approach*. This allows curriculum renewal to be well considered and cost effective, while also being an innovative and exciting journey involving many parts of the university and its key stakeholder groups. In Part 3 of this book, we enquire into each of these elements in more detail, providing guidance for academics on how to optimise their curriculum renewal experiences, depending on their local context.

TABLE 3.3 Summary of elements in the model for deliberative and dynamic curriculum renewal

Element	*Description*
Deliberative elements for curriculum renewal	
Curriculum renewal strategy	At the centre of the process a curriculum renewal strategy provides clarity and focus, potentially comprising program-, department- and institutional-level aspirations, goals and milestones.
Identify graduate attributes	Early in the process, identifying desired graduate attributes at a program level ensures that the curriculum renewal strategy is directly translated into meaningful knowledge and skill sets for graduating students that respond to a range of demands and anticipated trends.
Map learning pathways	Once graduate attributes are identified, mapping ideal learning pathways through a program provides an opportunity for an inclusive process to best present knowledge and skills to students as an integrated component of the program that can be easily tracked, and opportunities for synergies identified.
Audit learning outcomes	Auditing the current program for existing content that delivers on the graduate attributes provides a key benchmark to inform the development of new materials. This helps in identifying priority action areas for curriculum renewal in the program. A key aspect of the process is then to take time to check that the knowledge and skills designed into the learning path deliver on the intended graduate attributes before going ahead and creating such material.

TABLE 3.3 Continued

Element	*Description*
Develop and update curriculum	Following the previous steps, developing and updating courses becomes a prioritised, informed and creative process of designing a learning experience that targets particular graduate attributes assigned to specific courses, threaded through learning pathways.
Implement program	Courses are implemented with clear understanding about the roles of existing and new content, delivery methods and assessment. Evaluation takes place with intentionality, focused on whether the course fulfils expectations in developing particular learning outcomes.
Dynamic elements providing context for curriculum renewal	
Raise awareness and build capacity	Awareness raising (regarding the curriculum renewal process) and capacity-building activities (regarding the ability to participate in the process) are key dynamic activities undertaken with internal and external stakeholders. This may include faculty, staff, students and employers, through activities such as keynote lectures, public addresses, lunchtime seminars, media articles and promoting existing initiatives internally and externally.
Collaborate internally and externally	Internal and external collaborations are encouraged, to inform the curriculum renewal process. External collaboration is critical to ensure the relevance of proposed changes to potential employers, current and future students, and current and future legislative and market environments. Internal collaboration is critical to assist in embedding the process into the program design process. This aspect is of particular focus in Chapter 4.
Continually monitor and evaluate	Monitoring and evaluation encourages continual improvement, within and between each of the steps, and in whole of program delivery. This includes evaluating whether the intentions of the curriculum renewal strategy are being met through the implemented course-level changes and ensuring the curriculum renewal strategy is being adhered to.

PART 2

Strategic transformation through rapid curriculum renewal

4

A WHOLE OF INSTITUTION APPROACH (CURRICULUM HELIX)

Despite decades of calls to respond to environmental issues within higher education, particularly related to climate change, highly regulated disciplines such as engineering have been relatively slow to incorporate sustainability knowledge and skill areas, and are generally poorly prepared to do so. If this process is left to a 'business as usual' approach it could take another two decades for sustainability to be embedded within the curriculum, clearly beyond many scientific calls for action and emerging government and industry targets. Responding to this dilemma, this chapter provides a platform for taking a whole of institution approach to curriculum renewal where knowledge and skills are embedded across higher education institutions. At the heart of this approach is the curriculum renewal strategy (described further in Chapter 5) that can facilitate a collaborative and timely transition.

In the absence of such a strategy there is typically a lot of 'reinventing the wheel' between, and within, institutions over time while they wait for program-wide, department-wide and institution-wide strategies. Clearly, attempting holistic change as part of isolated projects is counter-productive, and also counter-intuitive. Although such efforts create experiences and build the capacity of those involved, this may actually foster an insular approach that involves few people internal and external to the university, hindering systemic efforts. In this chapter we explore the question of how a department transitions from occasional, *ad hoc* experiences in curriculum renewal to undertaking a strategic whole of institution approach that delivers multiple benefits across the university.

Answers to this question will differ for each institution depending on factors such as current processes, governance structures and preferences. However, a number of aspects will be reasonably consistent across institutions, which support the curriculum renewal strategy.

The curriculum helix provides a framework for the curriculum renewal strategy, to allow a strategic and collaborative approach to embedding sustainability into curriculum in a manner that delivers the strongest benefits to the institution.[1] The helix is arranged to focus on six activity streams in four stages including preparatory activities, piloting the curriculum renewal efforts, transferring learnings and achieving integration across programs.

The helical (oscillating) nature of the streams highlights that they are interconnected in undertaking various tasks within the strategy, delivering milestones within each stage. Furthermore, each activity stream will have a role to play as each of the elements of the deliberative and dynamic model are undertaken – indeed a whole of institution approach!

A missing dimension – timely curriculum renewal

This chapter was presented as an invited plenary presentation and paper at the 2012 World Symposium on Sustainable Development at Universities,[2] with thanks to the Conference Chair and Editor of the proceedings, Professor Walter Filho, for permission to take excerpts of the paper and adapt text for use within this publication. The chapter builds on Chapter 10 of *The Natural Advantage of Nations* (Hargroves and Smith[3]), applying this work to the higher education sector.

> Whole-institution approaches – which require the active engagement of multiple actors in the joint redesign of basic operations, processes and relationships – are increasingly put forward as a mechanism for making meaningful progress towards sustainability . . . visionary leadership, social networking, new forms of research and high levels of participation [are] key elements of such approaches.[4]
>
> 2012 Full-length Report on the UN Decade of Education for Sustainable Development

Considering the timeframe for curriculum renewal

> Professor John Franklin Bobbit, in his preface to the seminal 1971 publication *The Curriculum*, stated that 'As the world presses eagerly forward toward the accomplishment of new things, education also must advance no less swiftly . . . Education must take a pace set, not by itself, but by social progress.'[5]

The topic of 'timely' curriculum renewal is almost absent in the curriculum theory literature, with researchers considering the question of 'what' should be included and 'how' this should occur, rather than 'by when'. Looking back as far as the early 1900s, we can see that Bobbitt understood that curriculum renewal needed to keep pace with social progress. In the 1970s, Stenhouse noted that curriculum renewal is about keeping curriculum up to date with developments in knowledge, teaching techniques and teaching materials. However, using his cooking analogy, he did not explore how the ingredients or instructions might change for the recipe according to how much time is available for cooking! This presents a challenge for addressing the time imperative identified in Chapter 1, where there is not a structure for ensuring that curriculum renewal can be undertaken over set timeframes, with timeframes related to many of society's challenges being much shorter than the historic timeframe afforded to curriculum renewal efforts.

In the absence of guidance as to how to accelerate curriculum renewal, three patterns or 'phases' of curriculum renewal have been observed, each with varying timeframes for implementation. Perhaps one or more will resonate with your experiences:

- Opportunistic, faculty driven, *ad hoc* exploration ('*ad hoc*'): This type of curriculum renewal appears to comprise a number of self-initiated and -led activities involving one or more educators who have a particular interest in sustainability. This activity appears to be driven by individuals and small groups, continuing for the duration of their personal investment in such activities. Such initiatives often appear to be ceased or curtailed due to larger institutional programming or resourcing decisions that do not recognise a need to continue the innovation in this area (e.g. course consolidation or cancellation).
- Department-led, flagship initiatives ('flagship'): This type of curriculum renewal appears to be led by the department, where a decision is made to offer one or more new courses within the program on the topic or to undertake a review, etc. This is often referred to as undertaking 'flagship' initiatives. In many examples, this follows a period of *ad hoc* exploration, but some literature also suggests that it can be a new initiative. For example, this type of renewal towards education for sustainability may include developing one or more specific courses targeted at one or more dimensions of sustainability, such as 'sustainable energy systems', or creating a minor specialisation in the topic.
- A period of full integration ('integration'): This type of curriculum renewal appears to be evident only in examples where strategic approaches have been adopted for systematic integration of new content within every course in the program; few examples are available in the literature. In these examples, faculty within the department are tasked with assessing and advising how to proceed with implementing a full curriculum transition, resulting in a period of planned and systematic integration of sustainability content into the curriculum.

We have observed that these types of activities are operating largely in sequence, starting with *ad hoc*, moving into flagship, and in a few cases expanding to an integrated approach, which could be said to broadly define the current process of curriculum renewal. Over the phases, there is a greater focus on the linear and cyclic elements (Table 3.3). On occasion, the phases (particularly the third phase) may have specific timeframes and at times the phases overlap. However, while these phases of curriculum renewal have the potential to influence the pace of change, they do not appear to be 'accelerating mechanisms'. For example, even in urgent and challenging times, without direction the progression from *ad hoc*, through flagship, to achieve integration could theoretically still stretch over many years. Looking back to the linear and cyclic elements discussed earlier in this chapter, the deliberative and dynamic model for curriculum renewal could assist with a systematic approach, but with the current 'strategy' focus on how rather than by when, would not drive desired outcomes within a desired timeframe. In our search for methods to achieve 'rapid' curriculum renewal, we have not seen an emergence of planned consideration of how to deliver outcomes within certain timeframes. Clearly there is a missing dimension of 'timely' curriculum renewal, which requires a timeframe need to be created somewhere in the process, and then supporting mechanisms to tailor the three phases discussed above.[6]

Timeframes for curriculum renewal are determined by many factors beyond curriculum considerations. These include, for example, budget allocation, faculty resourcing, accreditation processes, institutional change processes, employer expectations, student demand and so on. Indeed, our experience in implementing the various elements of the model in Chapter 3 is that there is a need for involvement across the institution beyond the curriculum designers and the educators. Such a 'whole of institution' approach at first may seem like it will slow down the process, and if not undertaken effectively it can; however, we propose that it can be a powerful way to accelerate and embed the process into the institution.

With this context in mind, we now focus on a whole of institution approach to curriculum renewal that can facilitate a process of 'timely' curriculum renewal. Through this approach we discuss the possibility for 'rapid' curriculum renewal, a special case of curriculum renewal suitable for circumstances where there is a significant backlog of content that needs to be embedded within curriculum in a contracted period of time. This includes strategy considerations (expanded on in Chapter 5) and staging considerations (expanded on in Chapter 6).

Harnessing the curriculum renewal strategy

> Universities educate most of the people who develop and manage society's institutions. For this reason, universities bear profound responsibilities to increase the awareness, knowledge, technologies, and tools to create an environmentally sustainable future.[7]
>
> Association of University Leaders for a Sustainable Future, 2003

In Chapter 3 we presented the 'deliberative and dynamic' model for curriculum renewal to provide a scaffold for efforts to embed sustainability knowledge and skills into the curriculum. This chapter will focus on the pace with which such renewal is achieved, as the model could be used over any timeframe, running the risk of producing an excellent program that completely misses time imperatives. The key to the timing of the renewal process is the curriculum renewal strategy, and this chapter outlines an approach to developing and implementing such a strategy.

It is now clearly recognised by the higher education sector, and other education sectors, that current curriculum will need to be renewed to embed sustainability in order to continue to meet employer and societal needs. However, the question of what is an appropriate timeframe for such curriculum renewal is complicated by the fact that there is still little pressure for such renewal to be undertaken by the world's government and industry. As such, it is difficult to predict when the level of compliance and action on a range of sustainability-related areas will begin to strengthen, with the literature suggesting that the shift in market and regulatory requirements is likely to ratchet up within the next 5–10 years. With this timeframe in mind, for some countries the call to rapidly renew higher education curriculum for sustainability may be slightly pre-emptive of a mainstream shift and leading institutions are typically acting in relative isolation. However, for other countries, or even cities, significant demand already exists that has led to action, with a growing number of institutions now using one or more of the types of elements in the deliberative and dynamic model.

A process for timely curriculum renewal needs to address a number of issues. In particular it is important to acknowledge the periodic and highly integrated nature of substantial curriculum renewal, which involves academics across a department completing a variety of tasks, supported by staff across the institution, and in collaboration with partners and key stakeholders.[8] The key to such a complex approach is a central strategy to drive the process and ensure momentum is sufficient to meet the intended timeframes and stages. At the level of the institution, there is a need for strategies and structures to incorporate curriculum renewal efforts into existing operational frameworks across the many areas of the institution or department. Departments also need operational capacity to support the transition, including increasing internal professional capacity, and recruiting educators who can assist to develop and deliver the required curriculum. The strategy will also need to include a focus on promotion of such opportunities to potential students and on keeping tabs on anticipated shifts in student enrolment and employer preferences.

From breakdowns to breakthroughs

When discussing with colleagues their experiences of curriculum renewal and sustainability, the question often comes up as to why their efforts are not going as well as they had hoped. When they look at the literature, they see many examples of what can be done, so why are they experiencing problems? The following examples from our many discussions over the last decade (identities withheld) highlight the tenuous nature of new curriculum efforts that are project based, or which rely on individual champions without whole-of-department support:

- 'We used to run a great course in . . . but we just didn't get enough enrolments . . . prospective students just didn't understand what we were offering!'
- 'This initiative has been recognised internationally as a leader in the field . . . but last year during program restructuring it was removed from the curriculum.'
- 'This is an excellent example of integrating sustainability content . . . although the subject is not being run anymore as the convenor left.'
- 'The initiative was very popular with students, but wasn't continued after the project finished.'

Indeed, as we have previously mentioned, many if not most curriculum renewal examples in the literature are based around *ad hoc* trials and pilots driven by individuals or small groups in the absence of a strategy for systemic integration. As a result there is a lot of 'reinventing the wheel' between, and within, institutions over time while they wait for program-wide, department-wide and institution-wide strategies. Clearly, attempting holistic change as part of isolated projects is counter-productive and also counter-intuitive in a world that strives for efficiency. Although attempting such efforts creates experience and builds the capacity of those involved, it may actually foster an insular approach that involves a minimum of other staff internal and external to the university, and can actually hinder systemic efforts.

Our colleagues – no matter where they are around the world – also regularly express frustration about the lack of action. Otherwise energetic individuals express despair about how a range of barriers hinder efforts, such as reluctance to review existing materials by colleagues; a mindset that some academics may retire before the changes are really called for so it is avoided; the lack of time, support and demand for professional development of faculty in the new areas; the challenge of already 'full' curriculum and limited guidance as to how to prioritise current and new materials; a lack of expansion of successful new courses (units) or electives in the area; and the lack of a sense of urgency to update programs. If such barriers are not proactively approached, this runs the risk of academic burn-out, which can lead to academics reducing the effort invested in such activities, changing institutions or even leaving the sector for industry positions.

Typically, breakdowns in mainstreaming education for sustainability boil down to poorly supported isolated actions being undertaken by champions of the cause, in the absence of an overarching strategic plan across the whole of the university.

The Online Guide to Quality and Education for Sustainability, funded by the Higher Education Funding Council of England (HEFCE), through the national Leading Curriculum Change for Sustainability: Strategic Approaches to Quality Enhancement project, was intended to share practice and insights from the initiative about systemically addressing education for sustainable development, including consideration of some of the academic and leadership issues involved. It includes commentary and tools to help institutions develop strategic approaches to education for sustainable development as a cross-cutting curriculum enhancement agenda.[9]

The UK's Higher Education Academy has initiated a 'Green Academy change programme' that supports holistic sustainability change at an institutional level. The success of the first program has led to the recruitment of a further ten institutions for the second program which started in early 2013, resulting in 18% of higher education institutions in England and Wales participating in Green Academy. Initially focusing on education for sustainable development in the curriculum, the program supports innovation and strategically embedding sustainable development across the university experience for both students and staff, incorporating curriculum, research, estates and student activities. The Green Academy enables successful applicants to:

- Engage in transformative initiatives that directly impact on practice through innovative student and staff partnerships;
- Develop and improve institutional policy, practice and curriculum design in education for sustainable development;
- Enhance their understanding of the research evidence base to support institutional change for sustainability;
- Share their expertise, along with positive and negative experiences, with other participants through multi-institutional partnerships;
- Collaborate with the leading researchers and practitioners in ESD who act as 'critical friends' to develop innovative approaches towards sustainability in the higher education sector.

According to the program's coordinator, Simon Kemp,

> Green Academy was the first UK programme to attempt to bring tertiary education experts and universities from across the sector together in the pursuit of innovation and collaboration for education for sustainability. The initial focus on curriculum evolved to collaborative working between students and academics, and across institutions for holistic institutional change for sustainability. In the pursuit of truly sustainable universities the collegial yet competitive model offered by Green Academy is one that has great potential for successful replication in countries across the world.[10]

Open to institutional teams from across the UK, the program also includes a legacy of online resources for others around the world to benefit from, including case studies of participating institutions and program evaluation reports.[11]

The literature and our investigations and trials have highlighted three key lessons for those seeking to renew curriculum for sustainability in a timely manner:

a The process will require a dynamic and integrated approach that involves staff and stakeholders across the university to be managed to deliver tangible benefits.
b There is a need for a central strategy to drive the process and ensure momentum is sufficient to meet the intended timeframes and stages.
c The process will involve a number of actions across the institution involving various staff and stakeholders and needs to be undertaken in a collaborative and synergistic manner.

We have observed a number of strategies being trialled over the last decade, ranging from problem- or project-based learning, to inverted curriculum, and the use of sustainability-related threshold learning concepts.[12] While there is no 'right' or 'wrong' way of proceeding, it is apparent that an acceptance and embracing of integration and non-linearity provides a valuable way to inform efforts to embed sustainability knowledge and skills within the curriculum. So the question really is, how does a department transition from occasional, *ad hoc* experiences in curriculum renewal for sustainability that are hindered by regular breakdowns in momentum, to undertaking a strategic whole of institution approach that delivers multiple benefits across the university?

The answer to this question will differ for each institution depending on the current processes, governance structures and preferences. However, there are a number of aspects that will be reasonably consistent across institutions that will support the curriculum renewal strategy. An example of an approach to bring together such aspects is the The Management Helix for the Sustainable Organisation (sustainability helix) developed by The Natural Edge Project (led by Charlie Hargroves), Natural Capitalism Solutions (led by Hunter Lovins) and partners between 2003 and 2006.[13] The 'sustainability helix' was designed to provide industry with a platform to inform efforts to improve sustainability outcomes in a way that harnessed the strengths across the organisation.

According to Hunter Lovins, 'Companies that begin to engage with sustainability find that they can't do it all at once. The Helix helps them lay out a strategy and prioritize their actions.'[14]

This work with industry provides insight into how to support a curriculum renewal strategy within a university as, like a company within the university, there are significant groups of employees considering sustainable operations, which could potentially link in with curriculum renewal initiatives. After a successful effort to tailor the 'sustainability helix' to the needs of a university (part of a trial with the Griffith University Department of Information Services in 2006),[15] we have focused on creating a hybrid of the model to support the deliberative and dynamic model, as described below.

The curriculum helix for education for sustainability (curriculum helix)

Taking a whole of institution approach

The curriculum helix is intended to provide a framework for the curriculum renewal strategy to allow a strategic and collaborative approach to embedding sustainability into curriculum in a manner that delivers the strongest benefits to the institution.[16] The curriculum helix suggests six key activity streams across the university that are critical to the success of the curriculum renewal strategy, namely:

- Governance and management (G), including the Dean or head of the department or school, and any heads of research and teaching and learning.
- Operations and facilities (O), including any staff employed by the school, or allocated to managing the school's assets, who oversee building operations and maintenance, laboratory and workshop management.
- Teaching and learning (T), including all personnel who are involved in curriculum design, delivery and renewal within the school. This spans from program convenors to individual subject lecturers, sessional staff, teaching support staff (who assist with laboratories, workshops, tutorials, marking, etc.) and guest lecturers who may come from elsewhere in the university, or from outside the university.
- Human resources and culture (H), including any staff employed by the school, or allocated to managing the school's human resources (i.e. annual staff appraisals, contracts, appointments, etc.).
- Marketing and communications (M), including any staff employed by the school, or allocated to managing the marketing of the school's programs and internal and external communications regarding program content and delivery.
- Partnerships and stakeholder engagement (P), including any staff employed by the school, or allocated to managing the interaction with potential employers, industry, alumni, donors, patrons, primary and high schools and the community.

In Figure 4.1, we present the curriculum helix as an integrated model with a strong focus on collaborative and synergistic efforts, rather than the typical siloed organisational arrangements. At the core of these efforts is the curriculum renewal strategy to provide a structure to progress over the defined stages. The curriculum renewal strategy essentially becomes the centre-piece of the institution's transition towards education for sustainability, and will be outlined in detail in Chapter 5. In conversations around such a process, Dr Leith Sharp, Founding Director of the Harvard Green Campus Initiative, reflected that 'The key point of this process is that most of us have to start with the work of building buy-in and learning key lessons while amassing enough institutional support to organize it formally.'[17]

The curriculum helix is arranged to focus on six activity streams (in arbitrary order) along four stages to allow for preparatory activities, piloting the curriculum renewal efforts, transferring learnings and achieving integration across programs. The helical nature of the streams indicates that the various activity streams are interconnected in undertaking the various tasks within the strategy to deliver the milestones as part of each stage.

For example, each of the activity streams would be involved in informing the development of the strategy, which needs to capture the multiple benefits across the streams. Further, as each of the elements of the deliberative and dynamic model are undertaken, each activity stream will have a role to play. It is intended that, using such a staged and integrated approach, the model will support institutions to identify flexible and collaborative ways to transition to education for sustainability.

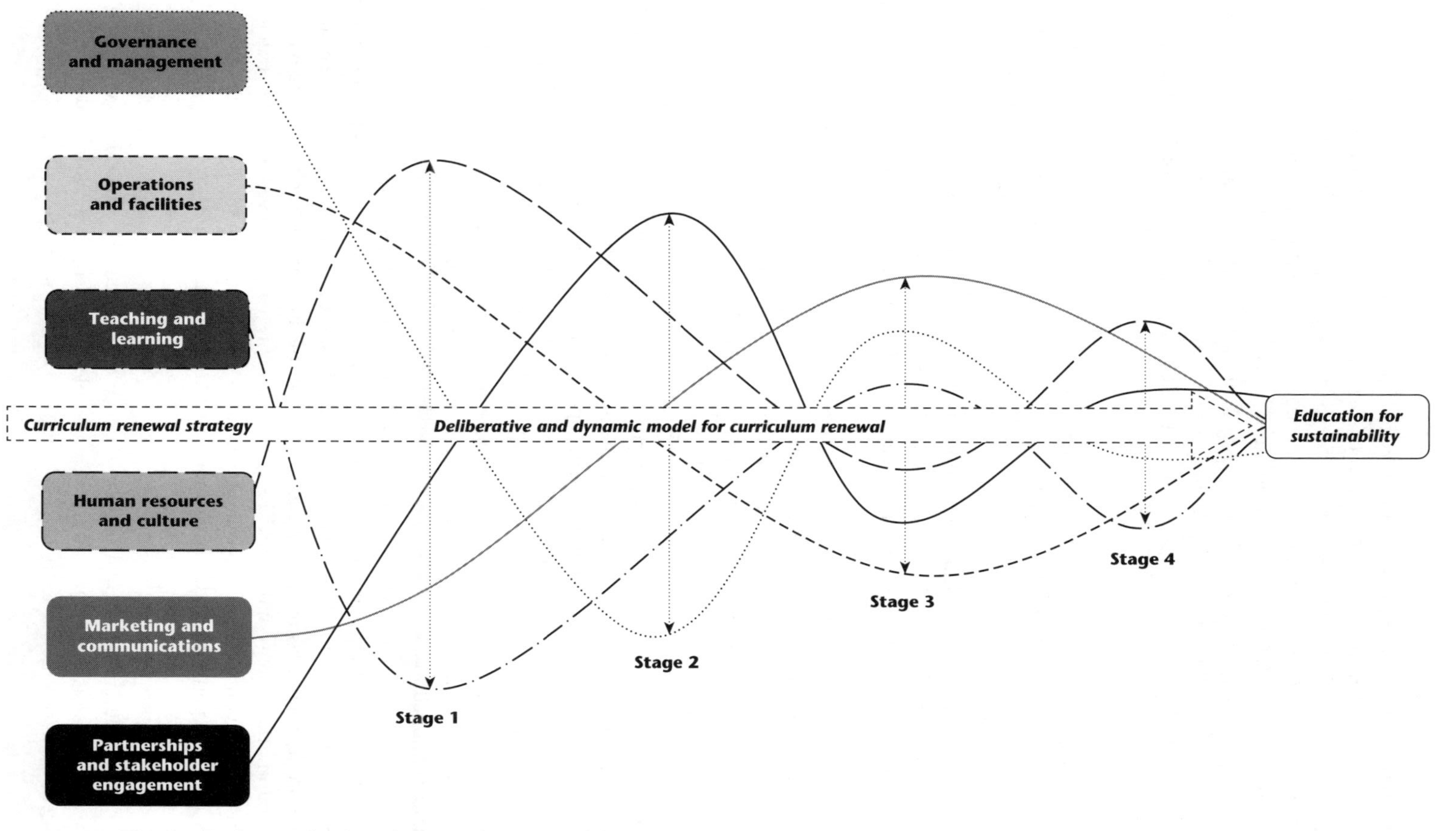

FIGURE 4.1 The Curriculum Helix for Education for Sustainability (Curriculum Helix)

Source: Adapted from 'The Management Helix for the Sustainable Organisation (Sustainability Helix)[18]

Chalmers University of Technology has a long history of education for sustainable development. We have found a whole of institution approach to curriculum renewal a critical ingredient to our continued success in upgrading our programs. Spreading responsibility and creating engagement, both vertically and horizontally in the organization, and efforts targeting both people's capacities and the structures in the organization are critical success factors.

Personal communication with Magdalena Svanström, Associate Professor, Chemical Environmental Science, and Director, Chalmers Learning Centre, Chalmers University of Technology

The curriculum helix provides guidance on how to approach the complex challenge of embedding sustainability into curriculum and provides a clear process that allows the various people involved to understand their role, gain support from others, and provide support to others involved in the process.

- Entering the model from the left along the central arrow, there is an immediate need for an overarching strategy to be developed, requiring some 'front-end' planning regarding aspects such as timing, resources, staffing, budget and risk management, before the whole of the institution is engaged in the process (see Chapter 5).
- Moving along the central arrow, there are a series of stages within the model with each of the activity streams making a contribution to each stage to ensure that curriculum renewal is achieved over the intended period. This includes regular review of the strategy to allow for improvements and adjustments. It is intended that the strategy will harness the deliberative and dynamic model for curriculum renewal as appropriate to the institution.
- Considering the intersections across the helix, there is a framework of actions and actors to use as the basis of the strategy to allow for clear assignment of tasks and performance indicators. Each intersection provides guidance to the various departments and groups across the university as to their role, along with guidance as to how to support and be supported by the other streams (see Appendix).
- Following the lines across the helix highlights how each of the activity streams will interact appropriately with each of the others. Some nodes will call for greater or lesser involvement. Hence, the various activity streams are interconnected in undertaking the tasks that deliver the milestones for each stage, providing the opportunity for streamlining activities, rapidly sharing learnings, and reducing duplication of efforts.
- Considering the overall helix process, it is important to undertake appropriate monitoring and evaluation so that efforts are well informed and those with limited benefits are targeted for either improvement or removal from the overall process. Such a process will also allow clear and transparent reporting of progress by the various actors across the institution. Such reporting will be important to inform the strategy and ensure the greatest value is being achieved from the strategy. This could include a mixture of qualitative and quantitative measures, from survey data through to focus groups and self-reporting on addressing audit recommendations.

The key to achieving education for sustainability in a manner that delivers maximum benefit is to strategically involve actors across the institution in complementary and synergistic ways. This involves the identification of key 'activity streams' and the assignment of a set of 'stages' as part of the curriculum renewal strategy. Within this framework various curriculum renewal efforts can be undertaken as outlined in the following chapters.

A note for early adopters of the curriculum helix

The challenge is to compress the timeframe for curriculum renewal through the use of progressive tasks and milestones, to ensure that the outcome is rapid (i.e. two accreditation cycles, or six to eight years) rather than standard curriculum renewal which can span up to four accreditation cycles (i.e. two decades). Even with clear staging there is still the potential for setting long time horizons for each of the stages, if a business-as-usual approach is taken. Whether or not the model will be useful for accelerating curriculum renewal depends on whether there are strong enough catalysts that set clear timeframes for the transition. In engineering, for example, although accreditation is currently one of the strongest potential drivers of change, the reality is that accreditation is still relatively weak in its expectations of education for sustainability.

The process relies on institutional leadership and support to establish timeframes for curriculum renewal within the institution, and to set the momentum and reduce the many organisational barriers to rapidly embedding new content into curriculum. The form of such support will be unique for each institution, including the provision of funding, marketing and resources, and flexibility in rules regarding developing new courses and modifying existing courses.

According to a study undertaken by the American Campus Sustainability Assessment Project of key characteristics of leading institutions,

> First, these 'sustainability leaders' have adopted serious strategies for systematically addressing the sustainability of the institution. They have policies stating their commitment to sustainability goals, and they have specific plans in place that explain how they intend to achieve them. Second, these institutions have provided the resources needed to implement their sustainability plans. They hire staff, form committees, allocate budgets, and show clear administrative support for sustainability initiatives. Third, these sustainability leaders know where they have been, where they are, and where they are headed in terms of sustainability. They measure and track their progress toward sustainability, and regularly meet and update goals and targets.[19]

Management will need to consider a range of incentive mechanisms to gain and maintain such leadership and momentum throughout the curriculum renewal process, highlighted in Table 4.1. Clearly, departmental leadership needs to be present to action the timeframes set by catalysts such as changes to accreditation requirements, shifts in regulation and policy that will affect the role played in society by graduates, and shifts in employer needs for graduates. A 2008 Australian report on addressing the supply and quality of engineering graduates for the new century observed four supporting actions that were common in institutions facilitating significant change, namely vision, leadership, stakeholder engagement and resources.[20]

TABLE 4.1 Leadership considerations for curriculum renewal

Leadership action	*Description*
Strong commitment to action	Making a strong commitment from department executive level to the process, including communicating the intent of the activities to adequately inform faculty of the implications of the process.
Internal champions	Supporting internal 'champions' to assist in communicating how the curriculum renewal efforts will affect faculty and the role that they can play. These champions are a critical part of the strategy as they create enthusiasm and are able to discuss various aspects of the process in appropriate terms and language.
Formal requirement for involvement	Including a formal request for faculty to participate in the curriculum renewal process, with a clear statement of the timelines and expectations. This may align with accreditation requirements or as part of a particular differentiation strategy by the university or department. This may include mandatory attendance at graduate attribute workshops.
Recognition of strengths	Recognising faculty and staff who already have experience in education for sustainability and encouraging them to share their experiences with others (e.g. service award, invitation to present, and representation on various boards and committees, internal and external).
Flexible workload allocations	Providing research assistance, teaching buy-out, or flexibility in appointments (e.g. research/teaching/service proportions) for staff to actively contribute to the curriculum renewal process.
Seed funding provisions	Providing seed funding opportunities (e.g. internal grants) to investigate research opportunities in this area to both inform the process and allow for academic recognition of the work. This topic is of growing interest in a number of international journals (as highlighted previously).
Flexible curriculum development	Permitting and encouraging faculty to appropriately use existing academically rigorous course materials available online under appropriate licensing arrangements.
Support for professional development	Providing opportunities and financial resources for professional development in the new area as a portfolio strengthening activity, potentially linked to specific requirements (e.g. becoming familiar with the topic area, identifying aspects that can be immediately incorporated into existing curriculum, identifying material in demand for postgraduate and professional development courses, or attracting international students faced with sustainable development challenges).

The intention of the development of both the deliberative and dynamic model and the curriculum helix is to support rapid curriculum renewal for sustainable development in all university programs. The models themselves will not ensure such a rapid process. Rather, there are a number of priority actions to create conditions for accelerated action. Institutional leadership and support is critical to ensure that momentum is maintained through implementation. Departments seeking to rapidly embed sustainability into curriculum could consider the model and its conceptual areas at the outset, to develop a strategy that is suited to its institutional and cultural context. Those institutions considering this process without strong external or internal pressure may find it more difficult to set substantial short-term goals and allocate budget for curriculum renewal such as graduate attribute mapping and curriculum auditing. Furthermore, these institutions may find it difficult to secure government and industry financial support for innovative renewal initiatives, relying on harnessing momentum from existing institutional projects.

In the remainder of this book we seek to guide those undertaking curriculum renewal informed by the two models with a particular focus on how such efforts can be accelerated. As each of the models is conceptual rather than prescriptive, they can be threaded within existing institutional frameworks and processes, providing flexibility to include the variety of cultural and organisational contexts in higher education.

Within this context, in Chapter 5 we discuss the challenge of creating a curriculum renewal strategy in the order of 6–8 years rather than the standard timeframe for such activity of 15–20 years. This strategy then becomes the centrepiece of the transition, where the aim is to align each of the activity streams to the implementation of the deliberative and dynamic curriculum model. Our conversation focuses on considerations to address the interwoven strands in Figure 4.1, where different activity streams within the department will be working on tasks together through the transition, each supporting others and being assisted to meet the milestone requirements.

5

INFORMING A CURRICULUM RENEWAL STRATEGY

Imagine for a moment that you are the head of an institution's executive and are convinced of the need to commit to education for sustainability. You understand that this will call for changes across each of your program offerings while retaining industry relevance, meeting accreditation requirements and continuing to attract students. Although your institution is worried about taking risks, it is also worried about not keeping up with competitors and potentially losing students and staff to other institutions.

There are also a number of significant operating challenges that demand your attention: research performance, student enrolment, capital works, infrastructure maintenance, faculty resourcing, workload allocations, and so on. As usual, these are all competing for priority. In short, although you have certainty that something has to be done, and clarity about the risks of not acting soon and the benefits of doing so, how will you develop a strategy for something as ambitious as achieving program-wide curriculum renewal? Perhaps it is without precedent.

Given the overwhelming evidence for the need to capacity build professionals to contribute to achieving sustainability in the coming decades, universities around the world are now posed with the question of 'how far . . . and how fast . . . do we proceed to transition to education for sustainability?'. This is a particularly daunting consideration, given the size of the scope, the short timeframe for action, and that there is little to no precedent or guidance.

The manner in which these questions are answered will heavily influence the quality of the response by the institution. Not only will it have wide-reaching budget and resourcing implications, it will also affect the level of risk and reward the institution achieves over time. In reality, a top-level commitment that provides a list of outcomes to achieve without a whole of organisation strategy is unlikely to compete effectively with other immediate, day-to-day challenges and bureaucratic pressures that face the institution, and may be overtaken by other agendas.

Education institutions are complex interconnected human systems that require a steady hand to steer towards comprehensive reform. When beginning to approach the questions of 'how far' and 'how fast' we present a number of considerations that will require some investigation, in particular related to external trends and internal capacity. This and the subsequent chapters provide guidance on key considerations for any institution seeking to develop a strategy to deliver education for sustainability, in a timely manner.

Shifting from 'business as usual'

The focus of this chapter is to consider how a strategic approach can be developed to achieve a transition to education for sustainability across entire institutions. Building on the deliberative and dynamic model presented in Chapter 3, and harnessing the curriculum helix presented in Chapter 4, in this chapter we focus on the cornerstone of each model, the 'curriculum renewal strategy'. Although the majority of examples are drawn from engineering education, the model could be applied to any discipline or professional field. The strategy is intended to set the timeframe for curriculum renewal to occur, support the institution to harnesses the existing strengths and rapidly develop new areas.

> Considering the evolution of school curriculum renewal processes in the UK in the late 1970s, Whitfield concluded that efficient curriculum reform cannot be carried out in the absence of clearly formulated goals. Without such goals, Whitfield reflects that difficulties can arise over establishing priorities, and frustrations and misunderstandings can develop between the agents of change, resulting in piecemeal and incoherent innovation. Whitfield was also concerned that effective mechanisms were not being set up to continuously appraise and modify 'new curricula of today' so that they would not become the 'old curricula of tomorrow'. As disciplines change in terms of content and insight, Whitfield surmised, so should educational objectives, and so the continual cycle of renewal must continue, otherwise curriculum will become stagnant.[1]

Understanding the typical approach

As discussed previously and shown schematically in Figure 5.1, typical forms of curriculum renewal can be grouped into three categories that despite intentions for them to follow on from each other, often don't in practice. The decreasing thickness of the arrows is intended to signify the reduced level of activity as the initiative moves from individual champions to wider involvement, with a very small proportion of initiatives ending up being fully integrated into the fabric of the programs offered. This is mainly due to the momentum developed by individual champions being dissipated due to the lack of a strategic approach across the department or institution.

In this case, the curriculum renewal process begins with a period of *ad hoc* exploration by one or more academics who have a particular interest in sustainability, until for one reason or another the initiative is noticed by the department or funders. Depending on variables such as budget and staff

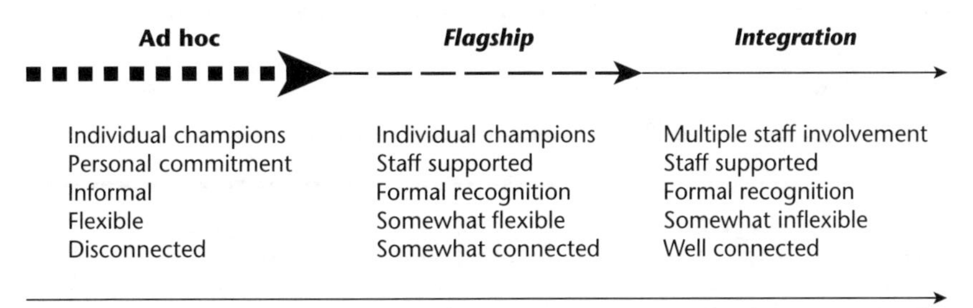

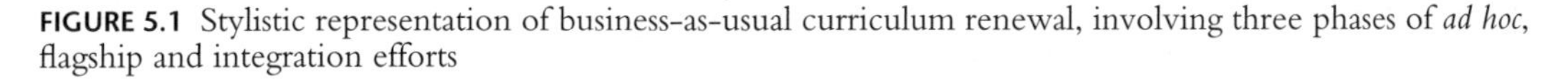

FIGURE 5.1 Stylistic representation of business-as-usual curriculum renewal, involving three phases of *ad hoc*, flagship and integration efforts

availability, this may then result in an expansion of the efforts of the champions to create, say, a new course (unit), a flagship course, on sustainability to be included in the overall program. For a minority of institutions where strategic approaches are adopted for systematic integration of sustainability content, or where individual champions persist, these flagship efforts may lead to a period of gradual integration. However, without a sense of urgency and a strategic approach that harnesses resources, experience and enthusiasm across the university, even in the most proactive examples this process can take in the order of one to two decades.

A paper comparing efforts by Chalmers University, Delft University and UPC-Spain to embed sustainability into their educational programs[2] concludes that the complex nature of changing needs requires strong and long-term strategies. The authors note that the three universities appear to be converging on developing comparable strategies, based on compulsory courses for all specialisation tracks as masters and minors. Moreover, they appear to be pursuing a progressive deep curriculum revision in order to embed sustainability in all programs.[3]

Accelerating the typical approach

The challenge for education for sustainability is to contract these three phases from a standard timeframe of one to two decades to within a window of six to eight years, while capturing and maintaining the innovation, creativity and passion of academics who are already engaging in such initiatives. To facilitate an accelerated program of curriculum renewal, the scope of the strategy is likely to be in the form of a multiple-track 'top-down', 'bottom-up' and 'middle-out' process, encouraging innovation and ownership while being clear about roles and responsibilities. Acknowledging current patterns of curriculum renewal (Figure 5.1) and considering the opportunity for a systemic 'plan-do-check-act' approach discussed in Chapter 4, we propose an alternative approach.

In the following schematic (Figure 5.2), we show how curriculum innovations in all three phases (*ad hoc*, flagship, integration) can be prepared for in Stage 1 and explored in Stage 2 of the Curriculum Helix. Initiatives with the potential for embedding into the curriculum are then tested and piloted in Stage 3, and – if successful – are fully integrated within the program in Stage 4. The thickness of arrows indicates the relative focus, shifting over time.

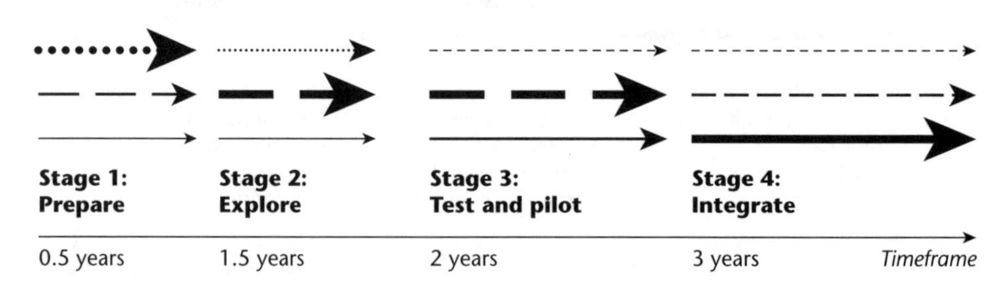

FIGURE 5.2 Stylistic representation of rapid curriculum renewal, contracting and re-prioritising the three phases of *ad hoc*, flagship and integration efforts

With this approach, assuming that the first year of renewed curriculum can be rolled out at the three-year mark and keep up with the student intake from that year (i.e. over their four- to five-year program), then the first graduates from the renewed program could theoretically be produced six years after commencing curriculum renewal. This is an ambitious – but we think possible – goal, which we discuss in the following pages.

As with any strategic planning there are a number of preparatory tasks to inform decisions about the form and scope of the strategy. This is particularly critical in developing a curriculum renewal strategy. Considering the literature, there is an overwhelming number of possible considerations and components of such a strategy, such as:

- Committing management support
- Including in graduate attribute list
- Recruiting faculty with expertise
- Providing training and capacity building
- Hosting topical event/s
- Understanding 'hot topic' areas
- Providing access to web-based courses
- Fostering interdisciplinary networks
- Providing financial assistance
- Creating a working party
- Setting future targets
- Creating a clear timeline
- Permitting workload discussions
- Providing seed funds for new research
- Providing seed funds for teaching research
- Harnessing other institutional overhauls
- Identifying and using modular content
- Investigating graduate career options
- Directly involving potential employers
- Engaging external support for advice

Given the steady increase in community awareness of sustainability issues over the last decade, a number of institutions have undertaken some or all of the components above to some degree. Perhaps some may need revisiting, or additional attention is needed in some discipline areas to provide a solid platform throughout the organisation to move forward. Working with partners over the last decade to assist them to incorporate the TNEP books and extensive freely available education resources into courses, we have often been asked how to approach the embedding of such materials across entire programs. What we have learned through informing and participating in these trials and initiatives is that a strategic approach is critical. Each institution will have staff who are very capable of creating the curriculum renewal strategy in line with the institution's own way of creating such strategies, with a range of differing formats and structures, hence our intention is not to dictate how this should be done but rather to inform efforts.

In this chapter we explore key considerations and provide samples of core strategic areas across the activity streams and stages of the strategy. The initial focus is on assisting institutions to consider 'how far' and 'how fast' it may proceed with the process of curriculum renewal, what the program offerings may look like and how this can be communicated across the departments.

Considering 'how far' and 'how fast'

What are the external trends?

Before investing time and resources into the potentially significant endeavour of creating a strategy for curriculum renewal, it is important to first get an idea of the 'lay of the land'. Getting an idea of things like the current level of student and employer demand, the likelihood of future shifts in accreditation requirements, the potential for shifting national research priorities in the area, and the level of achievement by competitors will allow greater clarity when considering the pace and scope of the strategy. This is best done as a quiet reconnaissance to clarify the reality of the situation across a number of key areas, to develop what we have dubbed the 'EfS Sit-Rep'. This can be done by asking a series of questions such as:

- Are there any current or likely shifts in legislation and policy towards sustainability?
- What is the current approach by competing institutions and programs (locally, nationally and internationally)?
- Are there current accreditation requirements related to sustainability?
- If so, are there signs that these will become more stringent in future rounds?
- Are there opportunities for new areas of research that may lead to top journal publications in this emerging area?
- Are there opportunities for access to new research grants and funding to support sustainability research or teaching?
- Are there signs that employers are increasing their interest in graduates with sustainability knowledge and skills?
- What new graduate employment opportunities are opening up in the various topics related to sustainability?
- What sort of media attention is being given to the field and its various topics?
- What is the level of existing student interest in the topic?

The findings of each of these questions have the potential to significantly affect the scope and scale of the curriculum renewal strategy. To ensure a rigorous approach to the development of the situation report we suggest the use of a systematic process that looks across society for important information to guide the strategy. The TNEP book *The Natural Advantage of Nations: Business opportunities, innovation and governance in the 21st century* presented such a model for a 'whole-of-society' approach, as shown in Figure 5.3.

Considering each of the major areas of society provides a valuable insight into the context that the institution operates in, with examples shown in Table 5.1. For instance, although accreditation requirements may be clear for the department, it may also want to ensure that decisions on 'how far' and 'how fast' the curriculum renewal process should proceed are in alignment with market and regulatory requirements. An advisory panel may be useful for this review process, which could be an extension of the role of advisory panels used for other functions such as strategic direction and accreditation reviews.

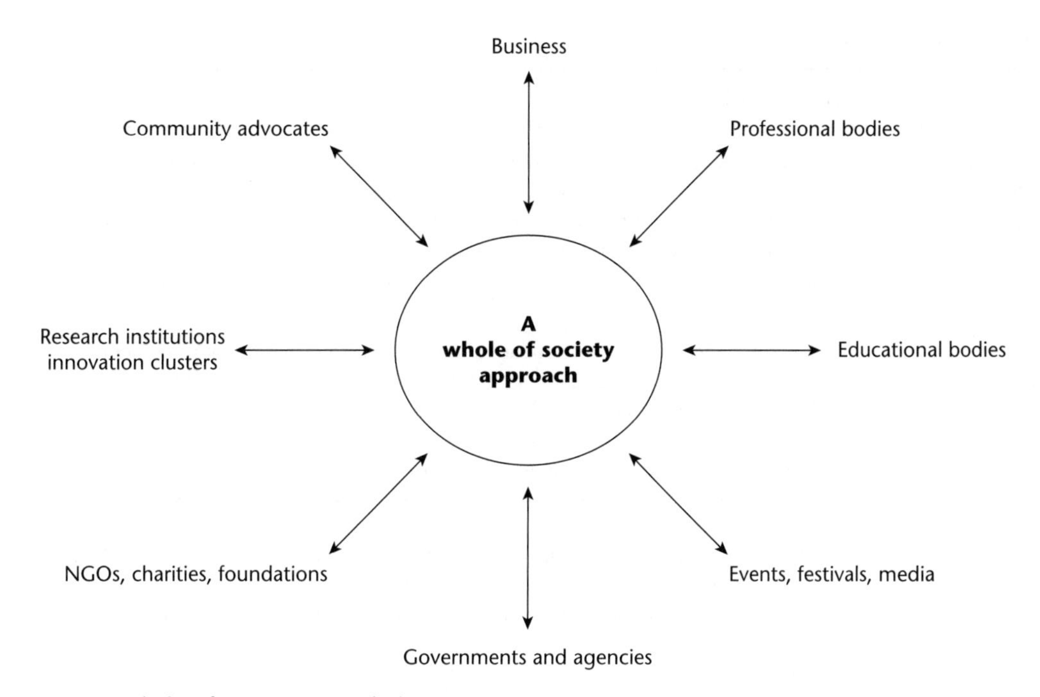

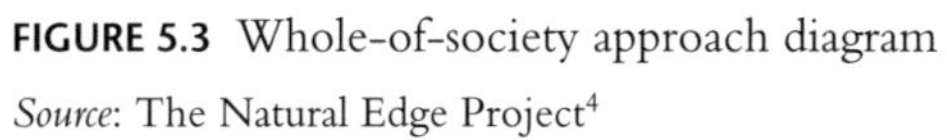

FIGURE 5.3 Whole-of-society approach diagram

Source: The Natural Edge Project[4]

TABLE 5.1 Sample questions related to external trends, using a whole-of-society approach

Sector	*Sample questions*
Business	• What current and future sustainability-related graduate employment opportunities are there, and are these anticipated to increase? • What topics are of the most interest to business and industry in the short term? • Are employers undertaking in-house graduate development in areas of sustainability? • Do employers include sustainability competency in recruitment processes for graduates? • What new areas are opening up related to sustainability in sectors serviced by our graduates?
Professional bodies	• What program accreditation requirements are expected in the upcoming rounds related to sustainability and associated topics? • Are there signs that such coverage will increase in future accreditation rounds, nationally and internationally? • Do professional body ethics statements (or equivalent) make mention of sustainability and associated topics? • Do professional bodies offer professional development in the area of sustainability, and are they well attended by industry?
Educational bodies	• What is being taught about this field, in primary and secondary school education? • Is there evidence that such coverage is set to increase in the future? • What are private training providers offering in this field, as professional development? • What are competing universities doing locally, elsewhere in this country and internationally?

TABLE 5.1 Continued

Sector	*Sample questions*
Events, festivals, media	• What coverage is there in the popular media about shifting requirements for professionals in sustainability and associate topics? • What coverage is sustainability getting at conferences, workshops and events that staff, employers and graduates may be attending? • What opportunities are there for refereed conference papers at events related to sustainability?
Governments and agencies	• Is there evidence that government policy will shift to include greater requirement for sustainability and associated topics in professional practice? • If so, what areas are anticipated to shift in the short term? • Are requirements related to sustainability and associate topics being included in qualifications for government grants and funding? • Are there new government grants available related to sustainability research and teaching? • How does sustainability feature in national research topics and among successful top-level research grant recipients such as the ARC programs? • Is there support for increasing education for sustainability available from national or state governments? • Is there a call for government to be advised by the research community on areas of sustainability? • Is there evidence of in-house capacity building in government agencies on sustainability? • Is sustainability a strong political issue locally, state and federally? • Do government agencies provide community education materials and programs in sustainability?
NGOs, charities, foundations	• What actions are these groups calling for from government, industry and academia related to sustainability and associated topics? • Are NGOs offering professional development courses in sustainability and are they well attended? • Are there specific areas of sustainability on which the NGOs are focusing lobbying efforts that may lead to a greater focus by industry and government? • Do charities and foundations require evidence of a focus on sustainability in grant applications, and if so, in what areas?
Research institutions, innovation clusters	• Are competing institutions gaining a reputation for advancing a focus on sustainability through research agendas? • What type of research institutes and centres are being formed to focus on sustainability and associated topics? • Are innovation clusters and other research collaborations being formed to focus on sustainability and associated topics?
Community advocates	• What are leading community advocates saying about future skill sets related to sustainability and associated topics? • What community programs are gaining support related to sustainability?

What is our level of readiness?

As well as developing an understanding of the external trends, it will be important to balance this knowledge with the actual internal strength in delivering education for sustainability. Getting an idea of things like the current coverage of sustainability in courses, the strength of sustainability teaching across departments, examples of successful new sustainability components of courses or new courses, will allow greater clarity as to the level of readiness of the institution to respond to the external situation explored above. This is best done again as a quiet reconnaissance to clarify the reality of the level of readiness across the institution. This can be done by asking a series of questions, some of which might interest your institution to develop what we have dubbed the 'EfS Loc-Stat':

- What is the general level of coverage across the various programs (without considering specific courses)?
- What is the level of senior executive and department head commitment to sustainability?
- Is this reflected in the university or department's plans or policy statements?
- What is the existing level of interest and capacity among the academic, research and general staff?
- What areas of expertise are there among existing faculty and elsewhere in the university?
- What gaps exist in capacity in existing faculty?
- Do we have the organisational capacity to undertake a department- or institution-wide initiative to renew curriculum?
- Are we losing faculty to other institutions with stronger efforts in this area?
- Are we attracting leading staff due to our efforts in this area?
- Are we considering such abilities in faculty recruitment?

To provide an example of this internal enquiry we use the six institutional 'activity streams' nominated as part of the 'curriculum helix' in Chapter 4, i.e. governance and management, operations and facilities, teaching and learning, human resources and culture, marketing and communications, and partnerships and stakeholder engagement. Samples of questions for each part of the diagram are presented in Table 5.2.

TABLE 5.2 Sample questions related to internal capacity across the 'Curriculum Helix'

Activity stream	*Sample queries*
Governance and management	• What is the level of senior executive and department heads' commitment? • Does the academic plan, or any institutional commitment/policy statement address education for sustainability? • Are suitable budgets available to support efforts? • What is happening elsewhere in this institution that may be leveraged? • Is there a process for managing curriculum renewal?
Operations and facilities	• What examples of sustainable development do we have on campus? • Do we have sustainability practitioners amongst faculty?
Teaching and learning	• What existing expertise do we have amongst our staff with respect to teachers and researchers? • What is the existing level of interest among faculty? • What current efforts and collaborations are underway? • Is there a process to respond to accreditation changes in curriculum?

TABLE 5.2 Continued

Activity stream	*Sample queries*
Human resources and culture	• Does the staff performance review process encourage action in this area? • Are there internal courses or capacity building available for curriculum renewal and/or sustainability? • Do the promotion pathways encourage actions?
Marketing and communications	• What sort of media attention is being given to the field, our programs and our competitors' efforts? • Have we communicated our achievements to-date in this area? • If so, what level of response did we receive from students and employers?
Partnerships and stakeholder engagement	• Are we engaging institutional stakeholders in discussing the level of integration at a degree program level? • Are we discussing the transition with current and former students? • Can this process cultivate research donors and grant providers?

It is important to check the alignment of resources and intentions for the proposed curriculum renewal with the institutional vision and initiatives underway to avoid issues as the transition occurs. Fundamentally, the scope of the strategy and budget need to align with the breadth and depth of the vision for curriculum renewal. In our experiences, at some point early in the process it has been enormously useful to scan institutional documents for signals that could align with the proposed curriculum renewal work. From declarations by senior management such as the Chancellor or Vice Chancellor, through to corporate plans and annual reporting tools, there are likely to be a number of 'hooks' on which the proposed curriculum renewal process can be attached, to ensure that it has a place within the institutional framework. Further, these hooks are often attached to budgets that may have funds to support the proposed activities.

In short, a strategy for curriculum renewal that is made in isolation of the broader institutional priorities and directions leaves itself vulnerable to competing staff resourcing pressures and budget constraints.

Once the external trends and internal level of readiness investigations have been completed, senior management will have a good idea of staff and budget implications and availability, and should be able to decide 'how far' and 'how fast' the transition will occur.

If consultants have been contracted to develop, or assist in the development of, the strategy, then the following considerations may increase the useability of the resultant documentation:

- The consultants should spend time within the department wherever possible for the duration of the contract, potentially sharing office and other facilities.

- Adequate time should be given within the contract to permit the consultants to become familiar with the institutional language, documentation and reporting processes.
- At key points during the strategy development, regardless of who develops the document, key staff need to be involved in formal review of the tasks, roles and responsibilities, to assist with buy-in and checking for cultural context and alignment.

Considering options for program renewal efforts

In any large-scale curriculum renewal process there will be institutional expectations regarding the 'look and feel' of the programs. For example, an institution may market itself as a leader in problem-based learning, mix-and-match programs, highly structured programs, and so on. Hence, any efforts to incorporate sustainability into programs will need to be undertaken within this context. At a more fundamental level, any curriculum renewal efforts will need to consider existing constraints and opportunities around budgets, approval processes and timeframes, rollout procedures and appointments.

It is therefore important to be clear on institutional expectations for what the renewed curriculum should look like. In the following paragraphs we reflect on our own direct experiences and those of our colleagues in the field, who have been trialling various forms of program offerings over the last two decades. In this chapter we focus on three structural considerations that have significant implications for curriculum renewal, namely, new niche degrees or updating existing degrees; the use of focused new 'flagship' courses (or subjects or units) within programs; and the use of 'armada' courses peppered across programs. Each of these has benefits and pitfalls and are discussed further in Chapter 8, when we consider learning pathways in more detail.

Considering the look and feel of course/program offerings, three key questions arise:

- Do we embark on creating a new niche program, or update our existing program?
- How much should we rely on single (i.e. flagship) courses included in programs to deliver these knowledge and skill areas?
- How much should we embed the new knowledge and skill areas into the various courses across the program (i.e. armada offerings)?

Developing new niche programs

There appears to be inertia arising from the mid 1950s to create new niche program offerings that will attract students.[5,6] This involves pulling new knowledge and skill areas into a focused program of delivery to cover the material, requiring students to enrol in that specialty area. Niche programs may be specialisations existing either within a larger discipline context (for example, a Bachelor of Sustainable Energy Systems as a major specialisation within engineering), or as 'hybrid' or 'trans-disciplinary' knowledge areas that cut across traditional boundaries (for example, a Bachelor of Environmental Management involving Environmental Science, Engineering, Business and the Arts).

In our experience, colleagues initially see niche programs as the preferred option for keeping up with the changing market place, as they appear to address emergent opportunities without upsetting the status quo of course supervision and content. They can also offer fundraising opportunities if offered as postgraduate or professional development courses. However, it is an urban myth that new programs are preferred over program updates. Documented audits such as that undertaken by the University of South Australia conclude that academic staff prefer that emergent topics – in this case sustainability – not be addressed in just one (or two) specialist sustainability programs.[7] Indeed, we have shown that 'education for sustainability' is about applying knowledge and skills for 21st-century living across programs within a discipline as appropriate and capturing the synergies between them, rather than defining a new discipline of study. The same could be said for other topic areas such as 'safety', 'professional practice', and 'information technology' that have permeated many traditional programs.

Where 'niche' sustainability bachelor programs have emerged and remained in the marketplace, it appears to be important to integrate core sustainability concepts with a base discipline training that is 'job-ready' (for example, engineering, science or law). These programs also seem to contain a number of elective options to cater for the variety of student interests, providing many non-linear pathways for students to articulate with their niche sustainability qualification.

Potential pitfalls to be addressed in creating niche sustainability programs that sustain enrolment numbers include:

- The absence of a discipline base makes it difficult to provide context for sustainability in practice which can lead to abstract and high-level understanding.
- As with any new program there can be significant resourcing costs to create additional programs, given existing academics are already teaching programs.
- Market risks in offering a new program with no track record of student employability.
- As with any new program it will need time to build credibility through its graduates to demonstrate its rigour and ability to create job-ready graduates that can bring value to employers.

> At TU Delft we have had many years of experience in integrating sustainability into our curriculum. We realised early on that niche programs are a niche option . . . TU Delft has therefore focused some of its efforts on providing some students with 'niche' experience opportunities. Most efforts have been devoted to upgrading every aspect of the students' studies to embed education for sustainability and to create the ability and support for that among the teaching staff.
>
> Dr Karel Mulder, Senior Lecturer and Head of the Sustainable Development Unit, Delft University of Technology

Developing new courses (subjects/units) – 'flagship approach'

As with so many management terms, 'flagship' has crossed over from the military vernacular into common use. The original meaning for a flagship is the lead ship in a fleet of vessels used by the commanding officer, which is the largest, fastest, newest, or the best-known. It is used to quickly identify the fleet to others, originally by flying a flag. It now has common derivations such as the 'flagship project' of a government, a 'flagship product' of a manufacturing company, or 'flagship store' of a retail chain, etc. In the context of higher education we have adopted the term to describe courses (subjects/units) that focus on a key area, with their inclusion in programs intended to distinguish them from other programs, such as advanced computer modelling. The flagship might be placed anywhere in undergraduate programs, co-offered as a postgraduate course, and convened by the resident expert or 'champion' within the department. It could be a first-year common course, an elective, or in the form of a doctoral study or a Master's.

Including flagship courses that focus on sustainability in existing programs, typically in early years, can influence the way students interpret and contextualise the other courses across the program.[8] For example, a first-year course called 'Sustainability and Professional Practice' can provide students with context for why and where they may embed sustainability thinking in their career paths. Such flagships can introduce the concepts of sustainability in relation to the overall discipline and provide a powerful shift in the mind of the student as to the contribution they may make to sustainability, or indeed un-sustainability. In particular, where champions exist in the absence of a community of practice in their institution, such flagships permit students to engage with the material and apply this new understanding elsewhere in their studies, regardless of whether other academics reinforce the acquired knowledge and skills. A significant benefit of the flagship approach is that it provides an opportunity for targeted intervention, involving minimal staff disruption and procedural issues. Over time, departments can integrate this material into relevant courses across the program to then make way for new flagships. In the past this process has allowed new material on topics, such as ethics and safety, to be vetted by staff, students and alumni to allow demand for the content to be established.

At an introductory level, a flagship introductory course might be developed for first-year students to 'kick-start' their learning in a certain topic area. At an intermediate or advanced level, flagship courses might be developed to cater for learning in a new content area previously not addressed in the program and offered as part of a minor or an elective. There are numerous examples of flagship engineering courses, often offered in first year, and which are used as the foundation for further learning. The University of Bristol's first-year 'Sustainable Development' course is one such example that won a 2007 UK Green Gown award for its design and interdisciplinary student enrolment achievements.[9]

However, there are a number of potential pitfalls to be addressed in using flagship courses:

- Although given context from the flagship course, students will be poorly supported to apply sustainability across their areas of professional practice, which they will be required to do once employed.
- Accreditation processes in the future may focus on program-wide incorporation of sustainability knowledge and skills and rank a singular course focus poorly.
- The presence of a flagship course can signal to other academics that sustainability is covered and they need not cover it.
- Flagship courses are susceptible to being terminated as they can be easily removed if they are not connected with other coursework or made an elective.
- Flagship courses that are convened by champions are vulnerable to disappearing should the academic move on or be unable to deliver the course.
- If this approach is used by itself, there is the potential to have too many flagships in a degree, which could result in a fragmented understanding of the program offerings.

Renewing courses (subjects/units) across the program – 'armada approach'

> The success of the work we have undertaken to embed sustainability into the first year engineering program at Monash University was really underpinned by an initial workshop we held which engaged external and internal stakeholders (facilitated by Desha and Hargroves). We then broke the implementation up into steps, which meant that staff were engaged in, supported through, and not overwhelmed by, the process of change. The development of a website, which supported the implementation, demonstrated the contribution that virtual learning environments can make to both staff and student learning about sustainability. That website served as a supply ship for the armada of first year courses where changes were made to embed sustainability.[10]
>
> Professor Geoff Rose, Professor of Education for Sustainability, Momash University

Continuing with the use of military terms, while a flagship flies the fleet's flag, the rest of the fleet, or the 'armada', is just as important in making an impression on their target, reinforcing the direction and protecting the flagship and its cargo. Within this context it makes sense that a flagship course would be complemented by a number of supporting courses that reinforce key messages and foster knowledge and skills development. In addition to flagship courses, the remaining armada courses are important components of a curriculum renewal strategy, reflecting the flagship course's key message to provide students with an integrated learning experience. In exploring options to harness courses across the program in an armada style we have discovered significant challenges to such efforts, perhaps why the step from '*ad hoc*' to 'integration' has been so difficult in the past. Strategic planning for renewing courses across programs to align with flagship content requires dealing with many faculty across the various components of a program, to discuss course outlines, assessment requirements, workload considerations and budget implications. Furthermore, without clear assignment of learning outcomes for each course and clear flagship objectives, academics may not incorporate appropriate content from the flagship into their courses, leading to a disjointed overall curriculum.

With this in mind, potential pitfalls to address in the armada approach include:

- If faculty are not committed to this approach, then there will be a disconnect between those courses that do support the flagship and those that don't, which could be confusing for students and detract from the program.
- An unstructured approach – e.g. without flagship objectives or clear learning pathways – could result in possible overlaps or gaps in knowledge and skills within the curriculum resulting in key knowledge and skill areas not being adequately developed.
- An approach that does not use a whole of institutional approach to finding expertise and opportunities for coverage in courses could result in lost opportunities for the program, and duplication of efforts in different schools or faculties.

Hence, the approach needs commitment from – and clear guidance for – a large number of faculty, to amend lectures, tutorials, workshops, assignments, laboratories and site visits, etc. The approach also needs to be steered carefully to avoid redundancy and a disorderly revision of content. Whereas faculty may teach with a 'unit-level focus', students need a 'program-wide focus' to ensure that the messages are woven into robust learning pathways that support the development of the required knowledge and skill areas. The armada approach also needs to be carefully coordinated to ensure that new content is aligned with regard to language, terms, definitions and overarching message, to provide a clear learning pathway for students. It needs to uncover academic strengths within and outside the department, potentially across a number of courses and programs. In the case of sustainability, the process requires a multi-disciplinary approach drawing upon the need to acquire knowledge in science, economics, law and engineering, so time needed to involve other departments also needs to be taken into account.

Strategic use of 'niche', 'flagship' and 'armada' approaches to develop sustainability-related graduate attributes has a number of potential benefits:

- It demonstrates a comprehensive 'whole-of-system' approach to curriculum renewal;
- It shows how graduate attributes are developed through clear learning pathways (see Chapter 7);
- It encourages prompt attention to changing needs, ensuring that faculty are involved in monitoring courses to align with desired graduate attributes.

So, while there may be hype in some parts of the sector about creating new, stand-alone offerings, or completely devolving all responsibility to faculty 'as they see fit', there does not appear to be a 'silver bullet' solution. Rather, our experience suggests that the answer is in blending the above three options to create programs that are resilient to changes in structure, faculty resourcing and budgets.

An example of such an integrated program is shown in Figure 5.4. Here there are 5 flagship courses, supported by 10 armada courses, within a program of 32 courses. There are also four additional

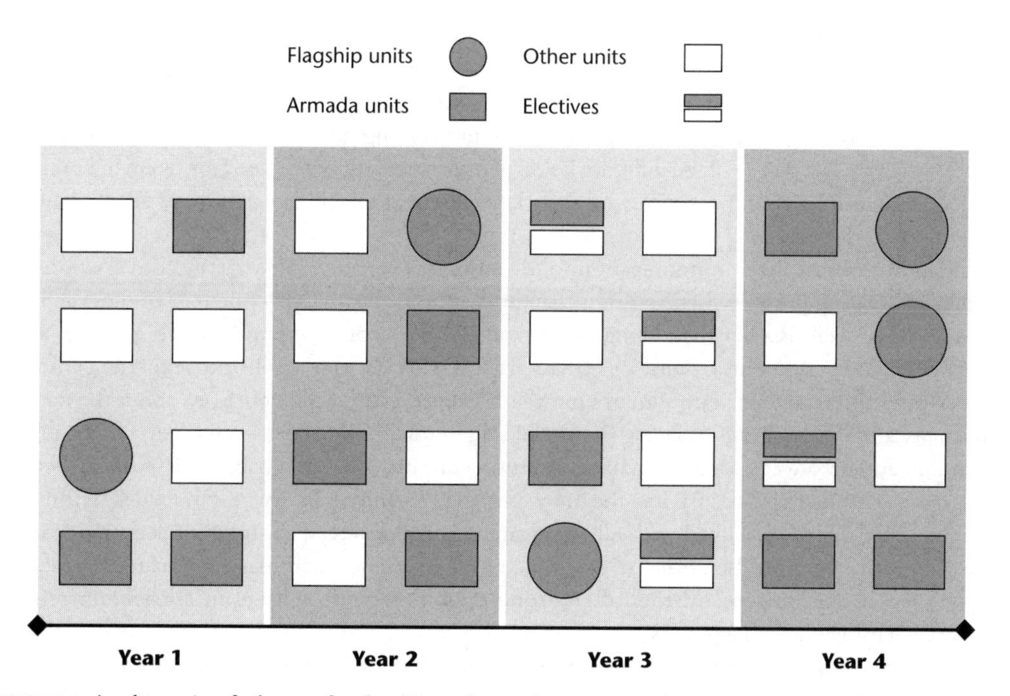

FIGURE 5.4 A schematic of a layout for flagship and armada courses within a program, over eight semesters of study

options for students to choose armada courses in their third and fourth years of study, through electives.

In this example, in first year there may be a common flagship course that targets the 'meta-narrative' (i.e. overarching discussion) about the attributes to be developed, with other courses supporting development of the learning outcomes. Alternatively, there might not be a flagship in first year, with learning outcomes assigned to all courses. In second and third year, common learning outcomes could be concentrated in common courses, and discipline-specific learning outcomes targeted in various electives. By the final year of study, the graduate attributes will be largely developed and one or more flagship courses may be used as a 'capstone' learning experience, where students are required to demonstrate competency across the spectrum of attributes including those related to sustainability. Electives may also be used to fully develop discipline-specific sustainability-related graduate attributes.

Internationally there are examples of leaders demonstrating this 'mix' of approaches towards education for sustainability within their programs, such as MIT, Delft, Carnegie, Tokyo University and UPC-Spain. This includes integrating, or 'embedding', these materials across the spectrum of courses on offer, combining integrated undergraduate bachelor degrees with postgraduate specialisations in sustainable development topics, offering flagship courses on leading-edge content and providing specialisations in various topic areas. Such an approach is certainly positioning them to be leading universities in the following decades; possibly the 'new world' ivy league!

Key strategic considerations

Communication and awareness raising

It is clear from the barriers highlighted in Chapter 2 that there are still few catalysts that are large enough to drive rapid curriculum renewal in the higher education sector. This is perhaps comparable to the primary and high school curriculum reform process in Australia, where it took almost three decades for the catalysts to become great enough (including politicians, policy makers and educators on both sides of politics) to trigger a period of 'rapid curriculum planning' in 2009 for national reform. In this absence of such drivers there will be differing levels of understanding of the need for renewing programs to include sustainability, and a process of clear communication will be critical to enrol faculty in embarking on the process.

Colleagues tend to list 'communication' and 'raising awareness' together in conversations and workshops discussing possible ways to drive change. Yet, in our experience, neither of these activities is typically done well. Rather than being given priority, they commonly get tacked on as an added extra to curriculum renewal activities, if there are any extra resources. In our experience, raising awareness is a critical factor in curriculum renewal; however, just raising awareness about the need for sustainability and the implications facing the world is not enough. Further, it can often stall the process as discussion and debate can focus on what constitutes education for sustainability. In reality, there are many considerations that are driving the new knowledge and skills to be embedded within the curriculum. Hence, closely linked with raising awareness, developing a common understanding of the context and expectations is also critical. Done well, raising awareness within the department creates the possibility for strong buy-in and thereafter strong commitment to engage in curriculum renewal initiatives, even the challenging parts.

As soon as the scope of the proposed transition is clear, it is timely to send an early signal to staff that something significant is about to happen. This might, for example, be a forum that includes an address by senior management members (depending on the institutional context). Depending on the culture of the institution, interaction with faculty may also include ensuring buy-in through feedback via focus groups and other presentation opportunities. Alternatively, it may be through a directive from the Dean or higher. As the strategy requires consideration of available resources, budget and staff availability, the person who writes the strategy document needs to be someone skilled in seeing the whole organisation and its potential interactions. This may be different to the appointee that will manage its implementation, who is skilled in organisational change and project management. Where there will be more than one person involved, the handover of documentation will be a critical factor in the success of the initiative.

The strategic planning phase provides an opportunity for some sort of up-front, high-level declaration of intentions, or 'announceable target', based on the findings of the initial reviews. For example, the department might declare, 'We are aiming to achieve full integration of sustainability-related knowledge and skills within our programs within six years'. However, with the best strategy in the world, the process is unlikely to be successful if faculty feel manipulated or resentful towards the process. Hence, this message should be accompanied by communication to the departments around expectations, and about the roles of key individuals in the institution. In the United Kingdom, the Higher Education Funding Council for England (HEFCE) published a report in 2007, titled *Strategic Review of Sustainable Development in Higher Education in England.*[11] The report concluded that higher education institutions (HEIs) have different perceptions of sustainable development and how it should be pursued (if at all) within the institution. It clearly defined the importance of staff engagement in achieving curriculum renewal, noting that in addition to the HEFCE's role supporting teaching committees within higher educations, HEIs will need to start teaching sustainability knowledge and skills because they want to, and they will decide how it is to be incorporated.

A number of professional bodies have been developing guidance documents over the last decade to clarify for engineering departments how the shifting requirements to education for sustainability may be met. For example, the UK Royal Academy of Engineering (RAE) produced a 2005 report *Engineering for Sustainable Development: Guiding principles,*[12] aimed at faculty within engineering departments or schools in the UK wanting to embed principles of sustainable development into curriculum. This report was also followed by a 2006 report on *Educating Engineers for the 21st Century: The industry view.*[13] Since 1998, the RAE has also operated a scheme for the appointment of visiting professors in Engineering Design for Sustainable Development at universities in the UK to assist in the generation of teaching materials to enhance the understanding of sustainability and sustainable development among academic staff and students alike.[14]

HEIs appear to be beginning to allocate resources to such awareness-raising endeavours. For example, in 2008, Oregon State University – one of America's top 25 'green colleges'[15] – employed a well-known expert in sustainable forestry and conservation issues as its new 'Director of Sustainability Programs' in its College of Engineering, to co-ordinate and expand sustainable construction and engineering concepts throughout the college's curriculum, collaborative research and outreach programs.[16] These days, despite fluctuations in national and regional economies, there is a continuous stream of job advertisements for sustainability coordinators, professors of sustainability and senior lecturers in sustainable development, across the spectrum of universities and colleges.

Globally there are a host of websites now that are geared towards assisting faculty understand what education for sustainability is all about. In addition to the TNEP website, here we have listed our top four hubs of resources for raising awareness and developing a common understanding amongst staff, which are user-friendly and which are regularly updated:

1 Association for the Advancement of Sustainability in Higher Education (AASHE): An American association of colleges and universities working to create a sustainable future through providing

resources, professional development and a network of support to enable institutions of higher education to model and advance sustainability in everything they do, from governance and operations to education and research. Their website provides a wealth of resources, from video links through to publications on education for sustainability and how it can be integrated within curriculum.

2 The Higher Education Academy Engineering Subject Centre (HEA-ESC): A national initiative funded by the UK government, servicing all engineering academics in the UK with subject-based support. Its strategic aims include sharing effective practice in teaching and learning among engineering academics, supporting curriculum change and innovation within departments, and informing and influencing policy in relation to engineering education. Their website contains a variety of sustainability-related resources, including a freely downloadable guide called *An Introduction to Sustainable Development in the Engineering Curriculum* produced by Penlington and Steiner.[17]

3 UNESCO Report – Drivers and Barriers: In 2006, UNESCO published proceedings from a 2005 Workshop in Göteborg (Sweden) on drivers and barriers for implementing sustainable development in higher education, edited by Holmberg and Samuelsson.[18] This publication captures a variety of experiences that have been attempted by higher education institutions around the world, including a number of papers focused on engineering education.

4 TED TV: A non-profit organisation based in America, which focuses on sharing information on the topics of technology, entertainment and design. This includes a 'TEDTalks' site where videos are released under a Creative Commons license, so they can be freely shared and reposted. It includes a rich supply of talks about sustainable development, which might provide a welcome break from curriculum-centred discussions, with staff in other activity streams from non-teaching backgrounds.

Identifying key tasks to be undertaken

As discussed in Chapter 4, we have used the concept of 'plan, do, check, act' to apply to the process of curriculum renewal. The resultant process comprises four stages and an endpoint whereby the curriculum can be evaluated for its intended goal. With these stages in mind, there are a number of ways we can generate a task list of actions, depending on how the organisation is structured. Again, the intention of this book is not to dictate such a process but rather to provide guidance to those undertaking the process. Given the number of parties involved across the institution there may be value in creating a stakeholder responsibility matrix to support the development of the strategy based on the Curriculum Helix, as suggested in Table 5.3.

> Once there is a commitment to undertake a period of focused curriculum renewal, where to next? However well intended, a commitment without a decent plan for action is destined for a rocky road. In particular, given the complexity that we have described in previous chapters, there are numerous hurdles to consider in implementing such a commitment, including pedagogical and organisational challenges. Furthermore, the interaction between members of faculty across these challenges is also critical to success!

TABLE 5.3 Potential stakeholder responsibility matrix

Strategy	*Stage 1: (S1) Prepare [Year 0–1]*	*Stages 2/3/4: (S2/3/4) Explore, test/pilot, integrate [Year 1–6]*
Faculty tasks	• External trends report • Internal readiness level report • Create curriculum renewal strategy • Monitoring and evaluation plan	• Update external trends report • Update internal readiness level report • Update curriculum renewal strategy • Update monitoring and evaluation plan
Governance/ Management (G)	**S1-G**1 Coordinate development of external trends report, including input from each of the other five activity streams **S1-G**2 Coordinate collection of data for internal readiness level report **S1-G**3 Develop strategy, including review by the other five activity streams	**S2/3/4-G**1 Coordinate review of external trends report, including input from each of the other five activity streams **S2/3/4-G**2 Coordinate collection of data for internal readiness level report **S2/3/4-G**3 Review strategy, including input from the other five activity streams
Operations/ facilities (O)	**S1-O**1 Identify point of contact/s for the strategy **S1-O**2 Report on existing student involvement related to strategy **S1-O**3 Report on sustainability initiatives/requirements on campus	**S2/3/4-O**1 Review point of contact/s for the strategy **S2/3/4-O**2 Report on existing student involvement related to strategy **S2/3/4-O**3 Report on sustainability initiatives/requirements on campus
Teaching and learning (TL)	**S1-TL**1 Identify point of contact/s for the strategy **S1-TL**2 Participate in strategy discussions with governance and management **S1-TL**3 Provide data requested **S1-TL**4 Review draft strategy	**S2/3/4-TL**1 Review point of contact/s for the strategy **S2/3/4-TL**2 Participate in strategy discussions with governance and management **S2/3/4-TL**3 Provide data requested **S2/3/4-TL**4 Review revised strategy
Human resources and culture (H)	**S1-H**1 Identify point of contact/s for the strategy **S1-H**2 Participate in strategy discussions with governance and management **S1-H**3 Provide data requested **S1-H**4 Review draft strategy	**S2/3/4-H**1 Review point of contact/s for the strategy **S2/3/4-H**2 Participate in strategy discussions with governance and management **S2/3/4-H**3 Provide data requested **S2/3/4-H**4 Review revised strategy
Marketing and communications (M)	**S1-M**1 Identify point of contact/s for the strategy **S1-M**2 Participate in strategy discussions with governance and management **S1-M**3 Provide data requested **S1-M**4 Review draft strategy	**S2/3/4-M**1 Review point of contact/s for the strategy **S2/3/4-M**2 Participate in strategy discussions with governance and management **S2/3/4-M**3 Provide data requested **S2/3/4-M**4 Review revised strategy
Partnerships and stakeholder engagement (P)	**S1-P**1 Identify point of contact/s for the strategy **S1-P**2 Participate in strategy discussions with governance and management **S1-P**3 Provide data requested **S1-P**4 Review draft strategy	**S2/3/4-P**1 Review point of contact/s for the strategy **S2/3/4-P**2 Participate in strategy discussions with governance and management **S2/3/4-P**3 Provide data requested **S2/3/4-P**4 Review revised strategy

Across these suggested stages the challenge is to compress the timeframe for curriculum renewal through the use of progressive tasks and milestones, to ensure that the outcome is achieved rapidly (i.e. 2 accreditation cycles, or 6–8 years) rather than in a typical timeframe of as long as 4 accreditation cycles (i.e. two decades). Such a successful and timely curriculum renewal transition will depend on faculty and staff from across the institution being part of the process and clearly understanding their role so they can remain actively involved. In addition to having a strategy with clear roles and responsibilities, the day-to-day tasks and timeframes need to be clear to encourage individual accountability and maintain momentum. The curriculum renewal strategy will provide the framework for a multitude of tasks to be planned for and scheduled to ensure that intentions are fulfilled.

Within each stage, each of the six activity streams of the Curriculum Helix could have a tailored work plan, describing how they contribute to the overall strategy. Critical to these work plans is the incorporation of tasks that direct action between the activity streams, as well as within the activity streams. Such interconnection provides a key framework for considering the various tasks involved in delivering the curriculum renewal strategy, namely:

- Direct: How can this activity stream contribute to achieving the strategy?
- Supporting: How can this activity stream help others achieve the strategy?
- Assisted: How can others help this activity stream achieve the strategy?

This framework can be used to consider each of the elements of the deliberative and dynamic model, for each of the activity streams in the curriculum helix, as shown in Figure 5.5.

An example of such tasks is shown in Table 5.4, with a sample staging table for each of the six streams suggested in the curriculum helix across each of the four stages shown in the Appendix. As you browse through the sample tables, perhaps various individuals occur to you, who might be responsible for ensuring that such tasks are completed in your institution. This may include academic faculty, administrative staff or students. Perhaps you will also identify tasks that involve individuals or groups outside the institution – you may need to create formal arrangements (for example including payment or an agreement) to ensure that the tasks can be completed in the desired timeframe. The sample tables have been included to provide a head start intended to reduce the 'time for preparation and planning' barrier to engaging in rapid curriculum renewal. We hope that this systemic overview of the process will also be a useful reference to revisit if the transition encounters difficulties along the way. The sample tables in the Appendix are intended as a provocation and template rather than a full 'to-do' list – we look forward to hearing of more examples.

How can efforts be continually monitored and improved?

An important component of any strategy is the mechanism for monitoring and evaluating progress, and using this to inform ongoing efforts. If using an approach like the curriculum helix, there will be a large number of tasks across the various activity streams that need to be undertaken as part of the overall strategy, with such tasks needing to be monitored and evaluated to ensure adequate support and achievement of objectives. There are a number of areas that will need to be reported on and this will vary somewhat in language and delineation across different institutions; however, typically at the end of the process management will want to report on:

- Whether the additional expenditure to carry out the strategy delivered suitable results to justify such expense;

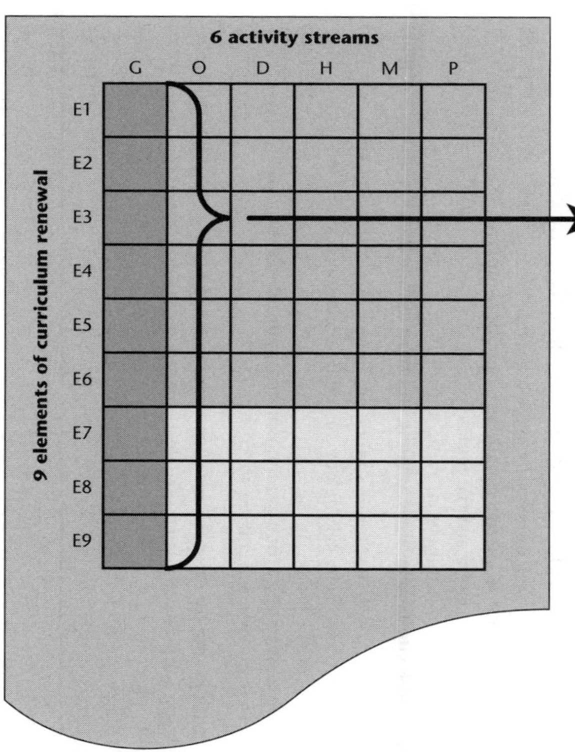

G – Governance:

Element	Role	Example task	Item ref
Curriculum renewal strategy	*Direct*		
	Supporting		
	Assisted		
Identify graduate attributes	*Direct*		
	Supporting		
	Assisted		
Map learning pathways	*Direct*		
	Supporting		
	Assisted		
Audit learning outcomes	*Direct*		
	Supporting		
	Assisted		
Develop and update curriculum	*Direct*		
	Supporting		
	Assisted		
Implement program	*Direct*		
	Supporting		
	Assisted		
Raise awareness and build capacity	*Direct*		
	Supporting		
	Assisted		
Collaborate internally and externally	*Direct*		
	Supporting		
	Assisted		
Continually monitor and evaluate	*Direct*		
	Supporting		
	Assisted		

FIGURE 5.5 Illustration of the staging tables that form the work plan for each stage

TABLE 5.4 Sample table – Governance and management – Stage 1 'Prepare' (G1)

Element	*Role*	*Example Tasks*	*Item Ref*
Curriculum renewal strategy (E1)	*Direct*	Drive the development of the curriculum renewal strategy	G1.E1.1
	Supporting	Communicate intentions to faculty	G1.E1.2
	Assisted	Seek contribution and comment from faculty	G1.E1.3
Identify graduate attributes (E2)	*Direct*	Budget and determine faculty resourcing for attributes work	G1.E2.1
	Supporting	Internally promote graduate attributes work as an important quality assurance measure	G1.E2.2
	Assisted	Become aware of upcoming processes to be undertaken	G1.E2.3
Map learning pathways (E3)	*Direct*	Budget and determine faculty resourcing for pathways work	G1.E3.1
	Supporting	Internally promote learning pathways work as an important quality assurance measure	G1.E3.2
	Assisted	Become aware of upcoming processes to be undertaken	G1.E3.3
Audit learning outcomes (E4)	*Direct*	Budget and determine faculty resourcing for audit work	G1.E4.1
	Supporting	Internally promote curriculum auditing as an important quality assurance measure	G1.E4.2
	Assisted	Become aware of upcoming processes to be undertaken	G1.E4.3
Develop and update curriculum (E5)	*Direct*	Gauge budget options for actions over the next 3–5 years	G1.E5.1
	Supporting	Internally promote the benefits of strategic rather than *ad hoc* content development and renewal	G1.E5.2
	Assisted	Understand the current situation regarding faculty resourcing	G1.E5.3
Implement the program (E6)	*Direct*	Gauge budget options for actions over the next 3–5 years	G1.E6.1
	Supporting	Internally promote the benefits of strategic rather than *ad hoc* content development and renewal	G1.E6.2
	Assisted	Review performance appraisal process to recognise contributions to this transition	G1.E6.3
Raise awareness and build capacity (E7)	*Direct*	Develop a vision with clear goals about the intended curriculum renewal process, to use internally and externally	G1.E7.1
	Supporting	Organise a series of internal seminars and public forums to introduce the context and the initiative	G1.E7.2
	Assisted	Investigate opportunity to create an annual awards event to recognise efforts in awareness raising and capacity building	G1.E7.3
Collaborate internally and externally (E8)	*Direct*	Communicate the institution's vision and intentions to stakeholders	G1.E8.1
	Supporting	Invite feedback from stakeholders regarding their expectations for this process	G1.E8.2
	Assisted	Become aware of curriculum innovation opportunities through internal and external collaboration	G1.E8.3
Continually monitor and evaluate (E9)	*Direct*	Identify funds for tracking this curriculum renewal process	G1.E9.1
	Supporting	Host a forum for academics, facilities management and local entrepreneurs to share experiences and lessons learnt to date	G1.E9.2
	Assisted	Become aware of low-cost opportunities to monitor and evaluate progress	G1.E9.3

- How the overall process has benefited the various programs in tangible ways, such as student intake, etc.;
- Whether the approach led to any disputes or issues between staff or various departments, and if so how they were dealt with; and
- Whether the process has achieved a curriculum that develops the desired graduate attributes.

Such reporting will require a mixture of quantitative and qualitative evaluative measures that take into account a range of complexities across the institution. This can be simplified somewhat through a focus on the activity streams of the curriculum helix. In this case the strategy would allow for budget and resources to track progress within each of the various activity streams chosen, across the selected stages (with these selected to suit the particular institution). Table 5.5 provides examples of key performance indicators that might be considered, for which targets and milestones could be set for each of the stages.

TABLE 5.5 Example monitoring and evaluation key performance indicators

Activity stream	*Example key performance indicators*
All streams	• Extent of completion on programmed tasks (%) • Proportion of invited staff attending seminars/forums (%)
Management/governance	• Proportion of policy statements incorporating commitment to sustainability (%) • Proportion of annual operating budget dedicated to sustainability strategy (%)
Operations/facilities	• Number of collaborative projects with teaching and learning staff (#/year) • Positive feedback from staff and students regarding collaboration (Q)
Teaching and learning	• Proportion of subjects where curriculum renewal for sustainability is taking place (%) • Number of staff hours spent on sustainability-related professional development (#/year) • The identification of preferred graduate attributes (#) • The auditing of current programs to identify such graduate attributes (% of courses) • The development of learning pathways for programs to develop graduate attributes (harnessing existing offerings) (% of programs)
Human resources and culture	• Proportion of successful job applications including sustainability competencies (%) • Proportion of staff performance reviews including sustainability activities (%) • Number of new staff engaged to participate in strategy items (#)
Marketing/communications	• Proportion of website pages about program offerings, which discuss sustainability content (%) • Number of media releases about new offering in sustainability (#) • Positive feedback from focus groups/survey regarding sustainability messaging (Q)
Stakeholder engagement	• Number of stakeholder communiqués regarding progress (#/year) • Proportion of responses to survey/focus group requests (%)

Note: 'Q' refers to a qualitative performance indicator

PART 3

Key considerations for each element of curriculum renewal

6

IDENTIFYING GRADUATE ATTRIBUTES

Graduate attributes are gaining popularity as a tool to inform curriculum renewal efforts. Not only do they help to define the kind of graduates programs seek to deliver, they provide a tangible way for accreditation bodies to communicate to universities what students are expected to develop by the time they complete their studies. Yet despite this increasing popularity, graduate attributes receive a mixed review among academics,[1] and the actual use of graduate attributes is still very low. As Professor King reflected in a 2008 review of engineering education in Australia,

> Few engineering education programs are underpinned by a comprehensive specification of program objectives and detailed graduate outcomes that provide a clear understanding of the knowledge, attributes and capability targets for graduates in the particular discipline.[2]

We often come across colleagues and educators in Australia and overseas who are not familiar with what the term means. Among those who are, there is clear understanding of the importance of graduate attributes; however, there is also reluctance to consider generating graduate attributes for their own programs, or interpreting what they mean at the level of individual courses. Often we hear complaints such as that the level of inquiry is too time consuming and does not seem practical. For those who have previously broached the concept of sustainability-related graduate attributes, their initiatives struggled due to the lack of a coherent process to reach agreement on what graduates should know in relation to sustainability.

In responding to this range of familiarity among educators of the use of graduate attributes and the mixed review of those who have some experience with their use, this chapter focuses on first explaining what graduate attributes are, and then outlines opportunities to use them to focus sustainability curriculum renewal efforts. The chapter also addresses the difficulty of reaching consensus in identifying sustainability-related graduate attributes, outlining a timely and effective facilitation approach that has proved invaluable for our team in facilitating many such workshops over the last decade.

Exploring the deliberative element 'Identify graduate attributes'

Introduction

So, what is a 'graduate attribute'? Curriculum literature tends to define the term as a particular ability that a graduate should have by the time they complete a given program of study, which may comprise one or more areas of knowledge and/or skill (i.e. 'competencies'). The language in this emerging field is somewhat mixed, with researchers also referring to graduate attributes as 'graduate competencies',[3] 'program outcomes',[4] and 'graduate capabilities'.[5] A set of graduate attributes can define the knowledge, skills and qualities that a department agrees its students should develop during their time with the institution.[6,7] Graduate attributes may reflect accreditation requirements, in addition to areas of focus that are particular to each institution (i.e. niche specialities), along with expressed employer needs. Some graduate attributes are at a high level and may be relevant across a number of disciplines, while others may be discipline specific. Examples of both will be discussed through this chapter.

This element 'identify graduate attributes' comprises the first step in the deliberative and dynamic model for curriculum renewal presented in Chapter 3, providing clarity with regard to expectations for what the program is trying to achieve related to education for sustainability, and subsequently the learning outcomes that need to be embedded within the curriculum to achieve this. When considering sustainability or any other new area of focus, a unique set of graduate attributes will evolve for each department from a variety of inputs as shown in Figure 6.1. This includes advice from the accreditation body regarding anticipated future requirements along with considering international 'best practice' requirements. There may also be clear market signals from advisory boards and market research, along with demands coming from current and potential students.

The UK Higher Education Academy has published two reports on *Student Attitudes and Skills for Sustainable Development*. The findings clearly demonstrate student demand for sustainability education, with more than two-thirds of 2012 first-year and second-year respondents believing that their university should cover sustainability. Furthermore, there is a continued student preference for reframing curriculum content rather than adding content or courses.[8]

Given the wide variety of external influences, and the preference of accreditation agencies for outcomes-based approaches to program design, institutions have significant flexibility in the type of sustainability knowledge and skills that are targeted in the curriculum. Indeed, the level of commitment may range from aiming to just meet accreditation requirements, through anticipating future requirements, to achieving sector leadership.

Creating a space to consider graduate attributes

In theory, the use of graduate attributes to focus attention sounds promising; however, in a field as emergent and wide reaching as sustainability, it can be difficult for a department to reach agreement on specific graduate attributes, and even more difficult to then develop them within a program.

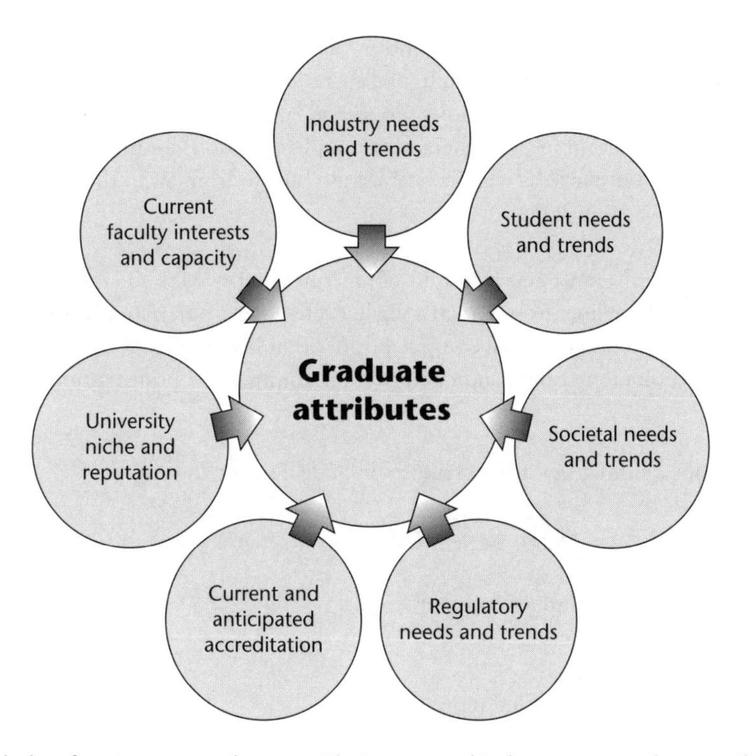

FIGURE 6.1 A whole-of-society approach to considering external influences on graduate attributes

Developing graduate attributes requires creative facilitation and careful preparation to distil information from faculty that is constructive and relevant. For example, all attendees will need to be appropriately briefed prior to the workshop to ensure that discussions can be based on a common understanding. Sustainability is a wide-reaching topic and there are many points of view that need to be considered constructively; however, this can sometimes become an interpersonal issue with faculty arguing for particular graduate attributes and being less inclined to compromise with others.

From our experiences, there are a few regularly occurring issues that stall discussions seeking to identify graduate attributes. The first is a lack of preparation – by the facilitators, organisers, and by the participants. Forecasting graduate attributes is in effect defining future graduates who may exist in the market in 5 years, 10 years, and beyond. The second is faculty reluctance to participate due to anxiety or concern regarding workload implications and availability of support going forwards. This might include potential requirements to participate in meetings and sub-committees to develop and update their courses, additional reporting and consultation requirements, and dealing with student enrolments and transitions. Clear acknowledgement and commitment to supporting these additional workload requirements during the transition can deal with this anxiety and facilitate quality discussions about moving forwards.

However, sometimes, even if these issues are addressed and the space cleared for a discussion about graduate attributes, faculty can still be reluctant to agree on specifics due to political or cultural issues that are well beyond the scope – and interest – of the curriculum renewal process. If this situation arises

– typically it can be foreseen as this type of cohort will likely have a track record in the institution – then we encourage a facilitation process designed to reduce the personal bias, such as the 'rotating control group' (RCG) facilitation methodology outlined later in this chapter. If for some reason face-to-face workshop time is not achievable, there are a couple of options for proceeding, although a word of caution that this may not work as well in establishing buy-in from all to the process and outcomes. Options may include:

- Faculty could be surveyed to generate a list of emerging knowledge and skills that they believe are important to their discipline, from which a smaller reference group may select specific attributes, or
- A set of graduate attributes could be compiled from other institutions (for example, competitors) or from related accreditation criteria and circulated for comment or prioritisation by faculty.

Considerations for defining graduate attributes include current and future:

- Industry and employer expectations
- Student demand
- Local and regional community context
- Legislation and agreements
- Accreditation requirements
- Market niche opportunities and reputation
- Faculty interests and capacity

Core activities to identify graduate attributes

In our work to assist colleagues identify preferred graduate attributes we have found two key activities to be very valuable, as described below. Each is designed to be an inclusive process that encourages consensus-based decision making within the institution, to maximise the potential for faculty buy-in. Each activity could be undertaken over one or more face-to-face interactions (i.e. workshops) with faculty, depending on the usual factors of timing, budget and availability.

Brainstorming a 'wish list' of sustainability-related graduate attributes

The first activity involves brainstorming a wish list of graduate attributes that span compliance through best practice to leadership. This provides an opportunity for the findings to be inclusive of all faculty ideas and aspirations by putting everything 'on the board'. It also provides a snapshot of ideas that can be revisited periodically, when reviewing program-level aspirations. It is helpful to ask participants to focus on sustainability-related attributes, and identify those that may be common to two or more sub-disciplines of interest. Once compiled it is helpful to break down the list of brainstormed attributes into three categories of increasing performance. An example is provided here for civil engineering. Sorting and prioritising can be done by participants, by the facilitator or a subset of faculty after the brainstorm:

- Compliance: The institution may focus on just meeting current accreditation requirements (i.e. required now). In this case, ensuring that attributes required for accreditation are appropriately mapped across the courses might be the extent of the process. The resultant list can be an important tool to reduce current accreditation risk exposure by demonstrating compliance. For example, an attribute for civil engineering might state that every graduate must have 'an understanding of engineering impact on the environment'.

- Beyond compliance: The institution may aim to go beyond existing accreditation requirements, planning for the development of graduate attributes to meet future regulatory, market, and accreditation demands and requirements (i.e. leading practice). For example it might be considered that having 'an understanding of the limitations and thresholds of environmental systems' is likely to become a graduate requirement for civil engineering within the next 5 years.
- Sector leader: The institution may want to identify graduate attributes that can 'get ahead of competitors' (i.e. best practice). For example, 'an appreciation of the interaction between carbon emissions and economic performance' may be considered a highly marketable competency that, while not required for accreditation purposes, will differentiate the program from competing institutions offering similar programs. In civil engineering this might comprise 'the ability to calculate carbon emissions from the construction and operation of infrastructure'.

In summary, the outcome of this activity is a comprehensive brainstorm of sustainability-related graduate attributes by faculty members, a grouping of these attributes into common and discipline-specific attributes, and a grading of the attributes by aspiration (in respect to accreditation requirement levels).

Developing a draft list of desired sustainability-related graduate attributes

The second activity involves prioritising graduate attributes from the brainstorm to create a 'desired graduate attribute' list for the program/s of enquiry, as summarised in Table 6.1. This involves selecting attributes that align with the level of aspiration for each program/discipline area (drawing on the curriculum renewal strategy). This exercise may need to be undertaken at the level of discipline, or in clusters, depending on the size and dynamics of the institution. There may also be a commitment to prioritise a number of common and discipline-specific graduate attributes for the program/s.

TABLE 6.1 Example desired graduate attribute set for an engineering department

Program (discipline)	*Aspiration*	*Sustainability graduate attributes*								
		'Compliance' (required now)			*'Beyond compliance' (leading practice)*			*'Sector leader' (best practice)*		
		Attribute 1	*Attribute 2*	*Attribute 3*	*Attribute 4*	*Attribute 5*	*Attribute 6*	*Attribute 7*	*Attribute 8*	*Attribute 9*
Chemical	Compliance	✓	✓	✓	✓					
Civil	Sector leader	✓	✓	✓	✓	✓			✓	
Electronic	Beyond compliance	✓	✓	✓	✓		✓			
Software	Compliance	✓	✓	✓						

Once the set of desired graduate attributes has been developed, they can then be prioritised with regard to when they will be incorporated (for example 'now', or 'in 2 years'). The language within each attribute can then also be refined to be clear about intentions as to the depth of coverage (e.g. 'knowledge of' versus 'ability to apply').

A facilitation model for identifying graduate attributes

Our experience favours these graduate attribute identification steps being undertaken as part of a facilitated workshop designed specifically for this purpose by our team. The following method has been developed through collaborations and support from Townsville City Council's Integrated Sustainability Services Division, the Australian Capital Territory's Land Development Authority, James Cook University, Griffith University, Queensland University of Technology, Monash University and the Australian Federal Government Department of Resources, Energy and Tourism. A full description of the method is available online.[9] The method assumes that participants have an understanding of issues such as existing accreditation requirements, current 'good practice', employer expectations and existing and anticipated future graduate needs in government and industry. If this is not the case, then preparation may be required (see 'awareness raising' in this chapter).

The 'control group' facilitation method is a reflective workshop tool providing participants with an opportunity to contribute both their specialised and cross-disciplinary knowledge. The tightly facilitated structured approach allows participants to voice their concerns and ideas, knowing that their input is being captured and considered. Such an inclusive approach reduces the potential for surprises later in the curriculum renewal process, where an issue might suddenly arise, for example, with competition with other institutions promoting similar graduate attributes that could have been foreseen.

The process allows for a group of 15–30 participants to collaboratively generate, review and prioritise graduate attributes related to sustainable development. The method is as follows:

1 Set-up – Participants and facilitator/s are selected and invited by an appropriately senior person. The facilitators do not need to be familiar with the specific program or discipline area, but should be familiar with the process, including the role of each of the participants and the required outcomes. To begin, senior management provide a short introduction to ensure a common platform of understanding and language.
2 Control group formation – Participants are asked to self-assign themselves to a table, which is designated one of a number of pre-allocated themes, which may be selected industry sectors, disciplines or key topic areas. This process allows participants to work with a theme of interest. Should the facilitators want particular representation on each of the tables (perhaps to ensure courses, gender or personalities are appropriately represented across the themes), then participants can be allocated groups (for example, through a number on their name badge, or a name list on each of the tables). These groups are called the 'control groups' and are responsible for the formation and final editing of notes for their theme.
3 Control group brainstorm – Each of the control groups is provided with a 'starter list' (i.e. a provocation) that introduces their particular theme and provides some initial possible options for relevant graduate attributes. The starter list may include a brief introduction to the topic area, and a quote from a relevant document that promotes creativity and innovation (i.e. a provocation) – it is intended to be a starter for the conversation and not a research paper. Participants may choose to accept or reject the ideas presented, and brainstorm additional items. Participants of each control group make notes directly on the starter list and on additional sheets. Depending on the desired outcomes, the participants may then be asked to work further with the list. Such a consideration could include, for example, prioritising each item by designating on the list a short- (S), medium-

(M) or long- (L) term priority. Clarity around nomenclature and acronyms is important. In this case an 'S' does not mean that the activity would be short lived, just that its development would start in the short term. Similarly, 'L' does not mean low priority, just that it would be developed over the longer term and may be left till later (perhaps for political/budgetary/cultural/logistical reasons).

4 Control group interaction – At this stage one of the following methods of control group interaction is used to allow for groups to comment on other groups' lists (providing comments in a different-coloured pen so as to be easily distinguished by the control group).
   - Rotating control group: Each group reviews each of the other groups' lists. Once each control group has generated a list of ideas using 'blue' pens, they then leave the documentation at their table and move to the next table, working through all of the tables over a series of iterations. At each table, each group considers the list and provides 'comments' in 'black' pen and 'cautions' in 'red' pen. For example, groups could add new items for consideration and/or indicate collaborative opportunities with their control group topic, along with providing cautionary notes as to potential risks or issues. Rotations continue until the groups end up back at their original 'control group' table to consider additions from other groups and compile the final list.
   - Scattered control group: Taking each table by turn, every group member goes to another table to present their group's considerations. This variation facilitates direct interaction between all group members on each of the selected themes for the workshop, and allows every member of the group an opportunity to present (and hence reflect on) the control group's ideas. By requiring each group member to present at some stage, it also acts as a driver for active participation and an eagerness to understand the content. In this variation, each of the control groups takes it in turn to divide up and visit each of the other groups to present their ideas. For this variation to work, the number of participants in each group needs to at least match the number of other groups (i.e. for a six-group workshop, there needs to be at least five participants in each group).
   - Representative rotation: Each control group identifies a representative to visit other groups and discuss the list of graduate attributes. The representative then rotates around the other groups, leaving the remaining group members at their theme table. It is important to ensure that the representatives are comfortable with presenting and receiving feedback from other groups (this could be quite confronting for some participants). It is also important to ensure that the environment is supportive, rather than confrontational. This process can be assisted if the rotating member of the group is able to take with them a photocopy of their control group's work.
   - Paired table rotations: Groups interact with other groups to see how their consideration of graduate attributes might better integrate with them. In this process, groups pair up with each other group and present their findings to each other, one followed by the other without discussion. The two groups then seek ways to incorporate the other group's theme and/or language into their findings and explore potential overlaps and conflicts. This option may assist in creating a systemic outcome that minimises double-up and the potential for gaps; however, it is advised to allow adequate time for discussion to build.

5 Control group presentation and discussion – Depending on the option selected above, each control group may then present their revised findings to the whole group for comment.

6 Control group review – At the conclusion of the process each of the control groups finalise their notes, incorporating considerations from each of the other control groups. The facilitator collects the documentation.

Workshop layout and other logistics

As this workshop method relies on group discussions, it is important that the workshop room is laid out to allow all participants on any one table to comfortably hear each other but not be overrun by discussions in the other groups. If possible, there should also be a space for presentations by the groups to the rest of the participants. There should be ample wall space or pin-board space for lists and/or drawings to be displayed. Occasional use of visual tools such as whiteboards and handouts may assist in stimulating the group atmosphere. Presentations at the front of the room are not recommended, as they can often create a lecture-style atmosphere, which can limit the group's stimulation and hence limit progress, except perhaps in the initial briefing.

It is recommended to have two facilitators and a note taker. Participant discussions should be recorded on butcher's paper/a whiteboard, and also electronically, through laptop note-taking and possibly the use of an audio recording device (provided that permission has been obtained and its use is clearly understood and adhered to). The real-time recording of discussion is also important to ensure that the dynamics of the participants are recorded; however, the knowledge that conversations are being recorded can affect the conversation and unless the recordings are intended to be studied should be avoided to allow open conversation. Where there is only one facilitator, this could also be achieved through the use of a video recording device; however, again, this will affect the conversation and participants may not want their contributions to the workshop to be recorded. Real-time note-taking also assists the facilitation team to pick up key language and ideas that can be further investigated, as well as giving a visual assurance for participants that their words/comments are being recorded by the facilitation team and are being incorporated into the summary. This role may be filled by advanced undergraduate students or postgraduate students. Participants need to be briefed on this role and prepared to be invited to restate items that may have been missed or require exact wording on request from the note taker or facilitator, and also to indicate if a pause is needed to capture a key note.

Sample Short Form Workshop Run Sheet (2.5 hour workshop)

1. **[0–15 mins] Introduction – workshop context and theme explanation**
   Describe the themes that have been chosen to represent the various disciplines for which programs are run within the department (maximum of 6). For example, this could comprise civil, environmental, electronic, structural, mechatronics and software engineering.
2. **[15–45 mins] Develop a list of graduate attributes**
   Participants develop the list of graduate attributes at a program level, using a 'starter list' distributed by the facilitator.
   Alternatively, the participants could answer the following questions:
   - What do you think the key sustainability attributes for [insert discipline] graduates will be over the next 5 years?
   - Considering your own understanding of the potential focus of sustainability in your discipline in the coming 5 years, what are the key sustainability attributes that you think would best prepare students for jobs?

   Alternatively, another structure may be preferred. For example, drawing on a workshop facilitation tool developed by Atkisson called the 'Compass Tool',[10] the following base structure could be used, depending on the nature of the program and the level of understanding among participants.

- *Nature*, which refers to environmental considerations;
- *Society*, which refers to local, regional and global community considerations;
- *Economy*, which refers to industry, government and international economic/supply chain/trade considerations; and
- *Well-being*, which refers to personal considerations which are often referred to as 'soft skills' with respect to each graduate's ability to work independently and in teams, communicate, write, time manage and project manage, etc.

3 **[45–70 mins] Prioritise graduate attributes**
Participants remain in their designated groups, considering for the brainstormed list:
- What should be the priority areas for students to learn about sustainable development (knowledge and skills)? Participants identify the top five areas, using a circle to mark.
- Considering the prioritised items, how are these currently addressed within your discipline's programs? Against each of the five items, allocate 0, 1, 2, or 3 stars for none to excellent existing coverage.

4 **[70–110 mins] Interdisciplinary discussion**
Each group presents a summary of their discussions from [2] and [3], highlighting the prioritised graduate attributes and their coverage within their discipline (10 mins each).
All other groups will listen for similarities with the presenting discipline, noting any changes that they consider relevant, and potential for collaborations.
The facilitator also listens for common attributes among the groups.

5 **[110–120 mins] Review of findings**
Participants spend 10 minutes in their groups discussing how these presentations have influenced their considerations, particularly with regard to identified common attributes.

6 **[120–140 mins] Final Summary**
In turn, the groups share their changes and final comments with the other groups (a shorter round of presentations to what just occurred in [4]).

7 **[140–150 mins] Workshop closure**
The workshop facilitator thanks all participants for their time and notes future actions.

Examples of this element in action

Example: James Cook University, Townsville

In 2010, James Cook University's engineering department was awarded an internal grant to undertake a 'curriculum refresh' project to review its curriculum for sustainability knowledge and skills.[11,12] The department, led by the Chemical Engineering discipline, subsequently collaborated with TNEP (led by the authors) to identify a set of preferred graduate attributes across five disciplines. The results of this process were then used in a pilot to map learning pathways for a particular graduate attribute in one discipline (see also mapping example in the following chapter). In the graduate attribute identification component of the project, faculty were initially provided with an invitation to two events on the same day: an onsite session that included an open-invite university-wide forum in the morning about education for sustainable development; and a 3-hour graduate identification workshop in the afternoon to brainstorm sustainability-related graduate attributes.

This format minimised disruption and helped to set the context for the afternoon workshop. The morning session also promoted the faculty who were already achieving in this area. The strong level of commitment to the initiative by the organisers and departmental managers resulted in 80 per cent attendance at the identification workshop, across five discipline areas. Moreover the list of potential and prioritised graduate attributes arising from the workshop was subsequently used by the department to distil five attributes common across the disciplines, for a department-wide focus.

Example: Queensland University of Technology, Brisbane

In 2005, the Queensland University of Technology's (QUT) Faculty of Built Environment and Engineering (FBEE) and TNEP (led by the authors) embarked on a process of curriculum renewal, focusing initially on a first-year common course to be offered to all first-year undergraduate built environment students (approximately 1200). QUT has one engineering bachelor program with a number of discipline 'majors'. This includes a common first year, and one of these courses is 'Introduction to Sustainability'. A significant part of the collaboration was the up-front engagement with key faculty to first generate a set of graduate attributes that address engineering and built environment professionals' requirements to deliver sustainable development, and to then identify the associated graduate attributes that would be addressed in the new first-year course. The course was designed to provide a platform to then extend the graduate attributes through various courses across the programs.

The generation, prioritisation and selection were undertaken in a workshop of 12 faculty members, over a period of 4 hours. The participants comprised the team of lecturers likely to be involved with the first-year course, the program convenor and lecturers from courses being amalgamated into the common course. The workshop format followed the facilitation method described above, using four 'themes of sustainability' – Nature, Society, Economy, Well-being – to develop the list. The resultant list was provided to the new course convenor responsible for building the course, and the Dean for consideration in the larger context of the faculty program offerings. This led to the creation of a set of supporting notes for the course.

Interaction with 'dynamic' elements of the model

Working with graduate attributes requires early and regular attention to the three dynamic elements of the deliberative and dynamic model, namely 'awareness raising', 'stakeholder engagement' and 'monitoring and evaluation'. In the following paragraphs we highlight key considerations to ensure that the graduate attribute identification process proceeds smoothly through the various stages of the curriculum renewal process.

Awareness and capacity building

Raising awareness is an important precursor to meetings or workshops about changes to the curriculum. As the title suggests, this part of the element comprises the preparatory role of raising awareness among faculty, from junior positions through to senior management and executive, and including faculty and staff. After all, everyone in the department will be interacting with students and the curriculum in some way and needs to be aware of the process and its value. Depending on the organisational culture this might include briefing sessions for senior management and executive on sustainability and graduate attributes, and an address from senior management on the same topics. Awareness raising may use senior faculty and bureaucrats, and take advantage of press releases, media attention and personal association with other leaders in the field. In the absence of this position, activities could also be initiated by other groups within or outside the institution, including students, the discipline's professional body and other special interest groups.

In addition to providing staff with new information, it's also critical to develop a common understanding about what education for sustainability means for the department and the curriculum. The term 'developing a common understanding' does not mean that everyone has to agree, or conformity needs to be achieved. Rather, management needs to be aware of how the proposed curriculum renewal affects the institution and its perspective on priorities. Awareness and capacity building efforts could involve:

- Clarifying graduate attribute expectations of competitors;
- Building faculty and staff capacity to create graduate attribute statements; and
- Building faculty capacity to discuss program opportunities in the context of institutional priorities.

One of many strategies for helping faculty come to terms with what might be possible for their courses is called the 'over the hump exercise' (with thanks to Dr Rowe, US Partnership for Education for Sustainable Development, and Jean MacGregor, Evergreen State College). After the basic introduction, participants are asked to take a big sustainability concept that they have just heard about, and a big concept from one of their courses that they have to teach anyway, and create a learning activity that achieves both – 'Go!' After the usual 'deer in the headlights' reaction, the room suddenly comes alive with staff talking about how this might happen in chemistry, physics, thermodynamics and so on. This group could then be given a wiki page or similar to share their ideas, and meet six months later to talk about their experiences in trialing these ideas. The key point is that raising awareness about possibilities does not have to be expensive to be powerful.[13]

Internal and external stakeholder engagement

Internal and external stakeholder engagement activities span all stages within the curriculum renewal process and play a particularly important support role in the process of identifying graduate attributes, as highlighted in Figure 6.1. It is critical for key parts of the institution, potential employers and key community members to participate in identifying priorities for program goals. Such engagement is also a strategic way for institutions to strengthen their identity, market their program/s and to raise awareness about educational capabilities among potential future students, both for undergraduate and postgraduate courses.

Monitoring and evaluation

Once identified, graduate attributes need to be reviewed each time the curriculum is reviewed. The purpose of this review will be to refresh the memory of faculty monitoring the program, or preparing for a major internal or external review (such as accreditation review), industry advisory panel recommendations, market changes in employer expectations for graduates and changes in legislation (for example, the introduction of a carbon trading scheme or the banning of certain substances in products).[14] Table 6.2 lists some examples of performance indicators across each of the activity streams for tracking the progress of this element within the process of curriculum renewal. The roles of each activity stream are discussed further in the following section.

TABLE 6.1 Performance indicator examples: identifying graduate attributes

Activity stream	*Example performance indicators*
Governance and management	Proportion of invited staff attending the graduate attribute identification workshop (number/event)
Operations and facilities	Involvement in the attribute identification process (number/event)
Teaching and learning	Level of faculty understanding of the sustainability-related graduate attributes, and the associated learning outcomes to be developed (qualitative)
Human resources and culture	Participant satisfaction on the extent of their contribution and the results (qualitative)
Marketing and communications	Level of comprehension of the graduate attribute identification process and opportunities for engagement (qualitative)
Partnerships and stakeholders	Involvement of key partners and stakeholders in the attribute identification process (number/event)

Roles and responsibilities across the curriculum renewal strategy

In the following we consider the deliberative element 'identify graduate attributes' with regard to how this translates across the various stages of curriculum renewal and the relative involvement by each of the various activity streams. For the purpose of illustration we have used a four-stage approach.

Governance and management

Stage 1: Prepare – This stage focuses on the exploration of the context for embedding sustainability into the curriculum, in order to inform the graduate attribute identification process to follow. Hence, the review of external trends outlined in Chapter 5 should include identifying current accreditation requirements, institutional commitments and requirements for curriculum, and industry and market signals (e.g. from advisory boards and market research). It should also include seeking advice from the relevant accreditation bodies regarding future directions, and international best-practice accreditation requirements.

Stage 2: Explore – This stage involves management tasks focused on ensuring that faculty participate in the required workshops and forums to identify and prioritise the graduate attributes. Once the attributes have been drafted, management will need to review the list and might seek external review on the prioritised graduate attribute list from stakeholders including the advisory board, current students and past students.

Stage 3: Test and pilot – This stage involves identifying strategic testing and piloting opportunities that address one or more of the graduate attributes. Management will be overseeing the selection process, which should include reviewing the proposed pilots for their ability to address the attributes. This may include meeting with program convenors about possibilities, and providing clear directive to faculty who will be part of the testing and piloting about timing and priorities.

Stage 4: Integrate – This stage involves management overseeing the integration of graduate attributes into programs and the monitoring of progress over the years of implementation. This could include occasional reporting to communicate progress with faculty and staff in other activity streams, and may include periodically reviewing the curriculum renewal budget for allocation of funds to projects that are targeted towards developing specific graduate attributes. Management will also need to be aware of any significant changes to the external context (see Figure 6.1), which may trigger a revision of the priorities, including, for example, legislative changes and changes to accreditation requirements.

Operations and facilities

For this activity stream, direct contributions to the process of graduate attribute identification will be through participation in awareness-raising activities and participation in graduate attribute identification workshops. In particular, if there are staff within operations and facilities who have relevant expertise, they might be specifically requested to be involved. In the later stages, operations and facilities may be involved in testing and piloting the development of graduate attributes, through collaborative curriculum–campus projects. Staff may also be on the lookout for changes to legislation and regulation which may feed into future revisions of the sustainability-related graduate attributes that affect the operation of the campus.

Teaching and learning

Stage 1: Prepare – This stage is critical in building capacity among staff to participate in the graduate attribute identification workshops that will occur in the next stage. This may include the nomination of key faculty who will be involved in the graduate attribute identification and follow course auditing. Such staff may undertake training or attend additional meetings to become familiar with the process and requirements. Faculty may be asked to consider prior to the workshops what sustainability means for their graduates.

Stage 2: Explore – The key activity is for faculty and teaching and learning support staff to participate in the graduate attribute identification workshop. Prior to the workshop they may also be provided with some key readings (related to sustainability in their field) and asked to brainstorm short-, medium- and long-term graduate attributes that would apply to their discipline. They might also be asked to consider this in light of the 'compliance', 'beyond compliance' and 'sector leader' considerations discussed at the beginning of this chapter. Essentially, more knowledge about opportunities at this stage translates to more detail in the identification process, reducing revisions and updates through the process.

Stage 3: Test and pilot – In this stage, faculty will be occupied with testing and piloting the development of graduate attributes that have been mapped to their courses. From these pilots, they will be reflecting on what worked and what didn't – perhaps the graduate attribute ambitions need to be reconsidered or refined in light of the pilot initiatives. This stage will also include consideration of how graduate attributes can be developed and expanded across multiple courses.

Stage 4: Integrate – This stage provides the opportunity for faculty to take learnings from the testing and piloting, and inform the integration of such content across entire programs, as per the mapping in the next element. This might include periodic review with regard to whether the graduate attribute expectations can be matched in practice. For example, faculty might consider whether the resultant composition of the subject is manageable, with regard to taught and assessed content. It may also involve program convenors periodically consulting with subject convenors collectively regarding any challenges to proceeding with integration.

Human resources and culture

The role of this activity stream across stages related to awareness raising and capacity building will be focused on ensuring that the concept of graduate attributes is embedded within the culture of the institution, facilitating use in formal teaching and learning documentation and performance monitoring. In the preparation stage this will involve human resource personnel becoming familiar with performance requirements directly relating to meeting the development of defined graduate attributes. The exploration stage will involve exploring opportunities to use graduate attribute language within documentation, including in the performance appraisal process and in faculty role descriptions. In the testing and piloting stage human resources will be drafting amendments to various documents with sustainability and graduate attribute language, for consultation with academics, management and other groups as appropriate. The final stage of implementation provides an opportunity for the documents to be finalised and the revisions become live within the system. For example, at this stage faculty may be formally required to report on curriculum renewal endeavours that meet the sustainability-related graduate attribute requirements.

Marketing and communications

The role of marketing and communications for this element will transition from internal to external over the four stages.

Stage 1: Prepare – During this stage the primary tasks will centre on assisting with internal documentation (led by teaching and learning staff and management) about the proposed graduate attribute identification process, and opportunities for academic staff involvement. There may also be some external communication about what is happening with the graduate attribute identification process to promote what the institution is doing in this regard.

Stage 2: Explore – In this stage the marketing and communications staff will be taking the results of the graduate attribute identification activities and using them to create a summary of proposed goals for graduate attributes that can be used to clearly communicate internally and externally the curriculum reforms underway. This summary will also provide a convenient reference guide for faculty, which can, for example, be put on a notice board or posted online.

Stage 3: Test and pilot – In this stage, the materials created in the previous stage can be checked with faculty and other stakeholders before being used in the institution's marketing campaigns. This stage should include a period of review of the initiatives undertaken to date to evaluate the effectiveness of the communication tool, and changes can be made as required in preparation for integration.

Stage 4: Integrate – In this stage, marketing and communications will have a clear understanding of the curriculum renewal underway, and will be able to clearly articulate this internally and to prospective students. As the renewal process continues, the communication documents may need periodic review to ensure that any changes made by other activity streams are incorporated into the communications. Marketing staff may also obtain feedback from stakeholders on how the curriculum renewal transition is perceived, with regard to value, etc.

Partnerships and stakeholders

Stage 1: Prepare – In this stage staff may need to do some capacity building to become familiar with graduate attributes and sustainability to be able to engage with partners and stakeholders in an informed way to encourage their input and involvement in the process.

Stage 2: Explore – Now familiar with what is required, staff will be identifying partners and stakeholders within the current network, who can participate in reviewing the drafted graduate attributes before the attributes are mapped across the course in the next step. Materials from marketing may be used in this process.

Stage 3: Test and pilot – In this stage, staff will be liaising with external contacts regarding progress, and may create further opportunities for faculty to check the progress towards achieving the graduate attributes with external stakeholders and institutional partners. There may be liaison with marketing and communications staff where potential articles/stories exist in the collaboration by stakeholders in this process.

Stage 4: Integrate – Staff in this activity stream will be clear about how departmental stakeholders and partners relate to sustainability within the curriculum, and their perspectives on the process as it has unfolded. Together with marketing and communications, there may be opportunities to share experiences, perspectives and possibly obtain endorsements of support for the transformed curriculum at this stage.

In the next chapter we consider the deliberative element of 'Graduate attribute mapping'.

7

MAPPING LEARNING PATHWAYS

The process of mapping graduate attributes across programs involves discussions about how such attributes can be developed over several years of study, before considering how this might be achieved in each course. With no course actually 'on the table', and no-one's course 'under threat', this level of discussion can be quite proactive and collaborative! However, when we talk to colleagues about mapping learning pathways, this element often gets confused with the previous element 'identifying graduate attributes'. Many workshop participants have not heard about the potential for mapping a series of learning outcomes within a program to develop one or more attributes, assuming it is sufficient simply to highlight program intentions for student capabilities at the end of their studies. Then, when conversation moves to what evidence could show how attributes are being developed, the discussion takes many turns. Layer after layer of confusion about terms and their meanings such as 'graduate attributes', 'competencies', 'learning outcomes', and so on are uncovered and we are all reminded that this field is as much another language as any foreign language, with many dialects!

In reality, most faculty probably 'map' learning pathways to achieve graduate attributes to some extent. Perhaps you have sat in assessment meetings or program planning meetings where topics such as 'design' are discussed with regard to 'who teaches what and where'. Perhaps this has been in the context of developing a particular skill such as computer aided drawing. Your department may have even been asked to show evidence of program mapping for attributes to do with 'professional practice', 'sustainability' or 'ethics' at a recent accreditation review. However, in practice, there are still few examples of detailed mapping for programs, across the full suite of graduate attributes that students are supposed to develop by the time they graduate. Further, this development appears to be more of an assumed 'accumulation' that will automatically occur, given the number and range of courses on offer.

With this in mind, this chapter begins by explaining what we mean by 'mapping', discussing a number of key terms and the evolution of the practice. Building on Chapter 6, which discussed the identification of graduate attribute statements, we explore how mapping fits into the process. For every graduate attribute that has been selected, there will be one or more component knowledge and skills – or competencies – that need to be embedded into student learning, across one or more courses, for that attribute to be developed by graduation.

The chapter then discusses how these competencies translate into learning outcomes with mapped pathways for developing graduate attributes in the curriculum. This includes a method for distilling competencies and learning outcome statements and mapping them, course by course. We also discuss why – in contrast to the process for identifying graduate attributes, which requires participation from the whole department – the mapping process can be undertaken by a small team including the program convenor, and teaching and learning expert.

Exploring the deliberative element 'Mapping learning pathways'

Introduction

Curriculum mapping is not new in the primary and high schools sector literature, although the interpretation of the term is somewhat different to what we are proposing here. In the late 1980s American educational leader and researcher Fenwick English was one of the first to develop the area of 'curriculum mapping' and also created a 'curriculum management audit' for schools and colleges. This early work has inspired researchers and educators for decades,[1,2] with authors such as Jacobs having written further about mapping processes for schools. This includes a curriculum-mapping model where teachers can produce a variety of maps through systematically collected information about the actual content taught in a classroom.[3,4] As Jacobs reflects in a foreword to another colleague Janet Hale's 2008 book on mapping, 'Curriculum mapping promotes a significant transition into 21st-century solutions to age-old problems of articulation and instruction.'[5]

Within the higher education sector, the literature is more recent and focused mainly on the medical and health-related disciplines. For example, in 2001 Harden defined curriculum mapping as 'a spatial representation of the different components of the curriculum so that the whole picture and the relationships and connections between their parts are easily seen'.[6] Essentially, in higher education, curriculum mapping involves identifying knowledge and skills taught in each course at each level, and then creating a calendar/year-based chart, or 'map' for each course that tracks the development of these skills across the program.[7,8,9] Over the last five years, Australia has taken a lead in applied research into mapping student learning in higher education, with many universities having some reference to mapping graduate attributes, or mapping competencies on their websites.[10,11,12,13] For example, in Western Australia, Curtin University's Office of Teaching and Learning developed a mapping tool as part of the University's Curriculum 2010 (C2010) Project, a 3-year initiative completed in 2009.[14] This included an Australian Learning and Teaching Council Fellowship on benchmarking partnerships for graduate employability.[15] Also in Western Australia, Murdoch University's Teaching and Learning Centre developed a commercial web-based mapping tool to facilitate the mapping of graduate attributes across programs in the university's Graduate Attribute Mapping Program.[16,17]

In 1997 an Australian Academy of Technical Societies and Engineering review of engineering education in Australia, concluded that[18] it is imperative that environmental issues are integrated into single branch engineering courses, such as electrical engineering, civil engineering etc. Within these courses, environmental issues should be integrated into existing course modules, as well as being taught in specialised environmental subjects.

Building on discussions in Chapter 6, graduate attributes reflect the knowledge and skills – or competencies – that a student should develop by the time they graduate, with the curriculum designed to develop such attributes within various courses in a program. Once preferred attributes are identified, the objective is to then map their development within a program across the year levels in a systematic way to create what we refer to as 'learning pathways'. Considering these learning pathways, a process can then be undertaken to identify the most suitable course/s to integrate each competency to develop the desired graduate attributes. These learning pathways and selected course/s may call for minor changes within one or more courses, through to the creation of completely new courses.

Learning pathway maps can become quite complex in their portrayal of graduate attribute development. Figure 7.1 provides an example of a summary that is tabulated, highlighting where competencies are learnt, practised and demonstrated across each year of study. In this example:

- 'Learn' refers to the presentation of the knowledge or skills related to the attribute to provide a basis for their practice and demonstration;
- 'Practise' refers to students repeatedly using the knowledge and theory; and
- 'Demonstrate' refers to the students being assessed as to their ability to apply the knowledge and theory to problem solving in a contextually appropriate way.

Component knowledge (K) and skills (S) [learning outcomes]		Year of Study			
		1	2	3	4
K1	Awareness of ...	**Learn**			
K2	Appreciation of ...		**Learn**	**Practise**	
S1	Ability to ...	**Learn**	**Practise**		**Demonstrate**
S2	Ability to ...	**Learn**		**Practise**	**Demonstrate**

FIGURE 7.1 Tabulated example of learning pathways for competencies for a particular graduate attribute

Considering Figure 7.1, a single course may be selected to introduce (at the level of 'learn') each of the three identified first-year knowledge and skill areas (i.e. K1, S1 and S2). This might result in a course outline that includes the following statement of learning outcomes:[19]

On successful completion of this course you will have:

- an awareness of, and ability to describe, the central tenets of sustainability and how these apply in the development of solutions;
- an understanding of what is required to communicate using a selected range of professional strategies;
- an appreciation of critical thinking, problem solving and reflection about sustainability in the context of realistic challenges or scenarios.

This process could be repeated for each of the knowledge and skill areas in the table, resulting in a suite of learning outcomes within a number of courses in each year of study that together form the learning pathways map for this graduate attribute.

Considering a civil engineering program example, a graduate attribute of 'proficiency in design of biological treatment systems for potable water' may be developed through a number of knowledge and skill areas that can be mapped to various courses across the year levels, including a course on 'Introduction to Design' in first year, 'Biological Treatment' in second year, and so on.

In another example, a graduate attribute of 'ability to evaluate engineering impacts on the environment' may be developed by addressing two competencies:

1. 'An understanding of the limitations and thresholds of environmental systems', through increasingly demanding learning outcomes in design courses such as 'Design 1', 'Design 2' and 'Design 3'.
2. 'The ability to calculate carbon emissions from the construction and operation of infrastructure'. In first year students may learn about carbon emissions and metrics, followed by exposure to calculations in second and third year, and practice in final year. A final-year course learning outcome might state that the student will be 'proficient in undertaking greenhouse gas emission calculations'.

While it is important to focus on key knowledge and skills needed to develop graduate attributes in specific disciplines, it is also important to focus on developing cross-disciplinary competencies. As shown in Figure 7.2 learning pathways can incorporate competencies that support both 'technical' and 'enabling' knowledge and skill development, which can be delivered as common courses across disciplines.

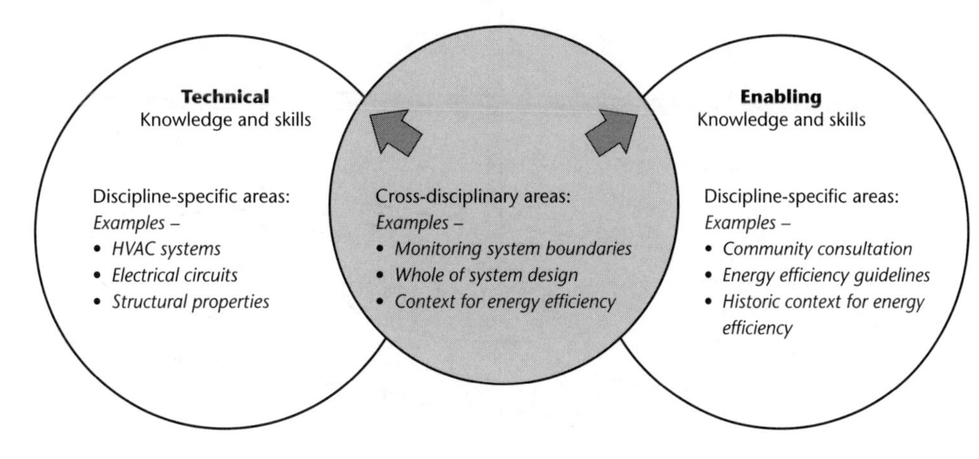

FIGURE 7.2 Illustration of the spectrum of knowledge and skill sets possible

Source: Adapted from a figure developed by the authors for the Australian Federal Government, with regard to energy efficiency education[20]

A consultation of engineering education institutions in Australia led by the authors and commissioned by the federal government identified numerous such cross-disciplinary competencies, including:[21]

- 'Big picture' thinking, whole systems design and whole-of-life analysis (in particular, ability to weigh lifetime costs and cost savings, versus initial expenditure).
- Ability to effectively communicate efficiency opportunities to key stakeholders, and to present a basic financial case for initial up-front investment in better technologies.
- An awareness and capacity to draw upon the specialist skills of other disciplines (industry representatives considered this to be a foundational tool for achieving the best outcomes through interdisciplinary collaboration).

Considering the range of technical and enabling knowledge and skills likely to be identified for any graduate attribute, the question arises: 'How do you teach this variety of requirements within an existing, highly regulated curriculum structure?' Perhaps you have experienced this frustration in the past, where there has been unnecessary duplication or omission of important enabling or technical knowledge and skill areas.

The following pages explore this challenge of optimising program deliverables through the process of mapping, including the potential for shared curriculum, where technical and enabling knowledge and skills are transdisciplinary in nature, as shown in Figure 7.3. Opportunities also arise here to reduce time and costs in shared curriculum development.

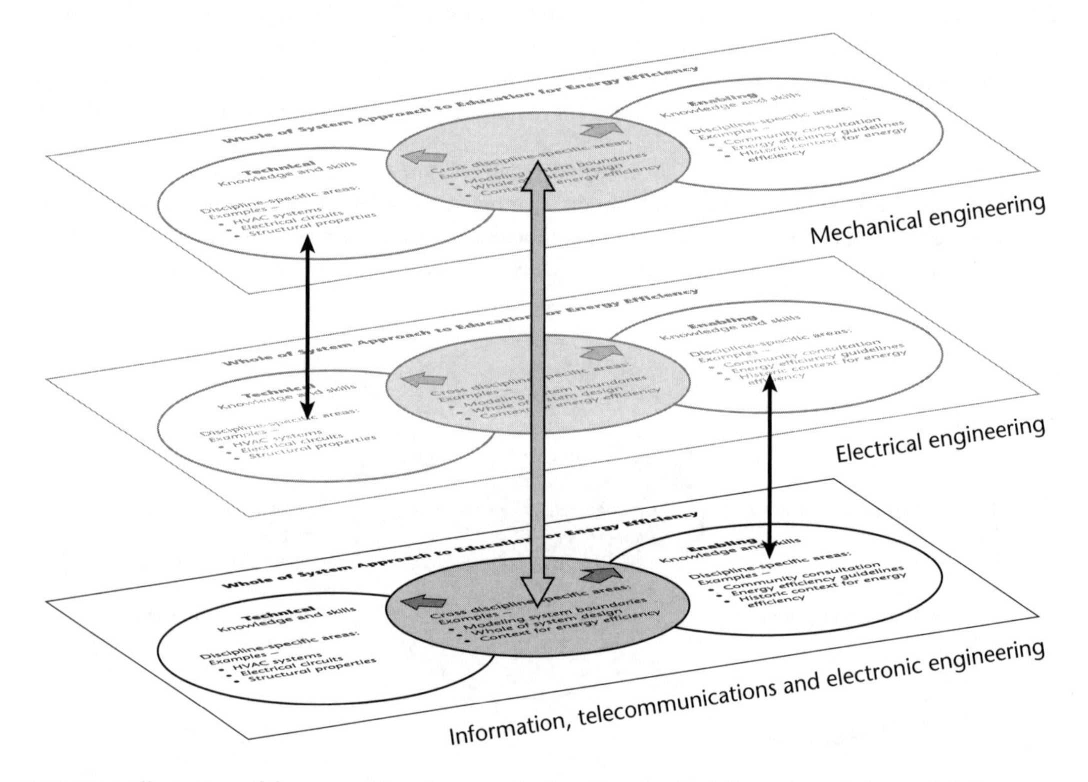

FIGURE 7.3 Illustration of the potential to share curriculum for transdisciplinary knowledge and skills

Source: Adapted from a figure developed by the authors for the Australian Federal Government, with regard to energy efficiency education.[22]

Assistance with mapping learning pathways

In the following paragraphs we outline a process to map learning pathways that has been developed and trialled in collaboration with a number of leading sustainability educators around the world.

Three key steps to mapping learning pathways include:

1 Identifying knowledge and skills (component competencies) for each graduate attribute;
2 Assigning those component competencies to one or more courses; and
3 Creating learning outcome statements for each course that integrates the competencies to clarify intent and scope.

Step 1: Identify competencies for each graduate attribute

This step involves the identification of specific knowledge and skills for each of the preferred graduate attributes. The process provides a tangible list of knowledge and skills to be used as the basis for mapping and also auditing (see Chapter 8). Unlike the graduate attribute identification process that involves a wide coverage of persons involved in curriculum design and delivery, determining the specific competencies for the prioritised attributes is best done within the program/discipline context. This may include the program convenor, to ensure consideration of various constraints such as staffing, budget, student numbers and so on; a selection of faculty knowledgeable about the program and its content and involved in its delivery; a person well versed in the sustainability context for the specific discipline (that may be external); and a learning and teaching expert to provide content, delivery and assessment advice. Once the competencies have been identified they may be peer reviewed by selected faculty and industry representatives or reviewed as part of a larger workshop.

An example of a graduate attribute and possible associated competencies is presented in Table 7.1. Extrapolating from the example, if there are several graduate attributes, there may be approximately 15–20 competencies associated with the subsequent map of learning pathways.

TABLE 7.1 Example graduate attribute and competencies

Graduate attribute	*Associated competencies*
1 Capacity to optimise/balance environmental, economic and social aspects into the design process.	1.1 Ability to identify and qualify impacts of the process. 1.2 Ability to identify and quantify economic impacts, short and long term. 1.3 Ability to identify and quantify social impacts, short and long term. 1.4 Knowledge of the ways in which systems interact.

Step 2: Map learning pathways

Once the list of competencies has been generated for each graduate attribute, consideration can be given to the appropriate location in the program, beginning with considering which year level they should be developed in, as shown in Figure 7.1. The next step is to identify specific courses that will provide knowledge and skills to develop the attribute, with some already existing and some needing to be developed, as shown in Figure 7.4.

The figure shows that, for example, in Semester 1 of second year, CH2051 has been assigned to incorporate 'Knowledge and Skill 1.1', with a high level of coverage. Further, in the first semester of fourth year, all knowledge and skills (1.1–1.5) will be covered, potentially within an integrated capstone course.

Using such a stylistic tabulated approach there are a variety of ways to communicate the location, level of coverage and emphasis in each of the courses, such as:

- Using a triangle to indicate a growing level of exposure across the year, and a rectangle to show a consistent coverage throughout the year. The height of the shapes can be used to indicate the relative emphasis compared with others.
- Using existing course codes or proposed new course codes. The level of coverage of the knowledge and skills for each unit can then be designated as either (L) for low, (M) for medium, or (H) for high.
- Using a dotted leader-line to indicate that component knowledge and skills may be referred to in this period (i.e. informally), without being assessed, in preparation for future coverage.

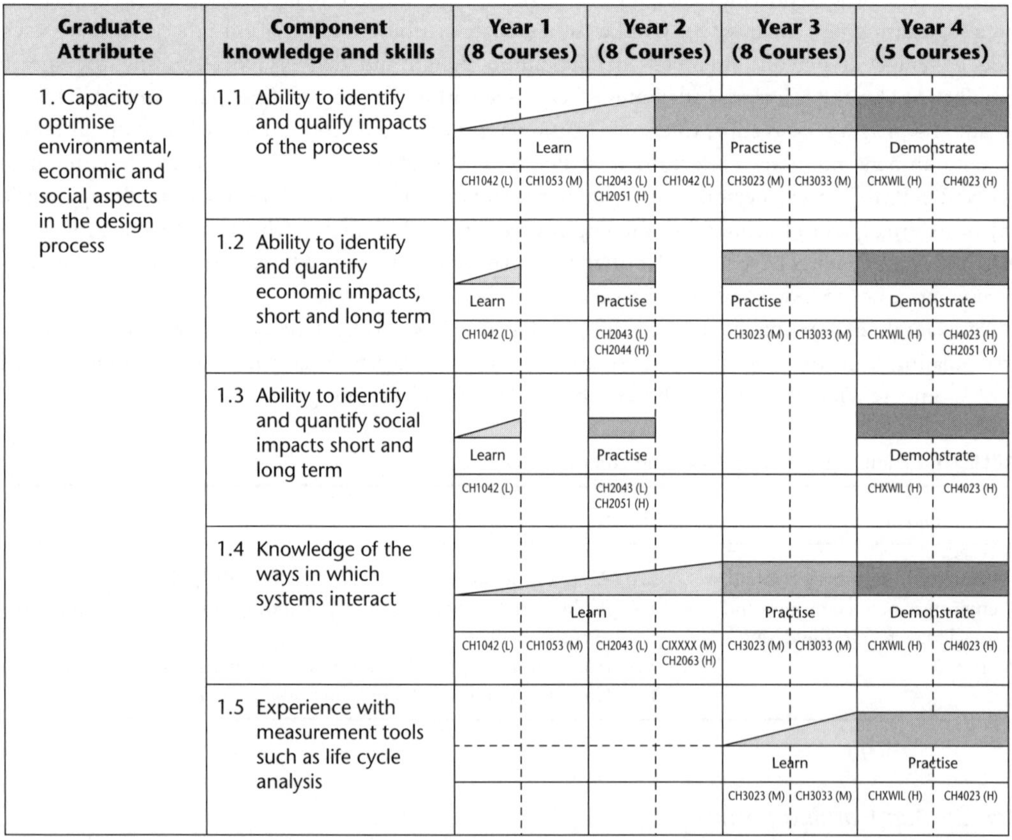

Graduate Attribute	Component knowledge and skills	Year 1 (8 Courses)	Year 2 (8 Courses)	Year 3 (8 Courses)	Year 4 (5 Courses)
1. Capacity to optimise environmental, economic and social aspects in the design process	1.1 Ability to identify and qualify impacts of the process	Learn CH1042 (L) CH1053 (M)	CH2043 (L) CH2051 (H) CH1042 (L)	Practise CH3023 (M) CH3033 (M)	Demonstrate CHXWIL (H) CH4023 (H)
	1.2 Ability to identify and quantify economic impacts, short and long term	Learn CH1042 (L)	Practise CH2043 (L) CH2044 (H)	Practise CH3023 (M) CH3033 (M)	Demonstrate CHXWIL (H) CH4023 (H) CH2051 (H)
	1.3 Ability to identify and quantify social impacts short and long term	Learn CH1042 (L)	Practise CH2043 (L) CH2051 (H)		Demonstrate CHXWIL (H) CH4023 (H)
	1.4 Knowledge of the ways in which systems interact	Learn CH1042 (L) CH1053 (M)	CH2043 (L) CIXXXX (M) CH2063 (H)	Practise CH3023 (M) CH3033 (M)	Demonstrate CHXWIL (H) CH4023 (H)
	1.5 Experience with measurement tools such as life cycle analysis			Learn CH3023 (M) CH3033 (M)	Practise CHXWIL (H) CH4023 (H)

FIGURE 7.4 Example map of learning pathways across a program for one graduate attribute

Source: The example learning pathways are drawn from the outcomes of the James Cook University, Curriculum Refresh Project, assisted by the authors and described in Sheehan, M., Desha, C. *et al.* (2012)[23]

Now that the graduate attribute has been mapped across the program, each course can be allocated the various knowledge and skills to be integrated, as shown in Table 7.2. The resultant map and summary table then informs the creation of learning outcomes for the assigned courses (Step 4). As the next stage in the deliberative and dynamic model is to 'audit learning outcomes', the map and table also provide a basis to audit the current curriculum with regard to how well these knowledge and skills are developed.

TABLE 7.2 Example table of competencies (knowledge and skills) assigned to courses

Course code	*Course name*	*Competencies to develop the graduate attribute and course emphasis*				
		K&S 1.1	*K&S 1.2*	*K&S 1.3*	*K&S 1.4*	*K&S 1.5*
CH1042	Introduction to Professional Practice	Learn (L)	Learn (L)	Learn (L)	Learn (L)	–
CH1053	Chemistry 101	Learn (M)	–	–	Learn (M)	–
CH2043	Introduction to Energy Balance	Learn (M)	Practise (L)	Practise (L)	Learn (L)	–

Course Emphasis: (L) Low; (M) Moderate; (H) High

Step 3: Create learning outcome statements for each competency

In preparation for the 'auditing learning outcomes' element, it is recommended that the final step in the mapping process is to create learning outcome statements for each of the competencies, in each of the respective courses. Considering the detail developed in Figure 7.4, the statements should relate to each of the three stages of 'learn', 'practise', and 'demonstrate', as well as reflecting the level of emphasis, either 'low', 'medium' or 'high'.

There is a wealth of information to assist with writing learning outcome statements, with universities often having a guide or template. In essence the short statements are about what students are expected to be able to do by the time they complete each course. Each learning outcome statement should include a verb (i.e. an action statement) such as 'know', 'understand', 'be aware of', 'recognise', 'use' and so on. Importantly, the statements should be able to be checked with regard to what the student actually learns, through course assessment. In this way the statements will also support the auditing process which is focused on identifying the current level of coverage in the particular units.

Once the learning outcome statements are compiled, they can be checked to make sure that unwanted duplication and/or omissions are avoided within the overall program of study. Further, learning outcome statements can be designed to reinforce each other in language and scope, to create clear pathways for the attribute being developed.

CREATING LEARNING OUTCOMES

Consider the graduate attribute, 'Have an understanding of existing and emerging legal requirements for the manufacture and disposal of electronic goods'. Several corresponding competencies to develop this attribute could include 'ability to identify environmental impacts from the disposal of electronic goods', 'knowledge of national and international electronic waste regulations', and 'the ability to monitor for emerging regulatory requirements'. Once identified, these competencies can now be developed into learning outcomes for courses. Take, for example, the third competency, 'the ability to monitor for emerging regulatory requirements' by graduation:

- The first year of study might include a compulsory course that addresses the knowledge competency with a learning outcome of 'knowledge of international electronic waste regulations'.
- In the second year of study another compulsory course may go some way to developing the ability competency, with a learning outcome of 'knowledge of national and international regulatory mechanisms for addressing environmental and social impacts of e-waste'.
- In final year, a design course or major project may include the learning outcome 'ability to track and evaluate the implications of regulatory requirements', where students can apply their knowledge of regulations and mechanisms to seek out for themselves emerging and potential future requirements.

Examples of this element in action

Example: James Cook University, Townsville

The previous chapter highlighted James Cook University's 'curriculum refresh' project, and the process used to identify graduate attributes was discussed. This project also included a pilot component to map learning pathways. Each discipline experienced the process for mapping learning pathways for one example graduate attribute.

The process used for mapping learning pathways was face-to-face discussions, following a faculty-wide workshop where desirable sustainability-related graduate attributes had been determined. On the day following the workshop, the authors and the department's program convenor met with discipline leaders one-on-one for 2-hour sessions. During the session the team considered, for one sustainability-related attribute, what the competencies would comprise, and then how learning pathways could ideally be mapped across the program to develop these knowledge and skills.

An observed benefit of this process included the ability of key faculty to make decisions during the workshop about changes to course content and to make commitments regarding the future swapping out of superseded knowledge and skills. Once faculty were familiar with the process, they were able to subsequently apply this to the rest of their graduate attributes without further assistance from our team.

Example: Queensland University of Technology, Brisbane

Another example of a map at the level of competencies is demonstrated in part of Queensland University of Technology's restructure of its Bachelor of Science program in 2012. The program included an environmental science major, which was earmarked for teaching by an interdisciplinary team of environmental engineering and science faculty. Furthermore, the timeframe for curriculum renewal required board consideration within six months.

Within this time-constrained and interdisciplinary stakeholder context, the curriculum renewal committee decided to undertake a learning pathway mapping process before commencing any detailed course review. Following a 2-hour graduate attribute brainstorming workshop, a second workshop was convened with the Head of School, teaching and learning expert, program convenor and a number of faculty who had an interest in the proposed new major. In the workshop, participants identified in which years and to what extent the priority graduate attributes should be developed. The teaching and learning expert then assisted the group by email and individual meetings, to establish learning outcome statements that needed to be embedded within the proposed major, across all semesters.

The outcome of this review included modifications to some existing courses taught within the school, and the addition of a number of new courses to address gaps. According to the Program Convenor Dr Ian Williamson,

> this whole of curriculum consideration of what was possible allowed us to put aside prior perceptions of 'what should be' and our own history with our courses, to imagine a new program that met the School's needs. Furthermore, we were able to get to updating and writing course outlines much more quickly than our colleagues in other minors where this mapping approach was not used.

Interaction with 'dynamic' elements of the model

Awareness and capacity building

The type of awareness and capacity building needed for successful mapping differs from that required for identifying graduate attributes. For this element, the key focus is two-fold:

- To empower those faculty and staff members who will be responsible for participating in the mapping process. This includes ensuring the team is familiar with the process, clear about the sustainability-related graduate attributes that have been selected and the associated competencies to be mapped (ideally involved in the process to select them), and understands staffing and budget constraints and opportunities that exist within the institution.
- To raise awareness among senior management, faculty and across the institution about the process, the value of the results and the implications for the curriculum renewal process. This includes creating appropriate communications to efficiently receive comments on drafts, and support buy-in to the results of the process.

Depending on organisational culture, activities that will increase the level of awareness of the need for curriculum renewal to embed sustainability and provide a capacity building base may include:

- briefing sessions for senior management and executive;
- an address from senior management to staff in the form of a briefing;
- keynote sustainability-related lectures by leaders in the field;
- profiling existing sustainability-related initiatives within the institution;
- profiling existing sustainability research or teaching champions within the institution;
- scheduling documentaries or current affairs coverage in lunchtime seminars; and
- distributing journal and media articles through an e-news list, intranet site or in the tea room/ meeting room.

Building on from such activities, particular faculty may be identified to receive capacity building in areas such as:

- the curriculum renewal process to be used;
- methods to interact with and enhance collaboration with other faculty; and
- the relevance and application of sustainability to specific disciplines and subjects.

Internal and external stakeholder engagement

Internal and external stakeholder engagement is an important aspect of the mapping process, with the goal to have 'the right people talk to the right people at the right time'.

Internal stakeholders may include those at higher levels that are not directly involved in the mapping process to identify constraints and opportunities at the institution level. These may include insights on available budget, faculty resourcing and accreditation requirements.

External stakeholders may include potential employers, accreditation agencies and future students who can inform the mapping process to ensure it is grounded in the kinds of competencies and learning pathways of interest.

Monitoring and evaluation

It is important to include periodic monitoring and evaluation of the mapping as part of the curriculum renewal strategy to ensure that the graduate attributes are being developed and that they are being developed in the appropriate location. Mapping can be influenced by a range of factors such as changes in policy and industry practices that affect employer expectations of graduates and can be advised on by advisory panels. As part of the curriculum renewal strategy a number of indicators can be considered to inform progress in this area, as suggested in Table 7.3.

TABLE 7.3 Sample performance indicators for mapping learning pathways

Activity stream	*Example performance indicators*
Governance and management	Proportion of invited faculty attending briefing sessions (number/event) Proportion of invited faculty participating in mapping (number/event)
Operations and facilities	Involvement in the attribute mapping process (number/event)
Teaching and learning	Level of faculty understanding of the sustainability-related graduate attributes, and where in their program they are intended to be developed (qualitative)
Human resources and culture	Participant satisfaction on the extent of their contribution and the results (qualitative)
Marketing and communications	Level of comprehension of the graduate attribute mapping process and opportunities for engagement (qualitative)
Partnerships and stakeholders	Involvement of key partners and stakeholders involved in mapping process (number/event)

Roles and responsibilities across the curriculum renewal strategy

In the following paragraphs we consider the deliberative element 'mapping learning pathways' with regard to how this translates across the various stages of curriculum renewal and the relative involvement by each of the various activity streams. For the purpose of illustration we have used a four-stage approach.

Governance and management

Stage 1: Prepare – This stage focuses on the identification of competencies for each graduate attribute, informed by the consideration of external trends as part of the curriculum renewal strategy development. It is important that governance and management staff involved in the process have access to guidance on the current and anticipated regulatory and market changes and expectations about graduate capabilities in the area. Decisions then need to be made as to how this guidance affects the graduate mapping process, such as the level of emphasis of various component knowledge and skills.

Stage 2: Explore – This stage involves identifying staff who will be involved in the knowledge and skills identification process, and subsequent mapping at a program level. Given the importance of this step in the curriculum renewal strategy, management will need to make sure that sufficient resources are available to these staff so that it can be prioritised in their workload. During the identification and mapping of the knowledge and skills across the program, management may also be required to make decisions regarding possible changes to courses and the addition of new courses, to meet the sustainability-related graduate attribute requirements.

Stage 3: Test and pilot – This stage involves identifying strategic testing and piloting opportunities to implement the mapped learning pathways. Management will be overseeing the selection process, which should include reviewing the proposed pilot projects for their alignment with the identified learning pathways, and ultimately the identified graduate attributes. This may include meeting with program convenors about possibilities, and providing clear directives to faculty who will be part of the testing and piloting about the focus on sustainability-related graduate attributes.

Stage 4: Integrate – In this stage management will be overseeing the expansion of the successful outcomes from the previous stage with the goal of implementing all learning pathways. As noted for graduate attribute identification, this could include occasional reporting to communicate progress with faculty and staff in other activity streams. It may also include periodically reviewing the curriculum renewal budget and prioritising the allocation of funds to projects targeting the development of specific graduate attributes. Management will also need to be prepared for a potential revision of the mapped learning pathways (stemming from a review of the graduate attributes discussed in the previous chapter), which may be prompted by legislative changes and changes to accreditation requirements.

Operations and facilities

For this activity stream, initially staff from operations and facilities will be primarily involved in participating in awareness-raising activities either to increase awareness or capacity to contribute to the process or to provide other staff across the university with guidance as to the potential contribution from operations and facilities in curriculum renewal activities (such as the potential for student involvement in actual infrastructure management projects, including the installation of renewable energy technology or bicycle facilities). Further staff in this area may provide valuable contributions to the process to identify the component knowledge and skills associated with the selected graduate attributes. In the later stages, as for the previous element, operations and facilities may be involved in testing and piloting initiatives related to the implementation of learning pathways, such as through collaborative curriculum–campus projects. Staff may also be on the lookout for changes to legislation and regulations that affect the universities' operations that may inform future revisions of the sustainability-related knowledge and skill areas.

Teaching and learning

Stage 1: Prepare – This stage will be critical to build capacity among the faculty who will be involved in the mapping process. This includes understanding the overall curriculum renewal strategy and the curriculum mapping process, in particular the need to identify specific component knowledge and skills for sustainability graduate attributes. These personnel will also need to have a thorough understanding of the current curriculum, which will be helpful when identifying possibilities for swapping or creating new courses for the program being mapped.

Stage 2: Explore – In this stage the selected faculty may undertake the mapping exercise for some or all of the graduate attributes selected. This process may be time consuming and will require support from management, and may need to be undertaken in consultation with other faculty as needed. The more knowledgeable faculty are about how sustainability relates to their areas of teaching and the opportunities it presents, the less potential for revisions and updates later in the process.

Stage 3: Test and pilot – This stage involves the selection of a learning pathway/s to pilot in a particular program. This will require those involved in the mapping process to assist, given their level of understanding of the intent of the map. Subsequent to the trials there may be a need to amend or refine the knowledge and skills, and potentially the pathways for their development. This will be valuable feedback to inform the wider implementation of learning pathways across various programs, and provide evidence of the impact of such initiatives.

Stage 4: Integrate – This stage involves the implementation of learning pathways for the graduate attributes across entire programs. This will include ensuring that the rollout matches a logical order for students in the transition, to minimise issues with split programs. As for the previous attribute, this may involve program convenors and academics discussing future potential challenges to the rollout with the map of learning outcomes, so that these can be addressed in advance.

Human resources and culture

The role of this activity stream across stages related to mapping learning pathways will be focused on ensuring that the concept of mapping learning pathways is embedded within the corporate culture of the department, which may involve incorporation into position descriptions, teaching and learning documentation, and performance monitoring processes. In the preparation stage, human resource personnel will need to be familiarised with the process to map learning outcomes across programs and represent them in a stylistic manner. In the exploration stage it will be important that staff, especially new staff, are familiar with the mapping diagrams and tables, and are able to contribute to this aspect of the curriculum renewal strategy. In the testing and piloting stage, the priorities for human resources may include assisting staff to amend various documents with the sustainability and graduate attribute language, consulting with academics, management and other groups as appropriate. In the final stage of implementation, human resources may be required to identify additional staff with appropriate skills to implement the learning pathways across various programs and to ensure that the associated protocols for doing so are appropriately developed or amended.

Marketing and communications

The role of marketing and communications for the element will be very similar to that of the graduate attribute identification element, transitioning from an internal institution focus to an external focus over the duration of the four stages.

Stage 1: Prepare – During this stage the primary tasks will centre on assisting with internal documentation (led by teaching and learning staff and management) about the intended mapping process that will be undertaken at the level of the program convenor, in consultation with faculty. There may also be some external communication about what is happening with the mapping process, to promote what the institution is doing in this regard.

Stage 2: Explore – In this stage the marketing and communications staff will be working with the teams creating the learning pathways to create a briefing of proposed learning outcome integration for courses within the program. This summary will also provide a convenient reference guide for staff, which can, for example, be put on a notice board or posted online. This is an important aspect of the curriculum renewal strategy as the greater the understanding of the process across the entire university and within its networks, the greater the level of participation.

Stage 3: Test and pilot – In this stage marketing and communications will play a key role in ensuring that the activities to embed sustainability graduate attributes across the selected programs are well known and key successes are well communicated. This stage should include marketing and communications staff assisting staff within the programs to communicate to other staff, tutors and students as to the changes being made and the rationale for doing so.

Stage 4: Integrate– In this stage marketing and communications will have a clear understanding of the curriculum renewal underway, and will be able to clearly articulate this internally and to prospective students. As the renewal process continues, the communication documents may need periodic review to ensure that any changes made by other activity streams are incorporated. Marketing staff may also obtain feedback from stakeholders on how the curriculum renewal transition is perceived.

Partnerships and stakeholders

Initially the primary role for partners and stakeholders will be to review the results of the mapping process, including the component knowledge and skills lists and their location across programs. Hence, it will be important for staff working with partners and stakeholders to become well versed in the curriculum renewal strategy and in particular the concept of curriculum mapping. In the later stages staff will need to identify and approach specific partners and stakeholders with appropriate experience and position to participate in the review processes. As part of the pilot and rollout of the mapping process, partners and stakeholders may be called upon to provide guest lectures, take student placements and to contribute to other curriculum aspects.

In the next chapter we consider the deliberative element of 'Auditing learning outcomes'.

8

AUDITING LEARNING OUTCOMES

Building on from the development of preferred graduate attributes in Chapter 6, the mapping of these attributes within programs to develop learning pathways and the identification of specific courses to support this in Chapter 7, this chapter focuses on gaining an understanding of the task at hand through an audit of learning outcomes at the course level. The chapter outlines a collaborative and non-confrontational process to systematically review existing courses in a program for the existence of knowledge and skills related to the preferred graduate attributes. The process seeks to both identify areas of strength and weakness, and consider threats and opportunities for the course. At this point the proposed audit approach addresses 'what's so' in the current curriculum, which then informs the curriculum renewal strategy.

Responding to the outcomes of the audit can deliver multiple benefits, not only in the development of the graduate attribute at the program level, but also to avoid re-inventing the wheel ('I didn't know you covered that too?'), missing key steps in developing an attribute ('I don't understand why my students can't do this by the time they get to my subject – I am wasting time covering this'), or missing key concepts altogether ('I thought someone else was covering that').

In summary, it is clear to us that timely curriculum renewal towards education for sustainability relies on an audit that produces a clear understanding about 'what's so' in the curriculum, versus 'what should be'. It is equally important that this process is undertaken in a collaborative, transparent and non-confrontational way (supported, or even led, by external facilitators). Rigorous assessment is critical in working out where the curriculum is performing well in developing the attributes, where it could be improved, and what priority it is for renewal. The audit also helps to define what expertise might be present or lacking in the department for potential future appointments and/or collaboration.

Exploring the deliberative element 'Audit learning outcomes'

Introduction

Within literature related to education for sustainability, there are numerous calls for existing programs to be urgently assessed as to the current coverage of sustainability-related knowledge and skills. The literature often cites students learning about sustainable development in *ad hoc* ways within one or more particular courses, rather than across programs – leading to limited application in professional roles.[1,2,3] However, while there are a number of methods available for assessing how well higher education institutions are addressing sustainability in general,[4,5] there is limited precedent for auditing programs at a program level for evidence of integrating sustainability knowledge and skills designed to develop particular graduate attributes. Considering the growing imperative for education for sustainability as discussed in Chapter 1, we anticipate that such detailed investigations will soon become mainstream as a matter of ensuring quality assurance in higher education offerings.

To systematically enquire into what is needed, the deliberative and dynamic model for curriculum renewal' next focuses on auditing learning outcomes. When we use the term 'audit' we mean to describe the process of reviewing existing courses to identify the level of coverage of knowledge and/or skills assigned to them, and assessing if they are to be created, updated or omitted. Audit materials include various components of the curriculum such as subject outlines, lecture and tutorial content, readings and assessment requirements.

As shown in Figure 8.1, this is the logical next step after identifying graduate attributes, mapping these attributes across programs and assigning them to particular courses.

> Critical determinant of success of the curriculum renewal strategy will be the candid and accurate assessment of coverage of sustainability knowledge and skills in particular courses and subsequently well-informed decisions around creating new content, updating existing content or not making any changes.

A valuable aspect of the audit process is that it provides an opportunity to identify efforts underway in various courses to embed sustainability and considers ways to build on these efforts to strengthen the overall program offering. Indeed, through formalising previously *ad hoc* approaches, existing champions and faculty with ideas gain increased access to management, to whom they can propose ideas in a structured environment.

Given that graduate attributes are often developed within more than one course, it is important that the audit process be mindful of potential inconsistencies in the language, terms and positions to ensure consistent messaging across the program. It is common for students to be exposed to sustainability by a faculty sustainability champion in their course, only to hear inconsistent and even contradictory materials in other courses. This is a serious risk to programs professing to include sustainability but delivering a fragmented approach as students now have access to worldwide instant communication to comment on and endorse programs.

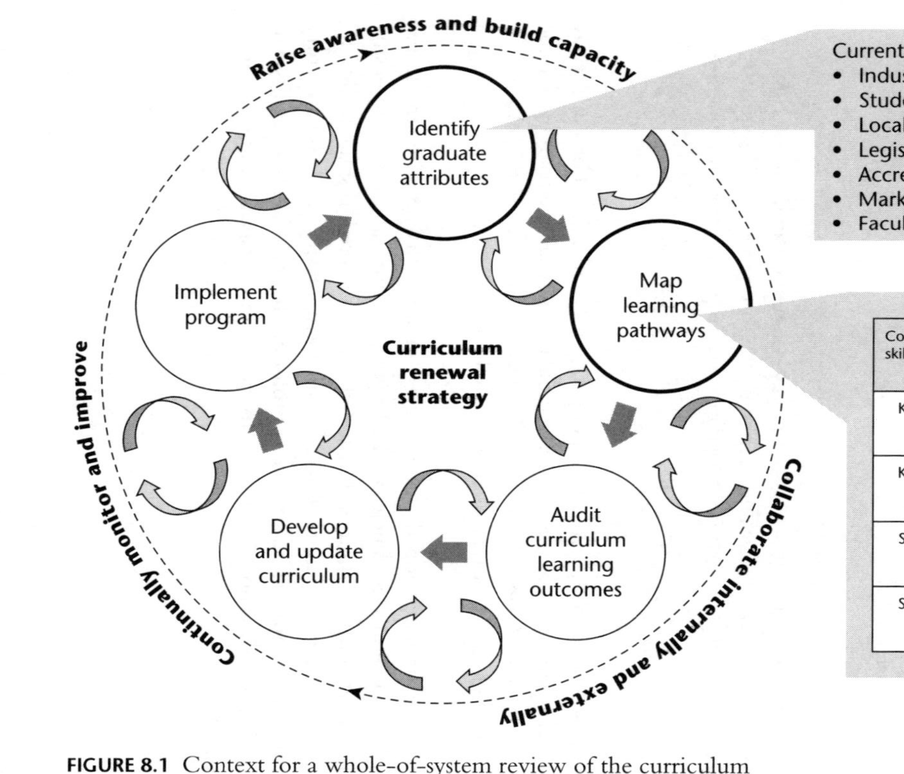

Current and future

- Industry and employer expectations
- Student demand
- Local and regional community context
- Legislation and agreements
- Accreditation requirements
- Market niche opportunities and reputation
- Faculty interests and capacity

Component knowledge (K) and skills (S) [learning outcomes]		Year of Study			
		1	2	3	4
K1	Awareness of …	Learn			
K2	Appreciation of …		Learn	Practise	
S1	Ability to …	Learn	Practise		Demonstrate
S2	Ability to …	Learn		Practise	Demonstrate

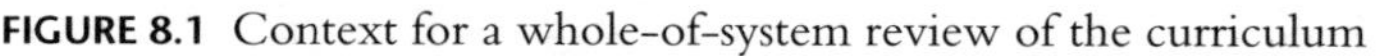

FIGURE 8.1 Context for a whole-of-system review of the curriculum

Once each course has been considered for coverage of its assigned sustainability knowledge and skills, the operational realities will be clear. The audit may recommend actions within courses, in addition to actions across courses at a program level, to be reflected in the learning pathways throughout the program.

Addressing faculty concerns

When discussing auditing with colleagues, there is usually great buy-in to the need for a transparent summary of what is happening within programs, but resistance to 'being the one' to manage an audit within their institution. Some of this resistance seems to be based on past experiences with quite confrontational processes that have diminished the standing of faculty members. Other colleagues have participated in detailed assessments only to find that their contributions regarding improvements were not included, or the reports didn't amount to the anticipated change. With such reflections from colleagues, no wonder there is hesitation in proceeding with another audit!

There is no 'short cut' option for auditing the curriculum. In response to colleagues who are eager to 'get on with it' and begin amending courses, we emphasise that the audit element can't just be skipped or rushed. After all, what would a student be told about retrofitting a green building for example, if they wanted to jump straight to solar panels as 'the solution'? Clearly we need to consider what the system – in this case the curriculum – has first, before we jump to conclusions about the best approach, especially by a single course in isolation from the greater program.

There will be those who insist that a cursory high-level review is sufficient; some may even resist the audit altogether. However, the reality is that effective curriculum renewal is dependent on having a detailed understanding of what is happening at the level of each course in the program. Using another analogy, we wouldn't want our mechanic to just replace the alternator because that's the part of the engine they understand rather than identify where in the engine the repairs need to be made. As with a program, each part of an engine is critical in performing a necessary task, and each needs to be examined with regard to its purpose and contribution to achieving the overall goal, in this case that of motion. Likewise, each subject within a program needs to be examined for its purpose in developing the selected graduate attributes, and its contribution to building a learning pathway through the overall program. With these concerns in mind and drawing on experiences over the last several years,[6] the following pages outline a straightforward, non-confrontational, collaborative and time-effective method for auditing the curriculum for learning outcomes.

A process for auditing learning outcomes

This process provides a risk management-based approach to embedding sustainability within the curriculum while preserving the program's diversity and innovation. The audit comprises assessment and classification of each course in a program with regard to whether the assigned learning outcomes are evident, and how much effort will be required to ensure sufficient coverage to develop the desired graduate attribute/s. A key outcome of the audit is a categorisation of courses ranging from those requiring 'no further action' to those requiring 'significant effort'. Using this information, priorities can be set within the curriculum renewal strategy to develop the learning pathways.

There is also the opportunity to use this process for benchmarking to provide a common reference point for discussing progress in future and to assist in evaluating the impact of various initiatives over time. With this in mind we have created the following process as a base for refinement, depending on variables such as scope, budget and timing.

Step 1: Form audit team

This step comprises two key preparatory sessions. The first is a session to create and brief an auditing team, which may include management, program convenors, external experts and other key persons. In this session the team seeks to identify the specific courses to be considered, agree on the particulars of the method to be used and identify related logistical considerations. The second session involves the auditing team and the program faculty, tutors, etc. responsible for delivery of courses in the audit, to explain the rationale, scope, process and potential implications. It is anticipated that many of these people will have participated in previous sessions as part of the overall curriculum renewal strategy.

Given the complexity of content to be audited, and of the pedagogy to consider, the selection of members of the audit team is a critical determinant of the success of the curriculum renewal strategy. An appropriate audit team can facilitate an enriched environment where faculty can discuss their curriculum innovations and hesitations, creating the opportunity for strategic discussion about opportunities for embedding sustainability-related competencies in a way that aligns with strengths and interests.

An audit team could comprise a lead auditor, the program convenor, an expert in sustainable development for the discipline area, and a faculty member trained in providing learning and teaching support.

It is important that the audit process is undertaken predominantly by the faculty for the faculty to encourage candid and direct contributions. This is because the process is intended to encourage reflection and dialogue about curriculum which is likely to include substantial tacit knowledge on how content is being delivered, in addition to standard curriculum documentation such as program and course outlines, assessment and lecture notes. However, there may be instances where some members of the audit team are colleagues invited from elsewhere such as another department or from outside the university. This might be appropriate where, for example, there will be more regard for an audit where external individuals are involved, particularly given the opportunity to involve national or international experts in the field. One or more external audit team members may be able to provide expert opinion on sustainability content, but may need briefing with regard to tacit historical or contextual knowledge about the program and desired graduate attributes. Ideally they would already have been involved in the mapping of learning outcomes.

Step 2: Review and classify courses

The audit team systematically works through key documents for each course (including the course outline, core content documents and assessment items) to identify either the presence of the component knowledge and skills, or the opportunity to include them (Chapter 7). The audit team bases its investigation on a checklist for each course, which specifies the targeted knowledge and skills for each graduate attribute (i.e. from the mapped learning pathway/s). Table 8.1 shows how both 'base' and 'general' knowledge can be identified along with the application of specific knowledge and skills.

TABLE 8.1 Categories of consideration in program evaluation

Category	*Description*
Base knowledge	How well does the *scope of the theory* and the *fundamental principles* in this course, form a foundation to develop the learning outcome/s?
General knowledge	How well does the *information* offered in this course relate the relevance and context of the base theory to develop the learning outcome/s?
Application skills	How well do the *examples and case studies* provided in this course demonstrate the application of fundamental principles and information to develop the learning outcome/s?

Source: Developed in accordance with the language used by the International Engineering Alliance in the Washington Accord, and the Institution of Engineers Australia in their Accreditation Management System, and aligned with curriculum domains identified by Barnett and Coate, of 'knowing', 'acting' and 'being'.[7]

Building on the categories and language described in Table 8.1 we suggest consideration of three key sets of materials as part of the audit, namely:

1 Course outline: How well does the course outline embed the assigned learning outcomes? This includes coverage in the course description, recognition in the aims, objectives and list of learning outcomes, and inclusion in the summary of topics and assessment.
2 Course delivery (content and structure): How well does the course (including what is taught and how it is delivered) embed the assigned learning outcomes? (i.e. including course outlines, lecture material and assessment briefs, and interviews with the program convenor and course convenors).
3 Course assessment: How well does the assessment for the course embed the assigned learning outcomes? This includes, for example, using sustainable development to set the context for assignment or exam questions, responding to an explicit sustainability-related query, and including sustainability-related measurement courses and conversions in calculation requirements.

Information gathered during the audit is used to evaluate each course with regard to how well it is addressing learning outcomes that develop the desired graduate attributes, and a priority is suggested with which updating needs to occur. It is recommended that the course classification used is scaled from 1 to 5, as shown in Table 8.2, where a higher classification reflects a higher level of effort required to transition the course to contribute effectively to the graduate attributes. Hence, a classification of '1' is the target and reflects full integration of knowledge and skills required to contribute to the graduate attribute selected with no further work required. A classification of '5' reflects minimal integration of such knowledge and skills and may result in their omission and relocation to another course or a significant change to the current course.

A useful way to structure the findings of the audit is in the form of a SWOT analysis, namely to consider the course's strengths, weaknesses, opportunities and threats (see Figure 8.2).

Step 3: Create course profiles

This step focuses on creating course profiles for each course being audited with direct input from the course convenors. This is undertaken through a series of semi-structured interviews or informal discussions, facilitated by the lead auditor with the faculty responsible for course delivery. It could be

TABLE 8.2 Example of potential course classification options

Level	*Description*
1	The course contains strong content and worked examples that embed the assigned sustainability-related learning outcomes. The particular knowledge and skills are well integrated, including representation in assessment. The course outline acknowledges the sustainability content and how this contributes to developing one or more sustainability-related learning outcomes.
2	The course contains sufficient content and worked examples that embed the assigned sustainability-related learning outcomes. The particular knowledge and skills are well integrated, including assessment, but there are limited examples of application in the content/assessment. The course outline acknowledges the sustainability content and how this content contributes to developing one or more sustainability-related learning outcomes.
3	The course contains some content and/or examples that embed the assigned sustainability-related learning outcomes, although the coverage is not completely aligned. The particular knowledge and skills are not accompanied by up-to-date worked examples or case studies. Sustainability content is addressed somewhat in assessment. The course outline may include limited mention of the sustainability content and how it contributes to developing one or more sustainability-related learning outcomes.
4	The course contains a scattering of content and/or examples that develop identified sustainability-related learning outcomes. The related knowledge and skills are presented in an *ad hoc* manner, are isolated rather than integrated, and may not be reflected in assessment. The course outline does not address sustainability or how it contributes to developing one or more sustainability-related learning outcomes.
5	There is little or no content or worked examples/case studies that develop identified sustainability-related learning outcomes. The course outline does not mention sustainability.

undertaken on an individual level with faculty responsible for the course/s, or in a group format, depending on the program size and preferences. It may also include formal sessions with student representatives from the course/s to cross-reference the intended learning outcomes with the students' perceptions. A format that we have found practical and concise is to sequentially address the course's strengths, weaknesses, opportunities and threats (i.e. a 'SWOT summary'):

- Strengths: This part highlights where learning outcomes were covered well, i.e. aspects of the course that embed the assigned learning outcomes. This could include mention of factors that positively affected the assessment, and which possibly strengthen the overall program. This can provide a source of ideas for improving other courses.
- Weaknesses: This part highlights where there are weaknesses in covering the learning outcomes. This includes parts of the course that need improvement, which adversely affected the assessment and which possibly weaken the overall program coverage of sustainability-related knowledge and skills.
- Opportunities: This part highlights where internal capacity could be harnessed to enhance the strengths and weaknesses. This includes identifying areas that could be addressed to improve the course performance profile and possibly other courses as well. This section may also include recommendations on possible sources of readily available content or expert support.
- Threats: This part highlights whether the identified gaps pose a threat to the program as it stands with regard to market and regulatory conditions. This could include areas within the course that could threaten the marketability or credibility of the program or institution. This may also include direct comparisons with other similar competing courses at other institutions that have made advances in embedding sustainability-related learning outcomes.

A sample profile is shown in Figure 8.2, informed through pilot programs with James Cook University (Townsville), Monash University (Melbourne) and QUT (Brisbane).

Course Name: **ENG1001 Professional Practice (First Year, Common Course)**

Audit Resources: 1-hour interview (with course convenor), course outline, mid-semester quiz, assignments (2), end-of-semester exam

Classification: 3 – Moderate renewal required

Audited learning outcomes (LO)	Intended emphasis	Course outline	Course content	Course assessment
Graduate Attribute 1: Ability to develop engineering solutions with regard for sustainability challenges and opportunities				
LO 1.1 Knowledge of sustainable development core concepts and definitions	Learn (High)	~ Room for improvement	✓ Clearly explained	~ Room for improvement
LO 1.2 Appreciation of complexity in natural and built environment systems	Learn (Low)	✓ Clearly explained	✓ Clearly explained	~ Room for improvement

SWOT Analysis Summary

Strengths: This subject appears to adopt a systematic approach to teaching the topic area. It incorporates an overview of what sustainable development means for professional practice, and some contemporary and recent examples of a range of emerging 21st-century challenges that are transdisciplinary, providing a platform for other courses within the program to build on. Sustainability lecture notes have been prepared by a member of faculty with experience across sub-disciplines and in sustainability. ...

Weaknesses: The subject lacks a 'meta-narrative' to contextualise sustainability learning within the program structure. For example the following or similar could appear in the Course Outline rationale/introduction: 'A critical component of professional practice involves understanding a number of core challenges and opportunities for sustainable development in the 21st century. This course will provide students with examples that provide a foundation for topics to be covered later in the program'. The lecture structure separates learning about key sustainability concepts within 2 of the 13-week course – a 'bolt on' measure, rather than being integrated. Currently these topics are represented in less than 5 per cent of assessment. The greenhouse gas emissions case study does not appear to be well integrated with the rest of the course. ...

Opportunities: There is a clear opportunity in this course to better integrate the sustainability component of professional practice within the 13 weeks of learning, and within assessment. For example: 2–3 of the other teaching weeks involving topics such as ethics and career planning could use sustainable development as the context for consideration; quiz questions could be reframed to have 21st-century challenges as the context; the assignment questions could ask students to search for a sustainable solution; the group project could also be reframed to address LO1.1 and LO1.2. Such actions would increase the level of coverage in assessment to 20–30 per cent. ...

Threats: The threat to successful renewal for this course appears to be low. The recommendations (above) would need some rewriting of 2–3 lectures, rewording of course assessment (including assignments and quizzes) and updating the course outline (major revision due to assessment changes). The course convenor has indicated they would appreciate support in the form of teaching and learning staff assistance to undertake these changes. The convenor has also identified two upcoming events (a seminar and workshop) that they would like support to attend, related to the proposed renewal of this course. See resources already available below, which could help to reframe the applied examples during class, and the assessment. ...

Suggestions for resources:
➢ (free, online) Case study support: ...
➢ (free, online) Student provocations: ...
➢ ...

FIGURE 8.2 An example course profile summary for one course

Step 4: Develop recommendations report

Once the course classification and profiles have been developed it is then possible to contemplate specific actions to be undertaken in particular courses and their priority, and develop a recommendations report to inform the curriculum renewal strategy. Following a first pass by the audit team, this could include one or more sessions with individual faculty or small groups of faculty responsible for course delivery to discuss the findings. This may include identifying the opportunity to shift a learning outcome to another course, and to streamline similar content across courses. The report will inform the scoping of resources and timing requirements to address areas of focus in particular courses across the program/s, such as existing course renewal and new course development/replacement.

At this point the audit team will need to refer to the curriculum renewal strategy, to focus the recommendations. The audit team can also return to the mapped learning outcomes to see whether there are any amendments that could address particular weaknesses or gaps within the overall program. This may be an opportunity for the learning pathway maps to be updated, setting the scene for a smooth process of program-wide curriculum renewal. Essentially, management will want to know what should be prioritised for curriculum renewal to provide an immediate 'critical mass' of sustainability content that sufficiently develops the selected graduate attributes for the program, and what can then be addressed in the medium to longer term. This could include, for example, increasing commitment from 'compliance' to 'beyond compliance' and 'market leader' positions depending on the external trends and internal readiness (outlined in Chapter 5).

Examples of this element in action

There are a number of examples of early development of assessment and classification processes based on sustainable development considerations generally, which can provide guidance in developing the audit of learning outcomes for sustainability-related graduate attributes. There are also emerging examples of institutions attempting their own assessment or classification tools.

For example in Wales (UK), the National Assembly Government required each higher education provider to undertake an assessment of curriculum with respect to sustainability by the end of 2008.[8] Cardiff University's Centre for Business Relationships, Accountability, Sustainability and Society (BRASS) developed a commercial-in-confidence tool called *STAUNCH* in 2007,[9] which evaluates course descriptions by grading them against a number of economic, environmental and social criteria and cross-cutting themes. In collaboration with the Higher Education Funding Council for Wales all Welsh institutions agreed to adopt a common approach to undertaking a sustainability audit of curriculum based on this work undertaken by Cardiff University.

In Australia, at the Royal Melbourne Institute of Technology (RMIT), the 'Beyond Leather Patches' (BELP) project (2005) included a course audit as a critical component in considering a framework for education for sustainable development in universities, in particular through the schools of management, property, construction and project management.[10] The University of South Australia reviewed their civil engineering program for sustainability content in 2008, deriving their own questions and analysis protocol to understand gaps and opportunities for improvements.[11]

Furthermore, this chapter has been informed by a collaboration between the authors and the Faculty of Engineering at Monash University, one of Australia's largest engineering faculties. The faculty offers a common first-year entry to the engineering degree program, where all courses are offered in both teaching semesters. Students select their courses and electives depending on their interests, across more than 14 discipline specialisations. In 2006, in an effort to understand the current level of coverage of sustainability, Professor Gary Codner (director of first-year teaching) invited our team to undertake an audit of the civil engineering undergraduate program, to determine how well the program integrated general sustainability language, content and application. Specifically, the department was interested in addressing its risk exposure to changing graduate needs in Australia from shifts in Australian and international legislation and industry demand. We therefore focused on auditing against what the institution perceived and understood to be likely future graduate needs over the next decade based on the team's research and experience (i.e. 'market leader' graduate attributes).

The audit identified that the program on the whole was achieving an integrated approach to embedding sustainability materials within the first year, rather than using a single first-year course as a flagship for sustainability issues. As each course catered for specific disciplinary needs to meet prerequisite knowledge for courses in subsequent years, the report included a recommendation that each course would benefit from using more sustainability-related case studies, introducing real-life applications that could be then expanded on in future years. The audit also identified the need across the first-year courses for a common introduction or a 'meta-discourse' in each course about why civil engineering plays an important role in addressing 21st-century challenges. The report concluded that embedding sustainable development language within each course outline, introductory lectures and examples would draw a common thread between the rather separate theoretical disciplines covered in first year, providing a foundation for future learning. All audited courses achieved a course classification of either 2 or 3, indicating minor targeted renewal.

Interaction with 'dynamic' elements of the model

Awareness raising and capacity building

Given that assessing and auditing courses program wide is often an unfamiliar exercise, there are three key areas for awareness raising and capacity building, namely:

- Non-confrontational approach: Highlighting and creating the capacity to undertake the auditing process in a non-confrontational way is crucial for the success of the process. This is especially important where past experiences from previous confrontational processes may have eroded morale and desire to contribute.
- Clarity on process: Creating clarity about the process and the importance of rigour can assist in circumventing discussions about skipping the process and 'just getting on with it'. The auditing process may at first appear negative; however, it can also be created as a crucial step to minimise time and effort needed to quickly embed learning outcomes in courses and achieve program-wide renewal.
- Collaborative process: Providing opportunities for faculty to contribute includes collecting ideas about how the process could run, and how they would like to benefit (e.g. course outlines, content resources and assessment). The audit process can be created as something that assists faculty to quickly identify tangible changes to implement to align to the graduate attributes, and each will have value to add to this process, and in many cases inform other faculty efforts.

Internal and external stakeholder engagement

As this process has the potential to be seen in a negative light it is important that internal and external stakeholders are strategically engaged in the process.

Internal stakeholders may include those that will participate in the audit team, such as a representative from the teaching and learning area of the university, along with inviting review of the audit process from appropriate people across the organisation. Key internal stakeholders can provide valuable advice on whether the audit findings resonate at an institutional level and any opportunities that may arise in addressing priority course/s. External stakeholders can be an important part of the process, with the inclusion of an industry expert on the audit team bringing valuable industry experience and creditability to the process. Further external stakeholders may be invited to review the audit process and inform the prioritisation process with regard to their urgency in receiving graduates with particular attributes.

Once the preliminary findings have been produced, there is an opportunity for review from both internal and external stakeholders, for example to nominate suitable resources to strengthen coverage of particular topics or to identify opportunities for weaknesses identified in one course to be offset by a strength in that area in another course with faculty working collaboratively to deliver the learning outcome.

Monitoring and evaluation activities

In effect, the process of auditing is a monitoring and evaluation activity, with the main areas to focus on being the delivery of the audit process itself, and the periodic review of the use of the audit findings. Considering tasks for each of the activity streams, Table 8.3 lists some examples of performance indicators for tracking the success of the audit process.

Following the audit, there will need to be periodic review of progress against the findings. This is important for two main reasons. First, as the audit report will recommend areas for improvement, the course convenors need opportunities to demonstrate that they have addressed these recommendations. Second, as the curriculum renewal process proceeds, each course can be reviewed to identify emerging challenges for implementation and opportunities for improvement to inform the wider implementation of learning pathways within programs. Depending on the timeframe for curriculum renewal, this might happen at the end of each semester of teaching for the courses that are run that semester. Alternatively it might take the form of an annual review.

TABLE 8.3 Sample performance indicators for auditing learning outcomes

Activity stream	*Example performance indicators*
Governance and management	The selection and setting up of the audit team (participants) Proportion of invited faculty participating in the audit process (percentage/event)
Operations and facilities	Number of existing campus–curriculum collaborations related to learning outcomes underway (number/year)
Teaching and learning	The extent of academic participant satisfaction with their input and guidance received (qualitative)
Human resources and culture	Proportion of relevant documentation that acknowledges sustainability-related audit component of faculty teaching workload (percentage)
Marketing and communications	Proportion of possible newsworthy stories evident in the audit that have been investigated (percentage/quarter)
Partnerships and stakeholders	Involvement of key partners and stakeholders involved in the audit process (qualitative)

Roles and responsibilities across the curriculum renewal strategy

In the following part we consider the deliberative element 'auditing learning outcomes' with regard to how this translates across the various stages of curriculum renewal and the relative involvement by each of the various activity streams. For the purpose of illustration we have used a four-stage approach.

Governance and management

Stage 1: Prepare – In this stage management will be exploring what an auditing process entails, with the aim of securing one or more faculty who will be responsible for leading the audit process, setting up the audit team and follow-up monitoring. There is the potential for an audit process, if not properly implemented, to be counterproductive to faculty engagement as faculty may feel threatened by the process. Hence, management needs to make a strong commitment to the audit and its findings and have a clear and consistent message about its intent. This includes clear communication of the focus on the non-confrontational, proactive and collaborative nature of the process, and to highlight the benefits to the programs and the university. In short, the intent of the audit is to ascertain the level of coverage of the selected learning outcomes in particular courses across programs, to then identify opportunities to increase coverage strategically as part of a program-wide strategy involving all relevant actors.

Stage 2: Explore – This stage involves management creating and supporting the audit team, and encouraging participation as required throughout the process. Management will also need to ensure that there is clarity about the audit process among faculty, and clarity about workload implications. This might include a formal request for faculty to participate and acknowledge the additional workload, which could include recognition through the faculty workload calculation of the time spent by faculty contributing to the audit process, potentially as part of the service load.

Stage 3: Test and pilot – During this stage, management is focused on supporting the use of the audit findings to determine the pilot projects in various programs. The audit findings also provide management with the opportunity to recognise faculty who have already embraced sustainability-related knowledge and skills in their course/s and encouraging them to share their curriculum renewal experiences with other faculty. This may be in the form of a service award, through invitation to present activities to formal gatherings and events within the university, or invitation to sit on boards and committees related to sustainability. It may include permitting and encouraging faculty appropriately using existing academically rigorous and cutting-edge course materials available online under open-source common attributes licensing arrangements to address the audit recommendations. Management will also need to strategically make available faculty resources and finances to undertake professional development where the audit has identified that sustainability-related capacity needs to be further developed. This could include providing seed funding opportunities (e.g. internal grants) for faculty to investigate research opportunities in this area. This has the added incentive of research recognition and also increases the likelihood that the course/s will be kept up to date with sustainable development theory, knowledge and application.

Stage 4: Integrate – In this stage management will need to ensure that the pilot evaluations are completed in a timely manner. Learning outcomes may need to be reviewed or refined in light of the pilot experiences (i.e. reviewing the map of learning outcomes), and work will need to commence on other courses across the program. Management will also need to consider a range of incentive mechanisms to gain and maintain momentum throughout the curriculum renewal process, and in particular to implement the recommendations of the audit. This includes continuing to make available the necessary resources and funds to facilitate actioning the audit recommendations. It also includes acting on review recommendations regarding emerging challenges and opportunities as they unfold during the integration process.

Operations and facilities

For this activity stream, initially the involvement in the auditing process may comprise assisting with any queries regarding existing curriculum collaborations in courses. Staff involved may need to participate in (and in some cases contribute to providing) awareness raising about sustainability and in auditing processes during this stage. Depending on staff familiarity with sustainability and the institution's commitment to on-campus collaboration, staff from this activity stream may participate in the audit panel to actively identify areas for potential campus–curriculum collaboration, under 'opportunities'. In the later stages this area will have an opportunity to offer, and participate in, pilot projects using on-campus projects and infrastructure. This activity stream may also be able to assist in finding industry or other partner funding to facilitate actioning audit recommendations that involve the expenditure of funds within the department, which can be a challenge for institutions in the increasingly competitive higher education market. This activity stream has the potential to learn from its pilot experiences in up-scaling the campus–curriculum collaborations attempted in the pilot/s.

Teaching and learning

Stage 1: Prepare – This stage requires participation by faculty who teach in the programs, in addition to teaching and learning personnel. In this first stage, once identified, there may need to be some professional development in education for sustainability and also in auditing, to ensure that they are prepared for the following stages, including the possible selection of sustainable development experts to build capacity. Sustainability champions may also do this internally by providing peer seminars or discussions. Further, there are a growing number of sustainability-related industry associations that provide presenters and materials. Throughout this process the data collected on the course can provide input into academic articles on curriculum renewal, which may help to expand the publication range of faculty.[12]

Stage 2: Explore – In this stage some faculty and staff from the teaching and learning stream will be involved in the team undertaking the audit of courses and documenting the results. This stage will also involve a revision of the mapped learning outcomes over the program in light of the audit findings. These actions will take significant time for some, who will need to be appropriately supported by management. For each course, faculty will need to gather documents for each of their subjects, and possibly meet with the audit team or complete a questionnaire or survey, reflecting on the extent to which their courses address the learning outcomes that have been assigned.

Stage 3: Test and pilot – This stage involves directly acting on the audit recommendations with regard to piloting the process of embedding learning outcomes into one or more courses. For the remainder of faculty, in Stage 3 they may be provided with opportunities to undertake professional development in the new area. This could be packaged with specific requirements, including becoming familiar with the topic area, identifying aspects that can be immediately incorporated into existing curriculum and identifying aspects needing significant new course development. It could also include a requirement for the faculty to identify material in demand for postgraduate and professional development courses or, for example, in attracting regional/international students faced with sustainable development challenges.

Stage 4: Integrate – This stage involves the full cohort of teaching faculty members to be engaged in actioning the audit recommendations, assisted through, for example, research support, teaching buy-out or flexibility in faculty appointments (for example, the research/teaching/service proportions). In this stage, there will also be a need to be revisit the audit process to check in on how the learning outcomes are being developed across the program.

Human resources and culture

Over the four stages, this activity stream will be engaged in embedding the element of curriculum auditing as an integral part of the institution's commitment to education for sustainability. Hence, in the early stages, one or more staff will need to be identified who are responsible for liaising with management and the audit team. These staff may participate in staff briefings and potentially some professional development in the areas of sustainability and auditing. Further, as part of the early stages, staff may contribute to the audit process by exploring existing policy documentation for references to curriculum renewal and academic staff requirements. They may also be working towards a situation where involvement in auditing actions will be considered an opportunity by staff, rather than 'another imposition'. In the later stages staff may draft amendments to existing documentation with references to the need for active review of the curriculum and actioning of audit recommendations. This could include requiring staff to address audit recommendations related to their subjects. It may also include reporting on progress with course renewal in faculty appraisals.

Marketing and communications

Internal communications will play an important role to ensure staff within the institution are aware of what the audit is about, how it provides them value, how they are expected to participate, what the results are and how they will be used, and expectations regarding responding to the recommendations. It will also be important for external marketing staff to be able to extract the 'good news' stories from the audit results and subsequent initiatives to enrol future students and potential employers in the curriculum transition.

Stage 1: Prepare – During this stage there will be a focus on identifying potential experts and key stakeholders internally and externally for the audit team that the marketing and communications staff can assist with. In this case staff may briefed on the role of the audit in the larger curriculum renewal process in order to understand its role in renewing the curriculum. There may also be a need to assist in preparing internal materials for academic staff about the audit, including its intention, intended usefulness and rigour. This stage may also be a good time to start marketing the fact that the program is undertaking a comprehensive review of its coverage of sustainability-related graduate attributes as this may pique the interest of potential students, high-quality academics in other institutions, and industry and government groups interested in supporting such efforts.

Stage 2: Explore – In this stage the marketing and communications staff will be focused on exploring opportunities to promote strengths and curriculum renewal activities that are identified in the audit. This includes highlighting efforts underway in curriculum renewal for sustainable development and identifying ways to build on these efforts. Marketing staff will also be providing assistance to academic staff adding the finalised learning outcomes to course promotional materials and course outlines.

Stage 3: Test and pilot – This stage presents an opportunity for marketing and communications staff to track and report on curriculum renewal initiatives as they are piloted, helping to maintain momentum around the pilots and the achievement of the curriculum renewal strategy.

Stage 4: Integrate – In this stage marketing and communications will participate in a number of additional projects rolling out as part of the integration phase. Within this context the primary role will be to increase the profile of the university in education for sustainability, maintaining a regular stream of news articles, industry briefings and marketing materials about integration within the curriculum.

Partnerships and stakeholders

Stage 1: Prepare – In the early stages it will be important to ensure that partners and stakeholders (i.e. the external network) are aware of how the audit process is reinforcing and strengthening the program's profile in education for sustainability. In this stage staff will focus on ensuring the intention of the audit and its potential benefits are clear to avoid the audit being seen as a surprise or in a negative light. This will be important to secure partners and stakeholders for participation in the audit team if required and for potential review of findings. This will involve staff involved in liaising with partners and stakeholders to be briefed on the curriculum renewal strategy, so that a consistent message is provided to align to the wider marketing and communications.

Stage 2: Explore – Staff in this activity stream may collaborate with the marketing and communications staff to cultivate opportunities to engage with partners and stakeholders to identify options to respond to the audit findings, such as suggesting curriculum–community initiatives, the development of industry case studies, the provision of scholarships, or the development and review of new materials. Staff may also be supporting the involvement of external experts on the audit team, taking care of liaison and logistical arrangements.

Stage 3: Test and pilot – This stage will focus on partners and stakeholders related to the particular courses being piloted being involved in the development of the course profile, through participation in the audit team, review of the audit findings and identification of potential resources or support to respond to the audit recommendations.

Stage 4: Integrate – Building on the lessons from the pilots, this stage will focus on the rollout of the process and will require significant engagement with a wide range of partners and stakeholders of each course being audited. As with the pilots, this engagement will include participation in the audit team, review of the audit findings and identification of potential resources or support to respond to the audit recommendations. Once entire programs are undergoing audits of its courses, this presents the opportunity for greater partner and stakeholder involvement, such as through industry course sponsorship, the appointment of funded 'sustainability chairs', and professional development bursaries.

In the next chapter we consider the deliberative element of 'Developing and updating curriculum'.

9

DEVELOPING AND UPDATING CURRICULUM

Our journey into this element was a surprise, requiring much more than a quick mention within the chapter on the model itself. It has taken many workshops and collaborations to understand the type and extent of the issues in developing and updating curriculum. In short, issues around integrating learning outcomes within courses are real and widespread, requiring immediate attention and support. Even with the three preceding elements (Chapters 6, 7 and 8) providing contex and direction, there are additional hurdles to overcome. No wonder, then, that there is such a high incidence of failure and frustration when faculty attempt to develop and update curriculum in isolation.

At this point in the curriculum renewal process there are a number of learning outcomes allocated to new and existing courses within a program, to achieve one or more graduate attributes. As a logical next step, one might conclude it to be a straightforward matter to address these new learning outcomes, then follow through with updating courses. After all, curriculum renewal is well established within higher education, and self-directed by faculty for the most part. The majority of student evaluations are above average, so why are we concerned? Unfortunately, in higher education, successfully embedding learning outcomes within courses is the exception rather than the rule. Not only do faculty struggle with emerging topic areas and knowledge and skills implications for their courses, they also struggle with the process of curriculum renewal itself, requiring greater support and strategic direction. Furthermore, there are few incentives to do curriculum renewal well, beyond 'passing' the next round of student evaluation and ticking the teaching boxes in performance appraisals.

With these considerations in mind, this chapter is not intended to be prescriptive in how faculty should shape courses to deliver intended learning outcomes. Rather, it provides a menu of opportunities and provokes thought about what may be possible for your institution, avoiding 're-inventing the wheel' and optimising faculty time on curriculum renewal. The chapter begins by addressing four common misconceptions to create a clearing for curriculum renewal to occur. The rest of the chapter then draws on lessons learned over the years to discuss four mechanisms for overcoming barriers to developing and updating curriculum, grouped under the following considerations:

- Harnessing existing resources;
- Engaging in institutional collaborations;
- Connecting with campus operations; and
- Connecting with community projects.

Exploring the deliberative element 'Developing and updating curriculum'

Introduction

Following the auditing process, there are now a number of learning outcomes assigned to various courses across the program that combine to deliver one or more graduate attributes related to sustainability. Some learning outcomes are allocated to existing courses while others need to be part of new offerings. The task now is to respond to the recommendations from the audit at the course level and integrate the new learning outcomes as needed. This process, usually starting as part of a trial or pilot as described in Stage 3 of the curriculum helix, needs to be approached strategically as it can often be *ad hoc*, reactive and laden with bureaucratic approval steps. A number of our Australian colleagues exploring curriculum renewal have concluded that faculty and managers experience the challenge of embedding graduate attributes in curriculum as 'a series of hurdles', where faculty are concerned about multiple barriers, which require substantial stamina and support to overcome.[1,2] On top of this there are a number of misconceptions which can hinder and even prevent action. We begin this chapter by addressing four common misconceptions to create a clearing for curriculum renewal to occur. The rest of the chapter then draws on lessons learned over the years to discuss four mechanisms for overcoming barriers to developing and updating curriculum.

Creating a clearing for curriculum renewal

In an Australian survey of energy efficiency education, nearly two-thirds of faculty surveyed (58%) considered the potential for course content overload to be an issue, while more than half (52%) considered having insufficient time to prepare new materials as a challenge to such curriculum renewal.[3]

Perhaps you have experienced colleagues discussing the following four misconceptions – they regularly appear in our discussions with colleagues about curriculum renewal. Here we provide some commentary that you may find useful to address and move past these statements, creating a space to develop and update the curriculum.

- 'Embedding sustainability content is risky' – This may have been the case a decade ago; however, given the significant advances in understanding of sustainability (and the availability of resources, papers and books on the various topics), it is becoming more commonplace. Much of the risk has been taken out of the process and those just starting to undertake the process are not trying something 'risky' or new, and they are not at the 'leading edge' in doing so. Rather, it is now a growing requirement in many accreditation systems to include at least some sustainability knowledge and skills. A study of the largest engineering employers in Australia found that 6 out of 10 provided in-house training to address gaps in energy efficiency graduate attributes, and 4 out of 10 incorporate energy efficiency expectations in graduate recruitment.[4]

- 'Programs already have a full quota of content that is all critical to learning' – This is an easy one to buy into; however, it is of course the case that the content that a program taught 50 years ago is likely to be different to the program taught today, all the while remaining 'full'. It is the very nature of the education system that the material taught will change over time in response to changing societal needs. For example, in many fields knowledge and skills that did not exist as little as 5 to 10 years ago are now included in programs, particularly in information technology and communications and medicine, and are regarded as essential. Similarly, knowledge and skills to address current and future issues such as climate change and environmental degradation will be incorporated into programs across all fields in the coming decades, and once in place will be seen as essential. The issue here is not that the program is full, as programs have to be full, but rather it is an issue of designing an appropriate strategy to streamline and resource the renewal process.
- 'The old program needs to be discarded' – Renewing programs to include sustainable development does not require 'starting from scratch', but rather identifying key areas for renewal in a systematic and integrated manner. This can result in a range of changes across programs from no change at all to major revisions. Much of the underpinning of sustainable development already exists as the core of a program, and the curriculum renewal process does not necessarily challenge first principles or core concepts but rather their completeness and application.
- 'Students should just be given the fundamentals, which they then apply throughout their career to any problem' – Some staff don't want to reduce fundamentals education to include another 'hot topic' or passing agenda. However, as discussed in Chapter 1, sustainability is not a passing agenda. In this context, curriculum renewal towards education for sustainability is about integrating with, rather than removing, these fundamentals or reducing their importance within curriculum.

Such misconceptions are important to address early in the curriculum renewal process, potentially as part of the early awareness raising and capacity building activities outlined previously.

There are a variety of options for curriculum renewal, which are regularly discussed by colleagues around the world, each of which has its own challenges for implementation. One way to gain a better understanding of the perceptions across the institution is to undertake a survey of perceived barriers and benefits to curriculum renewal for sustainability, and to consider how they may impact on the implementation of key aspects of the curriculum renewal strategy. For example, an Australian study led by the authors between 2007 and 2009 investigated university staff perceptions of the barriers and benefits[5] of a number of specific options for increasing the level of engineering education for energy efficiency in Australia.[6] The options included:

1 Include a case study on energy efficiency
2 Offer supervised research topics on energy efficiency themes
3 Include a guest lecturer to teach a sub-topic
4 Include tutorials that align with the energy efficiency theme in the course
5 Offer energy efficiency as a topic in a problem-based learning course

6 Include assessment that aligns with the energy efficiency theme within the course
7 Overhaul the course to embed energy efficiency
8 Include a field trip related to energy efficiency
9 Include one workshop (i.e. experiments) on energy efficiency in the course
10 Develop a new course on energy efficiency

As can be seen from Table 9.1, putting in place mechanisms to address one defined barrier can have benefits for addressing other barriers. For example, 7 of the 10 options have the barrier of 'insufficient time for preparation', which could all be addressed through setting up an annual allocation of teaching buy-out funds or altering workload allocations. With such information, budgets and human resourcing can be strategically allocated to enable maximum value for money and time.

TABLE 9.1 Issues for implementing curriculum renewal in energy efficiency education

Key issues for implementation	*Shortlisted options for curriculum renewal*									
	1. Case Study	*2 Guest Lecture*	*3 Research*	*4 PBL* Topic*	*5 Assessment*	*6 Tutorials*	*7 Overhaul*	*8 Workshop*	*9 Field Trip*	*10 New Course*
Common barriers										
Lack of available information	●	●		●	●	●	●		●	●
Lack of time for preparation	●	●		●	●	●		●		●
Prohibitive cost	●		●	●	●	●		●	●	●
Lack of knowledge	●	●	●	●	●		●		●	●
An overcrowded curriculum	●		●	●		●			●	●
Lack of value attached	●		●			●				
Lack of industry contacts		●	●					●		
Administrative coordination							●	●		●
Lecturer apathy		●					●			
Resistance to top-down directive			●				●			
Students' prior learning habits					●				●	
Common benefits										
Improved marketability	●	●					●	●		●
Improved pedagogy – PBL*				●	●	●			●	
Improved pedagogy – skills				●	●	●			●	
Cross-functionality of content	●						●			●
Networking for students		●	●					●		
Networking for lecturers		●	●					●		
Research opportunities		●								●
Professional development		●				●				
Experience in renewal			●				●			
Addressing the time-lag issue			●				●			

* Problem based learning.
Source: Desha, C., Hargroves, K. and Reeve, A. (2009)[7]

Assistance with developing and updating courses

The identification of such barriers and benefits provides a tangible approach to advancing curriculum renewal efforts. Our work with universities to review, pilot and implement curriculum renewal processes has led to a greater understanding of how to assist in the renewal process. Often there are specific areas that need dedicated support and can be relatively new to the university or faculty. The following part focuses on four such areas, two related to the curriculum helix, namely, 'Harnessing existing resources' (teaching and learning) and 'Connecting with campus operations' (operations and facilities), and two related to the deliberative and dynamic model of curriculum renewal, namely 'Engaging in institutional collaborations' (internal engagement) and 'Connecting with community projects' (external engagement).

Harnessing existing resources

Within education literature there is regular reference to pre-prepared content to quickly and cost-effectively increase the amount of sustainability content in programs. Indeed, there is a steadily growing wealth of suggestions available for faculty wondering how to proceed with embedding sustainability-related knowledge and skills in the curriculum, covering all teaching and learning pedagogies. As Dr Rowe, President of the US Partnership for Education for Sustainable Development, comments in her introduction to this book, it is critical to share these resources to leverage initiatives that have already occurred, fast-tracking the experiences of faculty who are embarking on significant curriculum renewal processes.

There are a number of advantages of using pre-prepared content resources. Many resources have been critically reviewed, are academically rigorous, and are based on core competencies that don't require constant updating (minimising resources for course building and updating). Pre-prepared curriculum can also contain relevant, rigorous introductory material that is readily accessible for undergraduate, postgraduate and professional development offerings. The content can also include discipline-specific and leading-edge materials that would otherwise take significant resources to develop in-house and can be altered in many cases to fit with the particular institution's preferred graduate attributes in the area. The pedagogy of such materials can be very straightforward and flexible, making the materials easily accessible by lecturers.

In the UK, the University of Nottingham's open courseware (U-Now) includes introductory sustainability modules for a number of disciplines including engineering, arts, humanities, business and geography. The U-Now project was sponsored by the university's e-learning strategy in close collaboration with the Information Services Learning Technology Section, with resources supplied by academics within the university. In addition to providing course notes and audiovisual materials, the website page for each topic area remains 'live' through the use of real-time smart external search engines such as 'Xpert' and 'YouTube' providing popular links to topics.[8]

> Web-based resources can certainly assist academics to develop [curriculum] and support their students learning. Resource networks . . . should certainly allow [faculty] to spend their curriculum development time allocation to greater effect.
>
> Addressing the Supply and Quality of Engineering Graduates for the New Century, 2008[9]

However, there are a number of potential issues with simply finding a report or set of lecture notes online and trying to use it in a course, particularly if it was not designed with such an end use in mind. For example, content may have been developed for programs in specific countries, meaning that examples may be country, hemisphere, language, metrics or currency specific. Given the rapid pace of change content can often be out of date even over short periods of time and may require updating. The cultural or language context may also be significantly different for departments engaging in curriculum renewal in different countries and with different student compositions.

Currently the majority of content resources are provided in English, which is not the first language for many countries. While organisations such as Japan for Sustainability (JFS) are constantly translating emerging technologies between languages, there is a need for the development, promotion and perhaps also translation of quality resources.

Cultural variety internationally also requires contextually sensitive content, or content that can be relatively easily amended to suit the student audience. For example, a university in Africa may consider using pre-prepared material on water scarcity and issues developed in Australia. This content may contain facts and figures that are specific to Australia, which could be enhanced for an African audience through including references to local legislation and context, local constraints and opportunities, and local innovations. Depending on the availability of knowledge about the local environment, the pre-prepared content might also be augmented by additional lectures/modules developed within the department, or parts of other pre-prepared material.

Bearing in mind the context for harnessing existing resources, Table 9.2 summarises a rich array of resources to assist faculty. Not only do they support course-level curriculum renewal, they are also an excellent professional development resource for faculty looking do some personal capacity building. There are also online courses available on a user-pays licence basis, including, for example, 'Chronos', a 3-hour e-learning tutorial on sustainable development that was developed by the University of Cambridge and the World Business Council for Sustainable Development.[10]

TABLE 9.2 Examples of free sustainability-related guiding and content resources

Resource	*Description*
The Natural Edge Project, online resources[11]	TNEP has developed more than 65 open source lectures covering a range of sustainability topic areas listed in the resource section of this chapter, as well as review and mentoring by numerous experts in the field.
The Handbook of Sustainability Literacy: Skills for a changing world[12]	This publication arose within a project on 'Soundings in Sustainability Literacy', funded by the HEA ESD Project and the University of Gloucestershire Centre for Active Learning. It is available in hard cover, or freely downloadable together with a companion online resource.
Second Nature 'Education for Sustainability'[13]	Second Nature is a US organisation helping higher education institutions prepare future professionals for increasingly complex environmental challenges. Second Nature's Resource Centre and website is a substantial and well-used repository of materials submitted by individuals from across higher education institutions.
United Nations Environment Program toolkits[14,15]	UNEP in collaboration with UNESCO has just released an online 'Higher Education Toolkit on Sustainability Communications', to provide interactive research, training and practical materials on issues that can be used in marketing and communications courses.

TABLE 9.2 Continued

Resource	*Description*
Online text books	There are a number of key sustainability books that are freely available online. For example: • *Plan B 3.0: Mobilising to Save Civilisation*[16] • *Natural Capitalism: Creating the next industrial revolution*[17] • *Whole System Design*[18]
Online lectures	Several organisations have emerged over the last five years to provide free online talks, interviews and presentations around sustainable development, including *TedTV*, *Big Picture TV*, and *Eco TV*.
The Future Fit Framework[19]	This UK-based introductory guide to teaching and learning for sustainability in higher education is a joint publication of Plymouth University's 'Teaching and learning sustainability' initiative, the Centre for Sustainable Futures and the Higher Education Academy.
UNESCO Teaching and Learning for a Sustainable Future Multimedia program[20,21]	A teacher education program published by UNESCO comprising 25 modules (approximately 100 hours) of interactive activities designed to enhance the teacher's understanding of sustainable development and related themes and practical skills for integrating themes into the school curriculum. Alongside this program two other publications were developed on 'Learning for a sustainable environment'.
Education for Sustainable Development toolkit[22]	A toolkit developed to help communities, schools and institutions take the first steps towards creating an education for sustainable development program.
playagreaterpart.org[23]	An online assessment match-making website for sustainability projects.
Massachusetts Institute of Technology Open Unitware (OCW)[24]	The Massachusetts Institute of Technology (MIT) open unitware (OCW) is a web-based publication of MIT course content. The site can be browsed by course, department or keyword to locate a specific course or topic.
Toolbox for Sustainable Design Education[25]	The UK Higher Education Academy's Engineering Subject Centre funded a mini-project for the development of a toolbox for teaching sustainable design, for lecturers looking for guidance and material to support the development of a module.
RAE Visiting Professors Scheme: Engineering Design for Sustainable Development[26]	The UK's Royal Academy of Engineering (RAE) has developed a set of engineering case studies, where each case study includes a set of documents and realistic data which describe the circumstances at each location, introductory slide presentations, and a set of guidelines and suggested exercises for lecturers.

In addition to the resources highlighted in Table 9.2, there are a growing number of online portals that support teaching and learning for sustainability, such as:

- The Natural Edge Project (TNEP) – see Table 9.3
- The Association for the Advancement of Sustainability in Higher Education (AASHE)
- The Environmental Association for Universities and Colleges (EAUC)
- The Disciplinary Associations Network for Sustainability (DANS)
- The US Partnership for Education for Sustainable Development (USP-ESD)
- The Higher Education Associations Sustainability Consortium

- The Higher Education Academy Engineering Subject Centre (HEA-ESC)
- The Australian 'Learning and Teaching Sustainability' portal

There are also a number of research institutes around the world whose websites include up-to-date resources (see also Chapter 1), including those with decades of experience such as the Rocky Mountain Institute in the United States, and the Wuppertal Institute in Europe, along with a new generation of teams such as The Natural Edge Project – here is a short summary of our group's work related to curriculum renewal.

Established in November 2002 with in-kind hosting from the Institution of Engineers Australia, The Natural Edge Project comprises Gen-X and Gen-Y engineers, scientists and business graduates hosted by several Australian higher education institutions focused on supporting sustainable development. Activities are not for profit and involve research, creating training material and producing publications, which are supported by grants, sponsorship (both in-kind and financial) and donations. Other activities involve delivering short courses and workshops, and working with partners and associates to test and improve the material. The team, led by the authors, has developed 4 books and over 2,000 pages of free access teaching and learning resources. The books have sold over 80,000 copies in 4 languages, having received an Australian Prime Minister's Banksia Award (*The Natural Advantage of Nations*), and being ranked 5th (*Cents and Sustainability*) and 12th (*Factor 5*) amongst the 'Top 40 Sustainability Books of 2010' by the Cambridge University Sustainability Leadership Program. The learning resources follow a standard format, comprising the educational aim, followed by a list of optional/required reading, key learning points, followed by background information on the topic being considered to assist faculty in knowing the emerging literature, and students needing extra tuition on the topic.[27] (See Table 9.3.)

TABLE 9.3 A summary of TNEP online curriculum resources – lecture materials

Content source	*Description*
Sustainability Education for High Schools (Introductory)	• Senior School/Higher Education/Professional Development • Supported by Griffith University and the Port of Brisbane Corporation as part of the national Sustainable Living Challenge hosted by the University of New South Wales • Twelve lessons comprising 'Technologies' – Grade 10, Senior School Physics, and Senior School Chemistry
Introduction to Sustainable Development (Introductory)	• Senior School/Higher Education/Professional Development • Supported by Engineers Australia, Society for Sustainability and Environmental Engineering, UNESCO • Twelve lessons comprising an overview of the context and opportunities for innovation across a number of emerging fields
Principles and Practices in Sustainable Development (Intermediate)	• Senior School/Higher Education/Professional Development • Supported by Engineers Australia, Society for Sustainability and Environmental Engineering and UNESCO • Twelve lessons comprising a discussion of the emerging principles of sustainable development, continuing the discussions begun in 'Introduction to Sustainable Development'
E-Waste Education Units (Intermediate)	• Senior School/Higher Education/Professional Development • Supported by Griffith University and DELL

TABLE 9.3 Continued

Content source	*Description*
	• Three lessons comprising an overview of the context for electronic waste production, implications for business as usual, and opportunities for sustainable solutions
Sustainable Information Technology (Intermediate)	• Senior School/Higher Education/Professional Development • Supported by Hewlett Packard • Five lessons comprising a discussion of opportunities for reducing carbon footprint and materials waste in the information technology sector
Whole System Design (Intermediate to Advanced)	• Higher Education/Professional Development • Supported by the Australian Federal Government Department of the Environment, Water, Heritage and the Arts • Ten lectures comprising a discussion of the context for sustainable design, 10 principles of whole system design and 5 worked examples
Sustainable Energy Solutions (Intermediate to Advanced)	• Higher Education/Professional Development • Supported by the CSIRO and the National Framework for Energy Efficiency • Thirty lessons comprising an overview of the context and opportunities for climate change mitigation through energy efficiency measures
Sustainable Water Solutions (Intermediate to Advanced)	• Higher Education/Professional Development • Supported by the Australian Federal Government Department of Climate Change, completed in collaboration with the Australian National University • Twenty-four lessons comprising an overview of the context and opportunities for climate change adaptation through addressing water supply and demand

* Resources are available as noted within each document or otherwise under a Creative Commons Attribution 3.0 Licence, at www.naturaledgeproject.net.

Engaging in institutional collaborations

Although collaboration across programs or universities is still relatively low with regard to education for sustainability, there are some emerging leaders, such as:

- In Japan, the University of Tokyo led the establishment of the 'Integrated Research System for Sustainability Science' (IR3S) in 2005, and involves an international Master's of Sustainability Science, taught in English. This builds on from the university's 'Intensive Japanese Program on Sustainability Science' (IPoS), a short-term experimental sustainability educational program, in co-operation with the Asian Institute of Technology (AIT), Thailand, and the Massachusetts Institute of Technology (MIT), USA.[28]
- In Switzerland, Zurich's Swiss Federal Institute of Technology's (ETH) Centre for Sustainability (ETHsustainability) collaborates with academics from a number of institutions internationally to develop the Youth Encounter on Sustainability (YES) course which is also focused on undergraduate and graduate students and which has been undertaken in a number of countries to date.[29]
- In Korea, the International Urban Training Centre (IUTC) was established in 2007 with the support of the United Nations Human Settlements Programme (UN-HABITAT), to build the capacity of cities and towns nationally and in the Asian and Pacific region in support of the goal of sustainable urbanisation. The Centre runs a number of short-course programs in collaboration with universities

internationally to bridge the knowledge gap between academia and local and central government in sustainable urban development.[30]

- In the United Kingdom, Forum for the Future is a non-government organisation that has been offering a Master's of Leadership for Sustainable Development since 1996 in collaboration with Middlesex University (UK), which validates it as a Master's in Professional Studies. Forum for the Future has now begun to develop national versions of the program internationally.[31]
- In Australia, Griffith University, University of Queensland, Monash University and the University of Western Australia have collaborated through the International Water Centre to offer a Master's of Integrated Water Management, combining expertise across four universities in a Master's degree that shares delivery costs.[32]
- In Victoria, Australia, Monash University has been running an academic program called 'Green Steps' for more than a decade, involving more than 500 participants and partnering with over 300 organisations Australia-wide, including a number of universities. In 2009 the program received the national Banksia Award in the education category.[33]
- Led by sustainability leader Gunter Pauli, the Global Zero Emissions Research and Initiatives (ZERI) network is a non-profit organisation originally formed within the United Nations University, which has partnered with Politecnico di Torino in Italy to offer a one-year Master's of Systems Design, which uses a new business model of open industrial systems.[34]
- In the UK, LEAD International is a non-profit organisation that delivers training programs to business executives, government officials, academics, non-government organisations, educators and media professionals on emerging issues relevant to leadership and sustainable development, often in collaboration with the higher education sector.[35]

There are a number of competitions internationally that have also received recognition for their ability to engage engineering students within their curriculum studies, while also providing a community benefit. We highlight three of them here:

- *Mondialogo Engineering Award:* Initiated through collaboration between Daimler and UNESCO in 2003, this international design challenge seeks to promote intercultural dialogue and exchange among young people, encouraging students and future engineers to think about new ways to develop intercultural learning and to achieve sustainable development.[36] Students from industrialised and developing countries team up to produce proposals for engineering applications that demonstrate excellence in applied engineering, help develop the technological infrastructures of developing countries and contribute to the Millennium Development Goals of eradicating poverty and achieving sustainable development. This challenge is not tied directly to curriculum, although there is the potential for students to seek academic credit for their projects.
- *Solar Decathlon Challenge:* The US Department of Energy Solar Decathlon Challenge sponsors this annual national challenge, which challenges college teams to design, build and operate solar-powered houses that are cost effective, energy efficient and attractive. The winner of the competition is the team that best blends affordability, consumer appeal and design excellence with optimal energy production and maximum efficiency.[37] Although not specifically tied within a particular college course, students are encouraged to apply and further develop their knowledge and skills to the design challenge with the support of their college. Following the success of this initiative in raising awareness about solar technologies in the United States, the initiative is intending to spread internationally.
- *EWB National Design Competition:* Focused on providing engineering students with third-world experiences, the Australian chapter of Engineers Without Borders International has developed an

annual competition around addressing disadvantaged community development challenges, including aspects of sustainability, which is embedded in the first semester of first-year engineering experiences. Through a collaborative partnership between several engineering professional bodies and an international construction company, the annual 'EWB Challenge' includes an introductory keynote lecture to students at each of the participating institutions, and a case study package related to the design challenge. Submissions consist of essays and preliminary designs (i.e. at low cost) which can also be used as an assessment item within a first-year subject.[38]

Connecting with campus operations

Universities are large consumers of water, energy and resources and are increasingly being expected to 'walk the talk' with regard to sustainable development, particularly where it delivers cost savings. However, such efforts are typically not linked to curriculum renewal efforts as encouraged in the curriculum helix. The good news is that there are significant existing campus greening initiatives underway. Institutions around the world are designing, constructing and retrofitting buildings that emulate green building principles of water, materials and energy sustainability. Furthermore, in what appears to be a natural progression that some institutions have already begun, the next step is for actions associated with these initiatives to be embedded within the curriculum. For education institutions to more deeply address sustainability, campus management needs to be linked to research, curriculum and administrative practice, so that sustainability can be embedded across every aspect of institutional operations in a synergistic way.

Campus greening efforts that involve students can yield educational dividends for the future, fast-tracking student experiences in real-life applications of the theory that they are being exposed to, and providing a supportive environment to address barriers surrounding dealing with new and emerging technologies and processes. Furthermore, where faculty may not have recent industry experience, on-campus initiatives can also provide practical experiences 'in the field'. Such experiences can also be important in providing professional development opportunities in their discipline, and also to build off-campus networks with industry, business and government who may also be interested in piloting new technologies on campus.

In the Australian federal government's 2009 *National Action Plan for Education for Sustainability*,[39] the government committed the Australian education sector to linking campus operations to research, curriculum and administrative practice. Similar national endorsement has also occurred for other countries (see Chapter 1).

A 2006 UNESCO publication on education for sustainability[40] outlined a systemic approach to connecting curriculum to campus greening operations. In addition to improving the environmental management of campus operations, this includes embedding education for sustainable development, developing partnerships with other organisations for mutual benefits and, either through focused research centres or the efforts of individual researchers, exploring the dimensions of sustainable development and its achievement.[41] The report concluded that given the wide range of activities that take place at a university, it is often actually a simple matter to focus the project on some aspect of the university's operations. In these situations, the university becomes an integral part of the learning experience for the students. There is also a direct local connection to their work, and from the university's perspective, there is the probability of receiving knowledge and ideas about the campus that can assist present and future plans.[42]

Despite the apparent ease of creating such possibilities, efforts to integrate campus operations with curriculum appear to be limited (i.e. as examples[43,44] rather than mainstreamed), *ad hoc* (i.e. driven by individual champions rather than embedded within institutional structures) and often initiated by students engaging with facilities operators, rather than faculty. Even in ivy-league institutions where world-leading campus greening initiatives have been underway for the last decade, connections with the curriculum are much more recent.

Harvard University established a central Harvard Green Campus initiative in 2000, supported by a US$3 million interest-free loan facility to finance environmental improvement, and a commitment of US$150,000 a year for 5 years to fund core staff. This has resulted in numerous achievements, including conversion of its fleet of diesel vehicles (including student shuttle buses) to bio-diesel, high-performing new 'green' buildings and renovations, reduced energy consumption and increases of over 50% in recycling rates in undergraduate student residences.[45] However, reflecting on Harvard University's highly successful 10-year Green Campus initiative, founder Leith Sharp concluded that while the university may be at least five or so years ahead of other higher education institutions in America and elsewhere, given the limited current cross-over between campus operations and curriculum, this is the next challenge for the university, which can only benefit from students living through such real projects.[46]

So, while facility managers are beginning to see faculty and student engagement as opportunities to engage in cost-effective campus improvement projects, there appears to be resistance to engage. The practicalities of connecting faculty with campus operations can be complicated by the fact that the operational activities of most universities are vastly different to the academic systems surrounding teaching and research, from financial accounts through to management committees, often only having common reporting structures at the level of university senior executive (i.e. through pro vice-chancellors to the vice-chancellor or president).

Encouragingly, there are signs that students are also keen to engage with campus operations, with examples of students undertaking water and energy audits of buildings, creating communal gardens, designing solar photovoltaic options for campus buildings, designing green buildings and retrofits, and exploring institutional consumption and procurement of goods such as electronic equipment, paper and furniture. There also appears to be an emergence of forums for faculty and campus staff to share knowledge about campus operations and curriculum areas identified for potential improvement.

Within the Australasian Campuses Towards Sustainability (ACTS) network, 2010 saw the first major conference focus on connecting curriculum and campus operations. Across the Pacific, the United States has made significant progress in connecting campus and curriculum, supported by the Association for the Advancement of Sustainability in Higher Education (AASHE) network, the US Partnership for Education for Sustainable Development, and significant online resources provided by both organisations. The UK has also used the Higher Education Academy (HEA) to encourage both campus greening and curriculum renewal towards education for sustainability, in particular through the subject centres. More recently it has placed education for sustainability as a core priority, which has stimulated a number of initiatives that are just getting underway.

There are a growing number of organisations which are actively promoting and supporting campus greening initiatives, such as the US Partnership for Education for Sustainable Development (US ESD), the US Disciplinary Associations Network (DANS), the UK Environmental Association for Universities and Colleges (EAUC), and the Australasian Campuses Towards Sustainability (ACTS). Other far-reaching initiatives such as the American College and University Presidents' Climate Commitment (ACUPCC) and the Sustainability Tracking, Assessment and Rating System (STARS) have also provided significant platforms for action.

Connecting with community projects

Higher education providers operate within a broader community and so have the ability to contribute to the sustainability of these communities. However, until recently there was not much literature describing curriculum and community interactions within higher education. In 2008, the US Association for the Advancement of Sustainability in Higher Education (AASHE) provided a significant leap forward, focusing its national conference on *Working Together for Sustainability – On Campus and Beyond,* for which all of the published papers and presentations are freely available.[47] The wide variety of topics from more than 55 curriculum-related papers and 12 papers on faculty and staff development provide an indication of the growing interaction between academics, students and the community, at least in the United States. A number of the papers also provide a sense of wider reach beyond one subject or one member of faculty. The Campus Community Partnership Foundation[48] is an international example (arising from the United States) of an initiative to develop student skills and knowledge needed for social responsibility, where institutions are encouraged to give recognition for such student participation.

From a review of these proceedings and papers, it is clear that in addition to taking advantage of sustainability activities already being undertaken on campus, a curriculum renewal strategy will substantially benefit from taking advantage of other change processes underway outside the institution. Indeed, as highlighted in one of the conference panel sessions, such interaction is a key way to improving the curriculum through connecting to the needs of the local community.[49] Interactions may range from local sustainable development infrastructure and building projects through to community capacity building in climate change awareness, regional planning initiatives and sustainability-related behaviour change programs.

It is interesting to compare this situation to primary and secondary schools around the world, which are readily engaging with campus and community sustainability projects in the classroom, perhaps partly due to the greater focus given to sustainability in primary and secondary education over the last decade, including a number of United Nations initiatives and programs by the World Wildlife Fund and others. In Australia for example, national initiatives such as the Australian Sustainable Schools Initiative (AuSSI), the Sustainable Living Challenge led by UNSW, and state initiatives such as the Queensland Environmentally Sustainable Schools Initiative (QESSI) and the state's sustainable schools policy have built significant momentum in school- and community-based action that is grounded in the curriculum. This includes portal access for teachers to upload (i.e. share) and download lesson templates, case studies and assessment items, professional development opportunities in action-research to renew their curriculum, and informal and formal peer mentoring initiatives.

With these types of efforts, on average it is now more likely that a school is undertaking such on- and off-campus sustainability projects than a university. This needs to be factored into institutional planning, as students from these schools will be keenly aware of institutional performance in the area of education for sustainability when selecting further studies.

Looking for successful examples of community engagement with the curriculum renewal process, we talked with two teams, one at Ithaca College in New York and the other at the Australian National University.

- Allen-Gil and colleagues from Ithaca College in New York have undertaken an innovative community partnership to enhance education for sustainability in their environmental science curriculum that began with an initial engagement with a local EcoVillage, and led to a variety of interactions between academia and the local community.[50] Student projects ranged from constructing solar photovoltaic systems through to designing and constructing a sustainable bus shelter. Staff professional development included the establishment of the annual sustainability curriculum workshop where local, national and international academic and community experts in various aspects of sustainable development are able to share their learning and experiences. Further to the initial engagement, the college formed an ongoing Partnerships in Sustainability Education team with representation from a number of academic disciplines, along with educators from the EcoVillage, to continue to advance sustainability-themed education on campus and further strengthen community ties to advance student learning through internship placements and class projects.[51] Ithaca's success subsequently attracted the attention of the HSBC Community Foundation, which awarded Environmental Studies and Sciences a US$500,000 grant to further enhance sustainability education. This included ongoing scholarships, fellowships and internships related to sustainability, support for the annual sustainability curriculum development workshop, support for student delegations from Ithaca College to attend the annual United Nations Framework Conventions on Climate Change, and assistance with creating the College's online professional certificate in Sustainability Leadership.
- The Australian National University (ANU) is an example of where an engineering department can strategically modify its existing undergraduate program to create a market niche and ensure sustainability-related graduate attributes are addressed for all students, avoiding the risks associated with a new 'niche sustainability degree'. ANU's Faculty of Engineering and Information Technology has embedded sustainability into the mainstream bachelor degree, focusing on systems engineering, or 'whole system engineering' as the foundation, based partly on the work of The Natural Edge Project.[52] Associate Head, Engineering (Coursework), Associate Professor Paul Compton reflects, 'Unique in the southern hemisphere, the degree produces graduates trained to become key members in teams of engineers that provide complete or 'whole system' solutions, rather than individuals contributing sub-systems to someone else's project.'[53]

Interaction with 'dynamic' elements of the model

Awareness raising and capacity building

At the point of developing and updating curriculum, it is important for faculty to have ongoing capacity-building opportunities that align with the expectation to embed sustainability-related topics within the program. As highlighted in the discussion of barriers to curriculum renewal earlier, this will not work in most institutions as a force-fed process. Rather, it will be a blend of professional development opportunities through seminars, research grants, course-focused workshops with experts and regular rewards for those who participate. In the UK, the Higher Education Funding Council for England's Higher Education Environmental Performance Improvement (HEEPI) initiative are creating opportunities for faculty and students to be rewarded for their campus efforts, including empowering staff and students around these innovations. In 2007, the winner of the initiative's 'Green Gown Award' was the Pershore Group of Colleges, which included integrating sustainability knowledge and skills into all programs for full-time students. Alongside faculty-focused activities, there is also a role for raising awareness in the institution – among staff and students – around how the courses are being updated to integrate the sustainability-related knowledge and skills. This will help to maintain profile around education for sustainability while the 'back-of-stage' work is being done by faculty, ensuring a receptive audience for the next step and providing feedback on the previous one.

Internal and external stakeholder engagement

This element highlights opportunities for curriculum renewal to be accelerated by taking advantage of existing project momentum created by on-campus greening initiatives and local community initiatives. As discussed earlier in this chapter, there are many effective and efficient ways that internal and external stakeholders can connect and collaborate with mutual benefit. An example is at the University of South Australia, which first commenced with sustainability research and campus greening operations, including the employment of institution-wide environmental officers. The institution subsequently embarked on curriculum renewal considerations to connect with green campus initiatives.[54] This included opportunities for faculty and students to be involved in auditing energy and water consumption of buildings on campus, calculating the potential costs and energy savings of onsite renewable energy options, water saving infrastructure and passive cooling initiatives (such as shading, or painting the roof white or a lighter colour).

Monitoring and evaluation

Of all elements, it will be critical to monitor and evaluate the actual progress of curriculum development and updating. After all, the programs themselves are primarily what students, potential employers and the community will see when they interact with the institution. Some examples are highlighted in Table 9.4, for each of the activity streams.

TABLE 9.4 Performance indicator examples: curriculum development and update

Activity stream	*Example performance indicators*
Governance and management	Proportion of departmental budget allocated to sustainability-related education initiatives ($/year)
Operations and facilities	Number of students accessing campus data as part of their sustainability-related course work (number/semester)
Teaching and learning	Proportion of courses completing the education for sustainability audit recommendations (cumulative percentage/semester)
Human resources and culture	Proportion of academic position descriptions including explicit reference to capabilities in sustainability (percentage/year)
Marketing and communications	Proportion of academic recruitment advertisements including explicit reference to capabilities in sustainability (percentage/year)
Partnerships and stakeholders	Number of partners and stakeholders in active collaboration with education for sustainability curriculum (number/year)

Of course, once the audit recommendations have been responded to within the designated timeframe, there will then be a 'watching brief' – an ongoing need to refresh courses based on shifting external requirements or advances in the field. Performance indicators can also be developed for this stage, which can help to ensure the department does not return to a static curriculum position – for example, the number of teaching and learning forums in curriculum innovations which include an education for sustainability focus (number/year), and the proportion of course outlines each semester which include education for sustainability explicitly within the learning outcomes (percentage/semester).

Roles and responsibilities across the curriculum renewal strategy

In the following paragraphs we consider the deliberative element 'developing and updating curriculum' with regard to how this translates across the various stages of curriculum renewal and the relative involvement by each of the various activity streams, understanding that much of the activity in this element will be dictated by the outcomes of the previous three deliberative elements. For the purpose of illustration we have used a four-stage approach.

Governance and management

Stage 1: Prepare – In preparation for Stage 3, this initial stage will focus on preparing for the fact that in the near future there will be the need to create new materials and teaching resources as part of the process. This may include a focus on the internal level of readiness of current faculty to respond to the need to renew courses, and consideration of short- and medium-term budgets and funding options to support such activity.

Stage 2: Explore – In preparation for Stage 3, in this stage it will be important to ensure that the level of readiness to create new materials and renew courses is carefully considered as part of the audit process and recommendations. Once completed, the audit will provide guidance as to options to embed sustainability that need to be carefully considered for selection as part of the piloting stage. The audit team may also be directly involved in this process, or contain a representative from management to ensure such consideration and prepare for pilot selection.

Stage 3: Test and pilot – In this stage management will focus on supporting and overseeing the testing and piloting of efforts to develop the desired graduate attributes in specific courses, working with departments to inform decisions around resourcing, timeframes and expectations of outcomes. Based on the results of the pilots, the curriculum renewal strategy will need to be expanded to create a framework for the integration stage to follow.

Stage 4: Integrate – In Stage 4, management will be overseeing the integration of successful curriculum development and renewal initiatives across the various programs. There will also be need for monitoring the roles and reporting of progress. This will be a significant task that will need to be managed through the implementation of the curriculum renewal strategy to ensure that efforts lead to the desired outcomes and that the value of the process is being achieved.

Operations and facilities

Operations and facilities may be closer to industry than academics as they deal in energy, waste etc. in the real world and are often very aware of sustainability-related opportunities on campus. Of all the elements, this is the one where this activity stream will have the most interaction, through the possibility of campus–curriculum innovation projects, as outlined above. In the early stages it will be important to identify staff who are interested in engaging with such projects, who will then participate in the departmental briefings about sustainability and the transition. These staff will have the opportunity, depending on their personal expertise and interest, to explore opportunities on campus for potential interaction with the curriculum. Once the audit is complete, staff in this activity stream will need to review the report to see where such opportunities might connect with curriculum needs, and then engage with academics, facilitated by the design and innovation activity stream. During the later stages staff may be involved in testing and piloting campus–curriculum collaborative projects that are coordinated by the design and innovation stream. Should the collaborations be successful, this may then evolve as part of the integration stage, to include the formal integration of these initiatives within the curriculum.

Teaching and learning

Stage 1: Prepare – In preparation for Stage 3, in this initial stage faculty may begin to reconsider any current and proposed curriculum renewal efforts, in light of departmental focus on strategic content development and renewal towards education for sustainability over coming years. Faculty may begin to put aside resources they come across related to sustainability in preparation for the update stage later in the process. Faculty will be briefed on future timeframes for testing, piloting and integration opportunities, following the graduate attribute identification and mapping, and curriculum audit.

Stage 2: Explore – In preparation for Stage 3, in this stage faculty will be informing the identification of graduate attributes, mapping of learning pathways and auditing of specific courses with consideration as to the follow-on impacts on updating the program. It will be important that such considerations are a part of these preparatory elements so as not to create an unachievable set of audit recommendations. Once the recommendations of the audit are complete, there will be opportunity for faculty to express interest in participating in the testing and piloting stage to follow.

Stage 3: Test and pilot – This will be a key stage for this element with courses being updated to align to the graduate attribute learning pathways and responding to the recommendations of the audit. Along with updating teaching materials this may also involve engaging with campus facilities management to determine opportunities for integrating curriculum renewal with campus operations and identifying opportunities for community engagement as outlined above. A key part of this stage will be learning from efforts to respond to the audit recommendations to inform the following integration stage.

Stage 4: Integrate – In Stage 4, faculty will be integrating the successful initiatives across the program, from high-profile projects that may involve more than one activity stream and have potential media opportunities, through to smaller initiatives within each subject. It will be important that teaching and learning faculty inform the curriculum renewal strategy to ensure that lessons from the pilot trials are incorporated and barriers to institution-wide implementation of the strategy can be well informed.

Human resources and culture

For this activity stream, the primary role is to ensure that the curriculum renewal requirements are not constrained by corporate documentation or bureaucracy. This may include incorporating tracking progress formally through staff performance reviews and ensuring that staff contracts and recruitment strategies align with departmental needs. In Stage 1, this activity stream will need to identify staff that will be the main contacts for the transition. In Stage 2, these staff will need to be briefed by management on the various types of curriculum renewal strategies possible, and what these mean for staff resourcing and potential recruitment opportunities. In Stage 3, staff will be drafting requirements for the annual staff review process, to incentivise action on curriculum renewal testing and piloting. This will continue in Stage 4, where the performance review process will be geared towards integration of the initiatives.

Marketing and communications

This element provides the marketing and communications activity stream with significant opportunities to learn about and share stories regarding curriculum innovations, internally and externally. Teaching and learning faculty will be able to extract key details that potential students and employers may find interesting and relevant to inform marketing efforts. For these reasons, it is imperative that the marketing and communications area use the initial stages to find a staff member who is interested in this element and these types of tasks, who can then participate in briefings and any necessary information sessions from the beginning of the process. In Stage 2, these staff members will review the findings of the curriculum audit for examples of existing strengths and innovations that can be reported. They will also be identifying, from testing and piloting decisions made by management, what stories can be developed during the piloting stage to follow. During Stage 3, this marketing and communications may also be responsible for running focus groups and surveys to test the language that has been developed to promote the renewed curriculum at the institution. In Stage 4, further media opportunities will develop as knowledge and skills are integrated throughout the curriculum and external partnerships developed with industry, government agencies and the community. Assistance will be needed to update program and course marketing materials to reflect the integration of sustainability and the opportunity it presents students.

Partnerships and stakeholders

This element carries opportunities to accelerate the process within the larger context of campus greening initiatives, and to also take advantage of momentum created by existing institutional change processes. This ensures that relationship-building activities are commenced with campus facilities staff and community stakeholders, and that future interaction opportunities are flagged and planned. This activity stream will be responsible for important actions in this element around ensuring that existing and desired partners and stakeholders are kept informed of the types of curriculum innovations that are being incorporated into the renewed programs. This will serve two purposes, including creating student – and staff – recruitment opportunities and creating curriculum collaboration opportunities such as keynote lectures, field trips and research projects. Hence, together with marketing and communications, this activity stream is critical in ensuring the successful implementation of this element.

In Stage 1, staff need to be identified who will lead tasks related to strategic content development and renewal throughout the transition. These staff will participate in briefing sessions which will inform subsequent briefing materials to the department's partner and stakeholder network – these staff will need to be able to clearly demonstrate the benefits for engaging in rapid curriculum renewal to network members. Towards the end of Stage 2, staff from this activity stream will lead meetings and possibly initiate workshops with the network and academics to explore opportunities for community–curriculum collaboration, in light of the audit findings. At this point, it will also be important to seek funding assistance from the network for testing and piloting the collaborations, and then potentially integrate the successful initiatives throughout the program. There may also be a need to expand the network, should there be significant gaps in the composition of partners and stakeholders in relation to needs. In Stage 3, these staff will continue to liaise between the network and academics, supporting the trials, dealing with any challenges and identifying further opportunities as they arise. In Stage 4, as the integration process proceeds, staff will continue to liaise with the network.

In the next chapter we consider the deliberative element of 'Implementing the Program'.

10

IMPLEMENTING THE PROGRAM

It is the equivalent of 'grand opening night'. The stage is set, casting complete, script prepared, the audience have paid for their tickets and are in their seats. The production – in this case your program – is ready to roll and perhaps there are some nerves back-stage! At this point, you may have several questions such as, 'what is the lowest-risk way to roll out the renewed curriculum?', 'will the students engage with the updated curriculum?', 'how will I know – and be able to communicate to others – whether it is successful?', 'are my colleagues taking action to address their new learning outcomes too?'. Perhaps there have been some 'reviews' and marketing of your sustainability inclusions in your course/s already, which has secured your student enrolment for the semester. Now it is time to implement them, and capture feedback for continual improvement. In this chapter we explore opportunities for minimising institutional risk and faculty time spent in course implementation, taking advantage of what has already been achieved from the previous steps in the process.

This chapter begins with a discussion of the 'reality' of course implementation, where, despite the best planning processes, new challenges are likely to arise that could threaten to derail the process. Through greater understanding of these it is possible to take actions that keep the process on track, with satisfied faculty, students and key stakeholders. The chapter then highlights a number of financially attractive opportunities that can be harnessed during this step, including suggestions for:

- Developing faculty strengths
- Catering for professional practitioners
- Sharing faculty and curriculum resources
- Reaching out to schools and the community

It has been tempting for us to use this publication to discuss the various types of teaching and learning options to engage with education for sustainability. Many papers on the topic blend content and pedagogy, signalling the importance of 'how' to teach the new knowledge and skill sets, as much as 'what' to teach. From blackboard and whiteboard lectures through to online interactive media and problem-based learning, there are myriad options available to interact with students, many of whom approach their studies with cultural preferences. Keeping this in mind, we highlight the importance of a whole-of-system approach to learning in the updated curriculum, through whatever pedagogy is chosen by faculty. In this approach, students are required to draw on their core training in a given field and interact with each other to apply knowledge and skills developed through their program, tackling increasingly complex and integrated challenges.

Exploring the deliberative element 'Implementing the program'

Introduction

Implementation is at the pointy end of the process. Students experience the updated program and the courses are fine-tuned to deliver the intended learning outcomes. However, without a systematic approach to curriculum renewal, faculty can feel unsupported and 'on their own', especially at this stage in the process. In the following pages we discuss the importance of approaching this step as the culmination of the planning steps described in previous chapters, rather than an *ad hoc* process of trial and error by faculty acting in isolation.

Planned and supported course implementation will ensure that new and renewed courses fit within a coherent structure of learning pathways for desired graduate attributes. Faculty can be confident that their courses contribute to program goals, be supported to do so, and be recognised for their efforts.

When discussing ways to accelerate curriculum renewal efforts with colleagues, the conversation quickly turns to discussing the issue of long-standing and often outdated teaching practices that are disconnected from professional practice, student expectations and progress by colleagues and other institutions. Despite the best planning processes that may have been undertaken, once in the hands of the course convenors, a new set of challenges appears.

A workshop on curriculum renewal typically includes the following conversation:

> Colleagues raise the issue of how departments are still offering degree programs in silos within the university that are disconnected from reality. There is a common complaint that today's curriculum is outdated, sometimes in the order of decades. Then someone will contribute that the broader community still doesn't understand what sustainability is about anyway, so how are academics supposed to know what to aim for in their graduates? The discussion subsequently moves to resignation about this being one of the reasons for lower than expected enrolments into sustainability-related courses. According to many workshop participants over the years, higher education will continue to be less than ideal, until . . . employers make their expectations clear, until the various professions sufficiently clarify the role of their profession in the community, until schools teach environment and sustainability better, until government stops using the term 'sustainability' to mean other things, and so on!

You may already be familiar with this type of discussion, and with areas in your institution where silos exist that impede the type of curriculum renewal being discussed in this book. While some of these concerns are real, and others perceived, many of them are outside faculty control. Moreover, by collapsing these issues into one another, it is difficult for colleagues to feel empowered to act.

We have subsequently focused attention on a question of 'at the level of each course, how can institutions create programs that are current and popular?'. In summary there are three 'classic' challenges to address, around faculty engagement with each other, with professional practice, and with potential students:

- Connecting with colleagues and institutions: An unsettling reality that persists in many institutions is the level of disconnect that exists between faculty within the very same department, with colleagues elsewhere in the university (for example other areas of the curriculum helix, such as operations and facilities or marketing and communications) and in other institutions. Despite today's access to email, web-conferencing, online learning, libraries and publications, an insular culture still persists in some institutions around the world among faculty in regard to teaching and learning practices. Given the scale of responding to education for sustainability across the higher education sector, such disconnect is a serious issue that should be addressed specifically as part of the curriculum renewal strategy. For example, not only does this create uncertainty in planning curriculum renewal (i.e. with regard to what is already being taught, by whom), it also stifles potential innovation in inter-institutional collaboration for teaching in areas with few experts.
- Connecting with professional practice: Many faculty still teach largely from the same notes that they have been using since they started the course, with only minor updates. In only a few cases is this appropriate, given the scale and pace of change around the world. There is now mounting advice from teaching and learning colleagues, advisory panels and accreditation reviews about the changing context of practice across the higher education sector. Accelerating curriculum renewal to respond to this changing context is hindered when textbooks and readings often remain the same – although perhaps the edition changes with minor updates. In some cases there has been no direct contact between the faculty member and practitioners for more than a decade, or the faculty member has never practised the profession that they are teaching into. This lack of interaction has two implications. First, where the academic has become an expert in their topic area, the knowledge and skill sets are not reaching practitioners. Alternatively, where the practitioner is the more experienced, there is little to no access for this to be transferred into the curriculum.
- Connecting with potential students: Many faculty do not engage with their local communities (government, business, industry and schools) with regard to their profession or programs on offer. Particularly within the school system, students emerging from senior studies now have a very different knowledge and skills base than students from a decade ago, often with a much greater understanding of environment and sustainability issues and opportunities. These potential customers are market-savvy and will not accept marketing that is not authentic about program offerings. In larger institutions, the gap in communication is wider, where 'recruitment' is given to a marketing or schools liaison officer – who often has not studied in the discipline and may not be familiar with emerging innovations.

Despite these challenges, there is more than enough evidence for improved communication and collaboration strategies that work, as discussed in the following pages. Indeed, towards the end of workshops on this topic, participants often begin to share what they are doing at their respective institutions and there is a lot of note-taking and swapping of contact details!

Dissolving silos on and off campus

Previous chapters outline a number of curriculum renewal activities that can engage faculty across the institution, and encourage the development of communities of practice and collaboration. This includes involvement in workshops, focus groups, sub-committees, research teams and supervision on sustainability-related research. However, considering various activity streams within the institution (described in Chapter 5), many are, for the most part, invisible to faculty, except when there are pay issues, graduations, or a room needs unlocking! Operations and facilities managers are even located at the other end of the campus, in a separate building that occasionally gets visited to pay parking fines or report missing property. Faculty interactions with support staff are minimal – reporting broken equipment, requesting a room-clean and participating in fire drills. Once in a while an email might arrive with news of a proposed new building on campus, to which faculty reply regarding space needs and perhaps with some suggestions on green building considerations – if there is time to spare!

Considering potential assistance from off-campus avenues, it appears that students and staff are most often engaged with their communities through extra-curricular activity. This might involve, for example, local community Lions and Rotary groups. For faculty, such activities are usually labelled as constituting part of an academic's 'service' load, which is only around 20% of their overall workload formula (the other 80% comprising research and teaching), and often already taken up by attending internal committee meetings. Students may under some circumstances get credit for such participation, but this is the exception rather than the rule.

> Many institutions' recognition of 'service' does not create much incentive for faculty, or students, to learn or even participate in life beyond theory and hypothetical applications of courses. Neither do they empower students with experience in making a difference through their coursework out in the community or profession before they graduate.

In reality, campus and community environments include professionals who are also being challenged by the implications of sustainable development. On any given campus or community project there will be professionals from all sorts of fields and disciplines, often acting in a mixture of technical and project management roles. With this in mind, there are a host of possibilities for interaction with curriculum (see examples on the following pages):

- On campus, operations and facilities management includes maintenance of the grounds, residential colleges, and office, theatre and library building operations (including the façade, structure, component equipment such as air conditioning, water, energy, etc.) and the management of new infrastructure projects, from lecture theatres to office space and central cooling and heating systems. Operations and facilities staff prepare contracts for tender, control campus procurement, manage the university fleet of vehicles and may also manage the leasing of retail and food outlets on campus.
- In the community, there are myriad initiatives underway to reclaim, revegetate, and reinvigorate spaces with a mixture of art, landscaping and functional infrastructure that may include play equipment, shelters, cooking and sleeping facilities and so on. There are also a host of initiatives related to reducing demand for energy and water through behaviour-change programs and energy-efficiency opportunities, and to increasing renewable energy supply through solar, wind and geothermal projects.

Using academic facilities as 'living labs' and 'demos' for sustainable solutions is not just a need but a must in order to lead innovative education . . . Campuses are becoming places where students, faculty, staff and the community can learn, teach, work and play together, and apply science in practical ways to address emerging and exacerbating environmental and social issues.[1]

Within this campus and community context, sustainable development opportunities are likely to be plentiful, including energy and water efficiency, materials productivity, centralised air conditioning, onsite energy generation and so on. Furthermore, with the university or funding provider for community initiatives (e.g. government) being an owner and user of much of the infrastructure, the financial feasibility of sustainable solutions are becoming increasingly attractive. Clearly, these environments outside the traditional classroom provide a significant opportunity to develop sustainability graduate attributes in a real-world setting where literally everyone is learning, including students, academics and the practising professionals. As highlighted in the following breakout box, there is an awareness that institutions can no longer afford to be seen to be teaching sustainability knowledge and skills, and then not practising it on campus (i.e. 'walking the talk'), given the significant potential for negative public relations, student intake and faculty recruitment ramifications.

In our discussions of opportunities around course implementation with community groups and operations and facilities personnel at various universities, the primary challenge appears to be interacting with faculty. Campus operations staff seem to understand that campus greening by itself will not be sufficient to 'green' the university, but don't have clear paths to engage the students through the curriculum. It seems faculty are collectively perceived as a rather reclusive bunch, and as somewhat reluctant participatants in 'real' projects, preferring the neat and controlled environment of theoretical assessment. In contrast, discussing on- and off-campus relationship building with faculty often creates excitement in the room about possibilities. One workshop even created the following possible motto, '*Many are talking about sustainable development, but at this university we practise what we teach!*'. There are, of course, a number of challenges that need addressing as part of the curriculum renewal strategy to increase access to 'real' projects. These include, for example, liability (i.e. regarding students giving advice), limited timeframes and scope for project work (i.e. within the teaching semester), quality variability (i.e. depending on the students' capabilities), and a lack of willingness to supervise or spend time assisting students working on such projects (i.e. providing data, answering questions).

Creating spin-off opportunities

In the following paragraphs a number of financially attractive opportunities are highlighted, which can be harnessed within this element. Each of these opportunities also provides additional 'anchors' for the renewed curriculum to be embedded within the program, making it increasingly difficult for the process to be inadvertently wound back or undone. There is the potential for learning to occur in a variety of ways between faculty and professional practitioners, schools and the community depending on where there is an interest in addressing knowledge and skills gaps.

Developing faculty strengths

Without formal requirements for professional development or associated accreditation requirements, academics who have been employed within the higher education sector for 10–20 years may well be teaching outdated information. In his keynote paper to the 2009 Australasian Association of Engineering Education conference, education researcher Malcolm Allan from the Glasgow Caledonian University in Scotland notes that educators need to be trained and developed to teach an understanding of competencies in the context of a global information society.[2] The 2008 Australian *Engineering Review* recognised this issue when they recommended increasing the uptake of academic positions by candidates with substantial and relevant industry experience, increasing the network of acknowledged expertise in engineering education, increasing the take-up of industry-based study leave opportunities, and increasing the sharing of resources between research and teaching.[3]

> Within international education literature there is a steady stream of discussion about academic educators not having recent industry experience, particularly in engineering and the built environment.[4]

This problem of currency in experience and knowledge appears to be compounded by the diminishing attractiveness of academic positions with workload and resourcing challenges and hence the difficulty, competing with the practising sector, to attract experienced engineers into academic roles. Where practitioners are experts in their field in an aspect of sustainability, there may be opportunities for academics within the department to engage in professional development through work placements, research projects, or conference or symposium gatherings. Such interactions with industry can strengthen relationships in regard to future collaborative research endeavours, highlight to employers that graduates are being exposed to such new curriculum, and also facilitate raising awareness about new opportunities for study. Depending on the knowledge and skill gaps within the department and the number of faculty who will be tasked with significant sustainability-related initiatives, there may also be a need to develop a short course that forms part of an induction process for these faculty. The increasing availability of open-source content and online support also appears to be an available mechanism for addressing this knowledge gap, although there is still concern that sufficient time and support is not available for faculty to make use of these resources.

Catering for professional practitioners

Where there are faculty who are experts in an aspect of sustainability, there may be opportunities to develop and offer postgraduate and professional development curriculum. Although few postgraduate programs require accreditation (and are therefore not under pressure by the accreditation organisation), there is strong and increasing pressure from industry and government for employee professional development in sustainability knowledge and skills. As the market is sending clear signals about demand for particular topics, the risk to departments seeking to offer such courses is reduced, particularly if they are offered in partnership with a number of departments such as engineering, business, humanities and/or science. This type of coursework could be undertaken on campus at the university, offered onsite, or offered online. Capacity building may be in the form of non-certificate short course training (on campus or onsite), or as a unit that has the option for assessment and credit towards an academic qualification.

When catering for such practitioners, although the content may stay the same, the mode of content delivery is likely to need some modification towards adult learning principles. Broadly speaking, adult learners share the characteristics of being self-directed learners who want knowledge that is immediately practical (i.e. problem centred), and who have a growing reservoir of experiences to draw on in learning.[5] Content delivery in these learning environments might immerse participants (both students and faculty) in predicaments and problems that participants would bring with them from their workplaces or other aspects of their lives. Such an approach makes access to online materials essential, together with faculty support.

A possible pitfall to consider is the potential for industry demand for short course training to undermine the department's capacity to deliver more formal postgraduate education (i.e. through faculty resourcing). This can be addressed by creating short courses with a pedagogy that is rigorous enough to contribute to a postgraduate qualification, providing the potential for articulation between informal training and formal postgraduate programs. Such a strategy may also encourage participants to consider undertaking further education such as a certificate, graduate diploma or masters studies. Institutions might also promote introductory and specialist courses internally to other parts of the institution and to international potential student audiences through collaborations.

Professor Michael Powell, Pro-Vice Chancellor (Business) and former Dean of the Griffith University Business School which offers a 'Master of Sustainable Enterprise', notes that the school is integrating corporate global responsibility principles across all programs because it thinks the business leaders of tomorrow should take corporate responsibility and sustainability as seriously as the bottom line, and also because many who show interest in coming to the public seminars subsequently engage in programs.[6]

Bournemouth University in the UK offers a flexible-delivery (distance learning) 'Green Economy' Master's, which combines work with studies, including the opportunity for professional placement. The courses of the program can be taken within formal Master's study, or as individual continuing professional development courses, where students can register for one course at a time. The initial two years had cohorts of 20 students each.[7] Chris Shiel, Associate Professor, Bournemouth University, shared with us that 'the students on the course come from a range of backgrounds, are passionate about "green" issues, and their engagement with the learning materials has been superb'.[8]

Sharing faculty and curriculum resources

For institutions that provide both postgraduate and undergraduate education, most postgraduate programs are offered under a different fee structure to undergraduate programs, as this gives institutions more flexibility in being able to fund the development of new courses. Postgraduate programs often have courses that can be co-offered to final-year undergraduate students. Postgraduate courses are often taught by faculty who also teach undergraduate courses and thus content from courses developed for industry could then be used to respond to the audit recommendations for undergraduate course renewal, while the Master's and other postgraduate content continues to be updated and extended.

Sharing faculty experience and teaching resources between undergraduate and postgraduate courses allows for sustainability education to be continually improved across university offerings.

An advantage of connecting postgraduate and undergraduate education is that institutions can fund and drive undergraduate curriculum renewal through accessing the significant and increasing demand for continuing professional development opportunities in education for sustainability across government, the industrial sector and licensing authorities. The timeframe for postgraduate programs varies between 1 year for a diploma to 3 years for a Master's certificate, meaning that students can develop sustainability knowledge and skills relatively quickly, then apply their new knowledge and skills in the workplace. As Master's students discover that developing a sustainability agenda in the workforce is far more effective when done as part of a multi-discipline team, this may increase demand for graduates with such abilities. In this way, postgraduate curriculum forms a key component in undergraduate curriculum renewal.

Reaching out to schools and the community

Most universities have existing programs to connect with their 'catchment' of potential students through bridging activities such as offering first-year courses for gifted final-year school students, and school outreach programs such as keynote lectures, competitions, open days and mentoring. Within this existing model, opportunities exist to engage potential students in sustainability topic areas taught by the institution.

Literature about education for sustainability includes increasing reference to institutional benefits of outreach activities with high schools and the community to both enhance understanding around sustainability and to recruit the brightest students who increasingly have high environmental values.

For example, first-year flagship course/s could be offered to high achievers in high schools as an accelerated year-12 course option. This could support existing sustainability challenges offered by local community groups, other institutions or government authorities by providing prizes and keynote speeches on sustainability topics at events. Universities that have not developed their own materials can also immediately begin outreach activities with resources already available online. Extracurricular education programs, such as intensive short courses, can also play an important role in bridging the divide to potential students and in introducing sustainability education in higher education. Finally, educational programs targeted at high school teacher professional development could also benefit both the institution with regard to student numbers and raising awareness among teachers.[9]

Universities will also need to continually adapt such outreach – and subsequent curriculum within university studies – to cater for changing incoming student knowledge and skills. Compared with the situation a decade ago, students already arrive at university with a greater awareness of sustainability. As such knowledge and skills become embedded in school education (i.e. from kindergarten through to year 12), and with increasing media exposure regarding the challenges and opportunities outlined in Chapter 1, student demands are likely to also change with regard to their preferred vocational education or higher education studies. This is therefore also an important consideration for universities considering outreach and subsequent modes of integrating sustainability in engineering education over the next decade.

Within the primary and secondary education systems there is literature discussing the need for kindergarten through to grade 12 (i.e. K–12) teachers to undertake professional development in specific areas such as engineering education as part of the plan to arrest the decline in tertiary enrolments.[10] Such professional development also provides an opportunity for teachers to expose students to new knowledge and skills in parallel with universities embedding it within programs.

Extracurricular programs such as intensive short courses can play an important role in successful curriculum renewal. For example, The Youth Encounter on Sustainability (a program within ETH Zurich) Program Director Michelle Grant reflects that:

> With over 900 alumni from 100 different countries, the course has played an important role in sensitising upper level university students to the complex issues of sustainable development . . . These initiatives have a spill over effect to other students and youth, which plays an important role in further raising awareness and increases demand for sustainability education.[11]

Examples of this element in action

A growing number of institutions around the world are seeking to address the issue of sustainability knowledge and skills currency, through initiatives such as the UK Visiting Professors Scheme,[12] and a number of others including:

- In China, Tongji University has been running a number of successful training programs over the past two decades through the Training Centre of Urban Planning and Management for leaders and professionals of urban management and construction, including a significant throughput of mayors.[13] For instance, once mayors complete their initial training, the college offers them 'service for life' through a 'sodality' within the Tongji alumni association, where program participants can continue to network and share experiences, and where the Tongji College of Architecture and Urban Planning[14,15] can assist them with planning issues.
- In Mexico the Monterrey Institute of Technology has been striving for education for sustainability since the mid 1990s, and since 2005 has been implementing an integrated, interconnected and multi-disciplinary approach for fostering sustainable development which includes a short course for sustainable development 'champions'.[16,17] This project also developed an assessment tool to track and assess staff progress in integrating sustainability within their courses.
- In Australia:
  - The University of South Australia's approach is consistent with the perspective that sustainability should be integrated into the policies, approaches and learning experiences of all academic members, professors and the students, as well as academic directors.[18]
  - RMIT University introduced a Master of Sustainable Practice in 2005 that was designed to embody adult learning principles. The structure was a mix of monthly whole-day workshop

classes where problems were discussed, combined with the traditional electives that provided supporting information and skills, including guest speakers, and outsourced content as elective options. The multidisciplinary program was built on adult learning principles with graduate capabilities focused on problem solving combined with change management, supported by pre-prepared modules.[19]

- Griffith University's Centre for Environment, Population and Health (CEPH) provides regular intensive training opportunities for middle to senior managers across industry and government internationally on areas including environmental health, public health and environmental management, which draw on the existing courses on offer through the university.[20] CEPH won the Queensland 2008 Education and Training International award for Best Practice in International Collaboration for outstanding achievements and positive collaborations with institutions in the Asia-pacific Region.[21]

- In Sweden, Chalmers University's academic staff have been writing about the implications for sustainable development in their curriculum and research since the mid 1990s. Internal awareness raising and inquiry have involved an organic process of individual champions creating momentum, which resulted in the formation of a specialist graduate-level education area called Engineering for Sustainable Development (ESD). Recognising the need for international peer support in this journey, in 2001 Chalmers joined the Alliance for Global Sustainability (AGS), a scientific co-operative venture with the Massachusetts Institute of Technology, ETH Zürich and the University of Tokyo. It also collaborated with AGS and the Delft University of Technology to create the Engineering Education for Sustainable Development Observatory to track European progress in curriculum renewal. These collaborations have resulted in a variety of papers on the topic of curriculum renewal towards ESD, which have helped to raise global awareness about the potential for science and engineering for sustainable development.[22]
- In the Netherlands, the Delft University of Technology was one of the early movers in transitioning curriculum towards education for sustainable development in the late 1990s, as previously mentioned. A critical part of the transition involved the formation of a 'sustainable development committee', which developed a case for such a transition. Subsequently, in 1998 the board took up the challenge of integrating sustainable development into curriculum and research, through an approach of 'learning by doing'. Ongoing awareness-raising activities have included the continuation of a lunch lecture program and the use of a platform called 'OSIRIS', a network of students and staff sharing a common interest for sustainable development, who organise activities such as lectures, discussions, excursions and workshops.[23] Ownership of various parts of the Delft strategy are shared across the university.
- In Spain, the Technical University of Catalunya has actively engaged in encouraging dialogue and research in sustainability, emerging from a strongly democratic approach to engaging with staff and students. The university's 2015 sustainable development vision is the result of 10 years of institutional maturation of the internal process. A key part of the engagement strategy has been to encourage staff to develop their own ideas for how they perceived sustainability and how they would like to engage in the process of curriculum renewal.[24] Institutional activities are coordinated through CITIES (Interdisciplinary Centre for Technology, Innovation and Education for Sustainability) which was formed in 1997, and which has a commitment to integrating sustainability into all areas of the UPC and providing opportunities for innovating in sustainability. This includes jointly organising the evaluation and updating of the Sustainable UPC 2015 plan and drafting its annual progress report.[25,26]

Interaction with 'dynamic' elements of the model

Awareness raising and capacity building

As faculty work to embed sustainability graduate attributes across the programs of the university the focus of awareness raising and capacity building will be on addressing silos within the institution. Further to the discussion about challenges earlier in this chapter, this will include a number of forms of communication to raise awareness and build capacity among faculty:

- Connecting colleagues – internally and externally: Creating opportunities for faculty to find out about progress and innovations in other courses and potentially in other institutions. This could include workshops, forums and seminars where faculty are encouraged to present on their initiatives, and peer-to-peer meetings with colleagues and teaching and learning experts to seek direct assistance with challenges or queries arising during the implementation processes.
- Connecting with professional practice: Facilitating faculty attendance at internal and external teaching and learning events about technical content and also about pedagogy. This could include a system of rewards for professional development in this area, where faculty can demonstrate subsequent efforts to renew their curriculum using their recently acquired knowledge. Professional development opportunities may also be created among industry and government to raise awareness among potential employers and students about new program offerings.
- Connecting with potential students: Supporting the development of marketing materials for recruitment that is authentic regarding the renewed curriculum intentions and offering. This may involve dedicated sessions between faculty and marketing and communications staff to develop the pitch for the program, where there is an opportunity for staff to learn about the approach and knowledge and skill development at the same time.

Clearly responsibility for these actions will be spread over several activity streams as we highlight below.

Internal and external stakeholder engagement

Wide stakeholder engagement is critical at this point in the process, where the success of the renewed program relies on student enrolment and satisfaction with their experience. To create a space for the curriculum to be promoted, the institution may engage in a number of stakeholder engagement activities that communicate the opportunities discussed earlier in this chapter. These might include short course and postgraduate offerings, innovative delivery options and articulation opportunities, innovative curriculum for future career pathways, and opportunities for university students to assist with off-campus projects in industry, school or the wider community. Avenues for this interaction could comprise:

- Industry engagement: 'hot topic' seminars, and short course 'tasters'.
- School engagement: school guest lectures and school visits to campus.
- Community engagement: community lectures and support of community events.
- Institutional (internal) engagement: internal showcases, seminars and displays.

Monitoring and evaluation

Of all elements, it will be especially important to monitor and evaluate the actual progress of implementation during this final stage of the curriculum renewal strategy. Management will want to be able to report on the initiatives and how they have delivered tangible benefits, and where they have been achieved. Literature is beginning to emerge that discusses possibilities for assessing the attainment of sustainability-related knowledge and skill development, which is problematic due to the need to test for systemic, complex and multidisciplinary skills in addition to straightforward knowledge about sustainability phenomena.[27,28] We look forward to monitoring how this research progresses with potential future application as an evaluation measure within the curriculum renewal model. Opportunities are also emerging – albeit slowly – for national incentives to monitor and evaluate success in embedding sustainability within the curriculum. For example, in the UK, the country's 'Quality Code for Higher Education' (which guides institutional teaching and learning policy as well as program-level curriculum development) includes education for sustainability as a cross-cutting theme within its learning and teaching chapter.[29]

Meanwhile, at an institutional level, some examples are highlighted in Table 10.1 for each of the activity streams.

TABLE 10.1 Sample performance indicators for implementing the program

Activity stream	*Example performance indicators*
Governance and management	• Number of awareness-raising seminars occurring within the department (number/semester) • Number of curriculum, campus and community collaboration initiatives that are able to leverage further research or teaching funds (number/year) • Number of staff who receive workload recognition for bridging and outreach accomplishments (qualitative)
Operations and facilities	• Number of curriculum, campus and community collaboration initiatives that are able to leverage further research or teaching funds (number/year) • Proportion of invited staff who participate in induction training (percentage/event) • Proportion of website pages related to sustainable development initiatives that include mention of institutional priorities (percentage)
Teaching and learning	• Proportion of campus sustainability initiatives that are linked in some way to sustainability-related curriculum (percentage of projects/year) • Proportion of invited staff who participate in induction training (percentage/event) • Level of understanding of key strategy actions, responsibilities and deliverables (qualitative) • The number of subjects that involve on-campus or off-campus community collaboration (number/year level)
Human resources and culture	• Level of staff satisfaction with their bridging and outreach workload recognition (qualitative) • Proportion of relevant induction documents mentioning sustainability as a priority departmental goal (percentage) • Proportion of staff performance review submissions where teaching achievements include curriculum–campus or curriculum–community initiatives (percentage/year)
Marketing and communications	• Number of internet hits on web pages related to bridging and outreach opportunities (number/month)

TABLE 10.1 Continued

Activity stream	*Example performance indicators*
	• Proportion of internal communiqué including reference to education for sustainability (percentage) • The number of published media articles on curriculum–campus or curriculum–community initiatives (number/semester)
Partnerships and stakeholders	• Extent of network awareness about bridging and outreach opportunities (qualitative) • Level of understanding of the strategy, intended curriculum outcomes, and opportunities for engagement (qualitative) • Proportion of invited partners and stakeholders attending events where curriculum collaboration projects are discussed (percentage/event)

Roles and responsibilities across the curriculum renewal strategy

In the following paragraphs we consider the deliberative element 'implementing the program' with regard to how this translates across the various stages of curriculum renewal and the relative involvement by each of the various activity streams. For the purpose of illustration we have used a four-stage approach. Most of the activity occurs in the later stages as part of the pilots and integration activities, with a range of preparatory activities in the initial stages.

Governance and management

Now that the work has been done to inform integration across entire programs, a number of benefits will start to emerge, such as the opportunity for additional income streams through short course and professional development programs. It may be possible to provide faculty with capacity-building experiences that are essentially cost neutral. It will therefore be important for governance and management to provide a clear platform through the curriculum renewal strategy and also to perform a strategic liaison role, engaging professional practitioners and the community in opportunities to collaborate with the department in the curriculum renewal process.

Stage 1: Prepare – In this stage, governance and management need to be aware of the goal of integration across the program/s when considering the early stages of the curriculum renewal strategy. This may include creating a list of existing professional contacts in industry, government and the community who may either be experts in some aspect of sustainable development that the department is interested in, or who may be seeking knowledge and skills from the department.

Stage 2: Explore – As the various staff are involved in graduate attribute mapping and audits, as part of this stage governance and management need to ensure that the reality of a program-wide rollout is considered to ensure a suitable aspiration that can be achieved in the final stage. A key part of this is to ensure that faculty are sufficiently supported with time and resources to undertake the various activities in this stage in preparation for the integration stage to follow.

Stage 3: Test and pilot – As part of the pilot, a number of courses will be implemented in response to the auditing recommendations, and governance and management need to ensure that responsibilities are delegated to appropriate faculty members. The lessons from this piloting will be crucial to the program-wide integration and need to be effectively incorporated into the curriculum renewal strategy and communicated internally and externally.

Stage 4: Integrate – In this stage, the lessons from the pilots will be drawn upon to systematically renew entire programs in line with the curriculum renewal strategy. It will be important that the pilot lessons inform the selection of the next round of courses, and that programs respond to the auditing and undertake integration of sustainability graduate attributes. It will be important to ensure that the successful collaborations are integrated throughout the curriculum, reviewing the revenue stream for potential support of other education for sustainability initiatives.

Operations and facilities

This element provides a significant opportunity for operations and facilities to engage faculty and students across the university in on-campus initiatives. This can be done by connecting contractors and suppliers with faculty. It can also involve activities through its own staff, some of whom will be practising professionals in building design, construction management, building maintenance, green campus initiatives, etc.

Stage 1: Prepare – In this stage, operations and facilities will identify staff who will be the point of contact for discussions about connecting with students and onsite potential projects. These staff will need to attend briefings to ensure that there is a clear understanding of the purpose of the collaborations and the overall curriculum renewal strategy in preparation for the piloting and integration stages to come.

Stage 2: Explore – A key aspect of this stage will be to inform the response to the audit recommendations; operations and facilities staff can contribute a list of contractors who may either be experts in some aspect of key knowledge and skill area that the department is interested in, or who may be seeking knowledge and skills from the department. Staff can also be identified who can contribute to course offerings (e.g. as guest speakers) or alternatively who would benefit from capacity building in one or more aspects of the topic area.

Stage 3: Test and pilot – In this stage, staff from operations and facilities could participate in a variety of piloted short courses or professional development activities coordinated by the design and process innovation activity stream.

Stage 4: Integrate – Depending on the success of these initiatives, more staff may participate in the curriculum renewal activities as they are rolled out across programs. This stage might also provide an opportunity for operations and facilities staff to develop relationships with contractors and suppliers who are participating in the various initiatives.

Teaching and learning

Teaching and learning faculty will potentially be involved in this element in two ways, depending on existing strengths. Faculty could be assisting with the dissemination of sustainability knowledge and skills – to practising professionals, schools and the community. They could also be learning these skills through professional development. This activity stream will also play a major role in ensuring that on-campus and off-campus collaboration opportunities are effective in providing professional development and at the same time improving students' experience in developing their graduate attributes.

Stage 1: Prepare – In this stage, all faculty attend briefings by management on the support and potential opportunities involved in curriculum, campus and community integration. The department will identify faculty who are experts in one or more aspects of sustainability and who can play a major role. The department can also identify faculty who have the potential to play a major role in the curriculum renewal strategy, but who require some professional development to do so.

Stage 2: Explore – It will be important in this stage that teaching and learning faculty involved in the graduate attribute identification, learning pathways mapping and course audits are keenly aware of the realities associated with implementing changes to courses. There may also be seminars or workshops hosted during this period, where academics, operations and facilities staff, and community project representatives can gather to share their ideas and existing curriculum innovations.

Stage 3: Test and pilot – This will be a key stage for teaching and learning faculty to build on from the process leading up to the development of audit recommendations and taking these and testing options in actual courses. This is the first time that students on the whole will be affected by the curriculum renewal strategy, and the pilot activities need to be aware of student responses to inform the integration stage to follow. It will be important for teaching and learning faculty and support staff to seek appropriate collaborations within and outside the university to support the pilots. These activities may include professional short courses for practitioners; for department-hosted forums where practitioners can provide professional development for faculty; and for department-hosted community/ school forums where both faculty and practitioners can share about the application of sustainable development to the program.

Stage 4: Integrate – This stage will involve the widespread renewal of courses and programs across the university providing both the opportunity for duplication and also for collaboration. As many topics related to sustainable development are multi-disciplinary, it will be an important part of the curriculum renewal strategy to identify synergies between courses and programs for leveraging new materials and teaching resources. It will be crucial for teaching and learning faculty to understand the graduate attributes and to have ideally been involved in their identification and the selection of the knowledge and skills to deliver them in the program. A key aspect of this stage will also be to prepare for accreditation reviews that call for demonstration of sustainability-related graduate attributes.

Human resources and culture

It will be important throughout the curriculum renewal process to create a culture of engagement between staff and with various internal and external groups. It will also be important to ensure staff morale is maintained through the process. Depending on the institutional culture, the transition may involve a mixture of voluntary and compulsory tasks, which both need to be managed in terms of staff expectations and recognition.

Stage 1: Prepare – This stage can involve key staff in this activity stream receiving a briefing from management regarding relative staff strengths within the department and intentions for proposing staff development and the identification of experts in preparation for the piloting and integration stages.

Stage 2: Explore – Staff may need to review corporate documentation regarding workload and role recognition to ensure that there are appropriate rewards for the varying levels of staff involvement (e.g. within the workload formula). During this stage, this activity stream could also review corporate documentation for opportunities to highlight the role of auditing in curriculum renewal.

Stage 3: Test and pilot – In this stage, staff can monitor the piloting of activities with regard to staff performance and satisfaction with the workload recognition. This stage also provides an opportunity for human resources and culture-related staff to trial the appropriate review processes and to gauge the level of staff satisfaction regarding workload in integrating real projects and sustainability into the curriculum.

Stage 4: Integrate – Following some refinement of review processes and activities, the findings from the pilots can be fully integrated across the human resources system to support the integration across programs. This will include staff recruiting requirements related to sustainability competency.

Marketing and communications

Now that the courses are actually being changed, those involved will be learning a great deal about the response from faculty, students and stakeholders that can inform marketing and communications. It will be important to ensure that the advances in this area are heralded clearly to capture the associated benefits.

Stage 1: Prepare – In this early stage it will be important for marketing and communications staff to advise on areas that are marketable and on the form of communications about the implementation of changes to the programs in the later stages. It will make it easier to receive and disseminate such material if staff involved across the other areas of the curriculum helix are prepared early.

Stage 2: Explore – Again, while faculty across the university are focusing on graduate attribute identification, learning pathways mapping and course audits, it will be important for marketing and communications staff to assist those involved to identify newsworthy outcomes later in the process.

Stage 3: Test and pilot – As Stage 3 proceeds, staff could collect examples of successful trials and student and staff experiences to further communicate the department's curriculum renewal initiative and the potential for the local community to become involved.

Stage 4: Integrate – As the results of the pilots are communicated and learned from the process to integrate, sustainability graduate attributes across the institution will require ongoing momentum to be maintained. This may be supported by marketing and communications staff by profiling staff that are making progress. This may include assisting in developing applications for awards that recognise leadership in education for sustainability.

Partnerships and stakeholders

This activity stream will have a substantial role to play, given the extensive interactions between the department, professional practitioners, community members and schools as part of implementing the recommendations of the audit process. It will be critical to ensure that the existing partner and stakeholder network is involved in considering ways for their projects to form part of the pilots and later the integration process.

Stage 1: Prepare – In this stage, staff will begin to consider the types of external partners and stakeholders that may be suitable for participation in the later stages of the process and sound them out for potential interest.

Stage 2: Explore – In this stage it will be important to strategically involve partners and stakeholders in the various activities related to graduate attribute identification, learning pathways mapping and unit audits. A clear intention, however, must be made on preparing such stakeholders for active involvement in the piloting and integration stages in the future. This will reduce grandstanding and allow for buy-in in the process at the early stages rather than approaching them to assist with unit renewal later in the process.

Stage 3: Test and pilot – Staff can be involved in liaising with the network as the pilot projects proceed. They could also engage with the network to become familiar with what is occurring in the community. Such activities may include inviting community group representatives to keynote lectures and staff workshops regarding sustainability. Once the test and pilot projects are confirmed, some partners and stakeholders may be involved in developing assessment tasks for students around their particular projects.

Stage 4: Integrate – As this stage involves the widespread renewal of courses across many programs, it presents numerous opportunities to engage partners and stakeholders, particularly the ones involved in the pilot initiatives. Given that there are many barriers to curriculum renewal, such support will be critical to the success of the curriculum renewal strategy. This may involve the funding to develop resources and materials, access to existing materials, provision of guest lecturers or problem-based learning opportunities, and potentially site visits to sustainability exemplars.

This concludes our presentation of the elements and their interactions. On the following pages we provide example tables for each activity stream over four stages, to support colleagues in implementing the model and the approaches that we have described.

APPENDIX: STAGING THE CURRICULUM RENEWAL STRATEGY

SAMPLE STAGING TABLES FOR 'GOVERNANCE AND MANAGEMENT'

TABLE A.1 Governance and management – Stage 1 'Prepare' (G1)

Element	*Role*	*Example Tasks*	*Item Ref*
Curriculum renewal strategy (E1)	Direct	Drive the development of the curriculum renewal strategy	G1.E1.1
	Supporting	Communicate intentions to faculty	G1.E1.2
	Assisted	Seek contribution and comment from faculty	G1.E1.3
Identify graduate attributes (E2)	Direct	Budget and determine faculty resourcing for attributes work	G1.E2.1
	Supporting	Internally promote graduate attributes work as an important quality assurance measure	G1.E2.2
	Assisted	Become aware of upcoming processes to be undertaken	G1.E2.3
Map learning pathways (E3)	Direct	Budget and determine faculty resourcing for pathways work	G1.E3.1
	Supporting	Internally promote learning pathways work as an important quality assurance measure	G1.E3.2
	Assisted	Become aware of upcoming processes to be undertaken	G1.E3.3
Audit learning outcomes (E4)	Direct	Budget and determine faculty resourcing for audit work	G1.E4.1
	Supporting	Internally promote curriculum auditing as an important quality assurance measure	G1.E4.2
	Assisted	Become aware of upcoming processes to be undertaken	G1.E4.3
Develop and update curriculum (E5)	Direct	Gauge budget options for actions over the next 3–5 years	G1.E5.1
	Supporting	Internally promote the benefits of strategic rather than *ad hoc* content development and renewal	G1.E5.2
	Assisted	Understand the current situation regarding faculty resourcing	G1.E5.3
Implement the program (E6)	Direct	Gauge budget options for actions over the next 3–5 years	G1.E6.1
	Supporting	Internally promote the benefits of strategic rather than *ad hoc* content development and renewal	G1.E6.2
	Assisted	Review performance appraisal process to recognise contributions to this transition	G1.E6.3

TABLE A.1 Continued

Element	*Role*	*Example Tasks*	*Item Ref*
Raise awareness and build capacity (E7)	Direct	Develop a vision with clear goals about the intended curriculum renewal process, to use internally and externally	G1.E7.1
	Supporting	Organise a series of internal seminars and public forums to introduce the context and the initiative	G1.E7.2
	Assisted	Investigate opportunity to create an annual awards event to recognise efforts in awareness raising and capacity building	G1.E7.3
Collaborate internally and externally (E8)	Direct	Communicate the institution's vision and intentions to stakeholders	G1.E8.1
	Supporting	Invite feedback from stakeholders regarding their expectations for this process	G1.E8.2
	Assisted	Become aware of curriculum innovation opportunities through internal and external collaboration	G1.E8.3
Continually monitor and evaluate (E9)	Direct	Identify funds for tracking this curriculum renewal process	G1.E9.1
	Supporting	Host a forum for academics, facilities management and local entrepreneurs to share experiences and lessons learnt to date	G1.E9.2
	Assisted	Become aware of low-cost opportunities to monitor and evaluate progress	G1.E9.3

TABLE A.2 Governance and management – Stage 2 'Explore' (G2)

Element	*Role*	*Example Tasks*	*Item Ref*
Curriculum renewal strategy (E1)	Direct	Update the strategy with actions arising from the attribute identification, learning pathways mapping and audit processes	G2.E1.1
	Supporting	Ensure that budget allocations are made available for the various processes underway	G2.E1.2
	Assisted	Seek feedback on priorities arising from Stage 2	G2.E1.3
Identify graduate attributes (E2)	Direct	Create the time and opportunity for faculty to participate in the identification process (e.g. workshop and review)	G2.E2.1
	Supporting	Invite experts as requested by the graduate attribute identification team, to be involved	G2.E2.2
	Assisted	Seek feedback from the identification process	G2.E2.3
Map learning pathways (E3)	Direct	Create the time and opportunity for faculty to participate in the mapping process (e.g. interviews and review)	G2.E3.1
	Supporting	Invite experts as requested by the mapping team, to be involved	G2.E3.2
	Assisted	Seek feedback from the mapping process	G2.E3.3
Audit learning outcomes (E4)	Direct	Create a formal requirement for participation in the curriculum audit process, stipulating the depth of investigation	G2.E4.1
	Supporting	Invite experts as requested by the audit team, to be involved	G2.E4.2
	Assisted	Determine the curriculum renewal short- and medium-term priorities	G2.E4.3

TABLE A.2 Continued

Element	*Role*	*Example Tasks*	*Item Ref*
Develop and update curriculum (E5)	Direct	Require content development/renewal proposals herein to identify how they assist with the curriculum transition	G2.E5.1
	Supporting	Encourage systematic exploration of innovation opportunities	G2.E5.2
	Assisted	Become familiar with leading initiatives underway elsewhere	G2.E5.3
Implement the program (E6)	Direct	Provide faculty incentives to review course pedagogy	G2.E6.1
	Supporting	Create opportunities for visiting experts to provide advice	G2.E6.2
	Assisted	Become familiar with leading initiatives underway elsewhere	G2.E6.3
Raise awareness and build capacity (E7)	Direct	Explore options for awards event to acknowledge achievements	G2.E7.1
	Supporting	Host a series of internal seminars on the findings of the external context and internal capacity reviews	G2.E7.2
	Assisted	Investigate opportunity to align staff position descriptions and performance review requirements with strategy	G2.E7.3
Collaborate internally and externally (E8)	Direct	Explore perspectives of key internal and external stakeholders, on findings from attributes, pathways and audit processes	G2.E8.1
	Supporting	Require actions in this stage to include consultation with campus facilities management and local community consultation	G2.E8.2
	Assisted	Become familiar with leading initiatives underway elsewhere	G2.E8.3
Continually monitor and evaluate (E9)	Direct	Review progress against the intended timeframes and scope of the strategy	G2.E9.1
	Supporting	Take remedial actions where necessary, to keep the process on track with regard to timing and scope	G2.E9.2
	Assisted	Become familiar with leading initiatives underway elsewhere	G2.E9.3

TABLE A.3 Governance and management – Stage 3 'Test and pilot' (G3)

Element	*Role*	*Example Tasks*	*Item Ref*
Curriculum renewal strategy (E1)	Direct	Allocate budget and resources for testing and pilot initiatives	G3.E1.1
	Supporting	Address requests for additional support as they arise	G3.E1.2
	Assisted	Check in with other activity streams regarding progress	G3.E1.3
Identify graduate attributes (E2)	Direct	Become familiar with the graduate attributes being targeted during this stage	G3.E2.1
	Supporting	Include desired attributes language in communications with faculty regarding curriculum renewal process	G3.E2.2
	Assisted	Blend emergent language into internal and external image	G3.E2.3
Map learning pathways (E3)	Direct	Become familiar with the learning pathways and learning outcomes being targeted during this stage	G3.E3.1
	Supporting	Contribute where possible to discussions around developing learning pathways	G3.E3.2
	Assisted	Seek briefing from key faculty regarding progress on actioning	G3.E3.3

TABLE A.3 Continued

Element	*Role*	*Example Tasks*	*Item Ref*
Audit learning outcomes (E4)	Direct	Become familiar with the learning pathways and learning outcomes being targeted during this stage	G3.E4.1
	Supporting	Reward faculty for creating a plan for actioning the audit recommendations related to their teaching	G3.E4.2
	Assisted	Seek briefing from key faculty regarding progress on actioning	G3.E4.3
Develop and update curriculum (E5)	Direct	Host an internal forum to present and discuss actions taken	G3.E5.1
	Supporting	Oversee process to coordinate test and pilot initiatives	G3.E5.2
	Assisted	Monitor testing and pilot results against strategy and timelines	G3.E5.3
Implement the program (E6)	Direct	Oversee regular reporting on progress	G3.E6.1
	Supporting	Reward faculty who action their intentions within their teaching	G3.E6.2
	Assisted	Encourage faculty to report on challenges and successes	G3.E6.3
Raise awareness and build capacity (E7)	Direct	Instigate a periodic awards event for faculty and stakeholders	G3.E7.1
	Supporting	Organise a series of internal seminars for all parts of the organisation, on progress	G3.E7.2
	Assisted	Organise a series of external forums incorporating faculty sharing regarding progress	G3.E7.3
Collaborate internally and externally (E8)	Direct	Encourage test and pilot initiatives that involve internal and external stakeholders	G3.E8.1
	Supporting	Create flexibility within codes and enrolment to permit professional development/short course/other innovative activities	G3.E8.2
	Assisted	Monitor efforts against strategy and timelines	G3.E8.3
Continually monitor and evaluate (E9)	Direct	Review budget and resourcing for testing and pilot initiatives	G3.E9.1
	Supporting	Allocate scholarship/grant for leading-edge pilot innovations around embedding new knowledge and skills	G3.E9.2
	Assisted	Recognise achievements in testing and pilot phase, through awards, certificates and so on	G3.E9.3

TABLE A.4 Governance and management – Stage 4 'Integrate' (G4)

Element	*Role*	*Example Tasks*	*Item Ref*
Curriculum renewal strategy (E1)	Direct	Develop a schedule of review dates for checking progress on transition	G4.E1.1
	Supporting	Provide update to all activity streams on successes of Stage 3 and intentions for Stage 4	G4.E1.2
	Assisted	Seek briefings from other activity streams on progress	G4.E1.3
Identify graduate attributes (E2)	Direct	Instigate periodic reporting on student attainment of attributes	G4.E2.1
	Supporting	Allocate responsibility and timeframe for periodic review of desired graduate attributes	G4.E2.2
	Assisted	Seek feedback from students and community on perceptions of curriculum renewal transition	G4.E2.3

TABLE A.4 Continued

Element	*Role*	*Example Tasks*	*Item Ref*
Map learning pathways (E3)	Direct	Instigate faculty periodic reporting on progress towards addressing learning pathways	G4.E3.1
	Supporting	Allocate responsibility and timeframe for periodic review of learning pathways	G4.E3.2
	Assisted	Seek feedback from students and community on perceptions of curriculum renewal transition	G4.E3.3
Audit learning outcomes (E4)	Direct	Develop a periodic schedule for auditing the curriculum against desired graduate attributes	G4.E4.1
	Supporting	Allocate responsibility and timeframe for review of audit findings	G4.E4.2
	Assisted	Seek feedback from students and community on perceptions of curriculum renewal transition	G4.E4.3
Develop and update curriculum (E5)	Direct	Use testing and pilot results to oversee work plan for integrating knowledge and skills throughout curriculum	G4.E5.1
	Supporting	Encourage staff to write education research papers on experiences in integrating content and institutional approach	G4.E5.2
	Assisted	Receive briefing from program convenors on plans for mainstreaming successful innovations from Stage 3	G4.E5.3
Implement the program (E6)	Direct	Oversee mainstreaming work plan for successful innovations	G4.E6.1
	Supporting	Review budget and other support mechanisms to ensure that faculty are equipped for this phase	G4.E6.2
	Assisted	Ensure that program offerings are synchronised with the integration process	G4.E6.3
Raise awareness and build capacity (E7)	Direct	Set up the structure for an ongoing awards process to recognise achievements in this area of curriculum renewal	G4.E7.1
	Supporting	Organise a series of internal seminars for all parts of the institution, on learnings and achievements	G4.E7.2
	Assisted	Host an external forum (education sector, community) to share learnings and achievements	G4.E7.3
Collaborate internally and externally (E8)	Direct	Approach alumni, research partners etc. to consider engaging their organisation in curriculum-related activities	G4.E8.1
	Supporting	Encourage faculty and staff to expand collaborations in teaching and learning, as renewed curriculum is rolled out	G4.E8.2
	Assisted	Request periodic briefing on initiatives aimed at increasing enrolments around the renewed curriculum	G4.E8.3
Continually monitor and evaluate (E9)	Direct	Regularly report to faculty on progress of this integration stage	G4.E9.1
	Supporting	Actively seek opportunities to maintain momentum for integration	G4.E9.2
	Assisted	Request periodic briefing on how curriculum is incorporating real projects with campus/community	G4.E9.3

SAMPLE STAGING TABLES FOR 'OPERATIONS AND FACILITIES MANAGEMENT'

TABLE A.5 Operations and facilities management – Stage 1 'Prepare' (O1)

Element	*Role*	*Example Tasks*	*Item Ref*
Curriculum renewal strategy (E1)	Direct	Review draft strategy	O1.E1.1
	Supporting	Communicate interest in contributing to the transition	O1.E1.2
	Assisted	Investigate opportunities for benefiting from involvement	O1.E1.3
Identify graduate attributes (E2)	Direct	Identify staff who could participate in process	O1.E2.1
	Supporting	Offer workshop participation	O1.E2.2
	Assisted	Become familiar with intended process	O1.E2.3
Map learning pathways (E3)	Direct	Identify staff who could participate in process	O1.E3.1
	Supporting	Offer workshop participation	O1.E3.2
	Assisted	Become familiar with intended process	O1.E3.3
Audit learning outcomes (E4)	Direct	–	O1.E4.1
	Supporting	–	O1.E4.2
	Assisted	–	O1.E4.3
Develop and update curriculum (E5)	Direct	Identify staff who could participate in process	O1.E5.1
	Supporting	Determine the extent of opportunities to engage on campus	O1.E5.2
	Assisted	Become familiar with intended process	O1.E5.3
Implement the program (E6)	Direct	Identify staff who could participate in process	O1.E6.1
	Supporting	Determine extent of on-campus interaction that could be offered	O1.E6.2
	Assisted	Become familiar with intended process	O1.E6.3
Raise awareness and build capacity (E7)	Direct	Delegate responsibility for coordinating actions between OFM and other parts of the organisation	O1.E7.1
	Supporting	Identify staff with experience/interest in proposed topic areas	O1.E7.2
	Assisted	Encourage interested staff to attend seminars in the institution	O1.E7.3
Collaborate internally and externally (E8)	Direct	Contact suppliers/design and construction contractors with news of the initiative and the potential for benefit through involvement	O1.E8.1
	Supporting	Put partnerships and stakeholder engagement personnel in contact with design and construction contractors	O1.E8.2
	Assisted	Become familiar with curriculum innovation opportunities through internal and external collaboration	O1.E8.3
Continually monitor and evaluate (E9)	Direct	Contact existing community partners regarding the potential for involvement in the curriculum renewal initiative	O1.E9.1
	Supporting	Put partnerships and stakeholder engagement personnel in contact with existing community partners	O1.E9.2
	Assisted	Become familiar with greening campus initiatives	O1.E9.3

TABLE A.6 Operations and facilities management – Stage 2 'Explore' (O2)

Element	*Role*	*Example Tasks*	*Item Ref*
Curriculum renewal strategy (E1)	Direct	Confirm the extent to which OFM can participate in next stages	O2.E1.1
	Supporting	Respond to any queries or invitations for involvement	O2.E1.2
	Assisted	Understand expectations regarding the role of on-campus experiences within the curriculum	O2.E1.3
Identify graduate attributes (E2)	Direct	Create opportunity for interested staff to participate in graduate attributes identification process	O2.E2.1
	Supporting	Have interested staff connect directly with organiser/s	O2.E2.2
	Assisted	Use findings to see how OFM can contribute to next stages	O2.E2.3
Map learning pathways (E3)	Direct	Create opportunity for interested staff to participate in mapping process	O2.E3.1
	Supporting	Have interested staff connect directly with organiser/s	O2.E3.2
	Assisted	Use findings to see how OFM can contribute to next stages	O2.E3.3
Audit learning outcomes (E4)	Direct	Create opportunity for interested staff to participate in audit process	O2.E4.1
	Supporting	Have interested staff connect directly with organiser/s	O2.E4.2
	Assisted	Use findings to see how OFM can contribute to next stages	O2.E4.3
Develop and update curriculum (E5)	Direct	Explore the kind of roles OFM can provide in future curriculum, including for example online/onsite tours regarding campus innovations	O2.E5.1
	Supporting	Be in communication with faculty who are exploring on-campus opportunities	O2.E5.2
	Assisted	Become familiar with leading initiatives underway elsewhere	O2.E5.3
Implement the program (E6)	Direct	Explore options for making data available, relating to building design characteristics, electricity use, water use and procurement (e.g. fleet management, information services)	O2.E6.1
	Supporting	Explore ways to provide academics with access to design and performance data on campus	O2.E6.2
	Assisted	Investigate the type of data that could be sought by academics	O2.E6.3
Raise awareness and build capacity (E7)	Direct	Host a forum between OFM and academics to discuss campus–curriculum opportunities	O2.E7.1
	Supporting	Explore role descriptions and performance review process, to value liaising with academics about curriculum collaboration	O2.E7.2
	Assisted	Receive briefing on the findings of this stage and consider implications and opportunities	O2.E7.3
Collaborate internally and externally (E8)	Direct	Continue to promote the opportunity to be involved in the initiative, to contractors and suppliers	O2.E8.1
	Supporting	Keep partnerships and stakeholder engagement activity stream informed of contractor discussions	O2.E8.2
	Assisted	Explore options to create a course induction requirement for all contractors, collaborating with academics	O2.E8.3

TABLE A.6 Continued

Element	*Role*	*Example Tasks*	*Item Ref*
Continually monitor and evaluate (E9)	Direct	Evaluate the extent of contractor engagement regarding the curriculum renewal process	O2.E9.1
	Supporting	Monitor the number of faculty approaching OFM regarding collaboration opportunities	O2.E9.2
	Assisted	Seek feedback from other activity streams regarding OFM willingness to engage in curriculum discussions	O2.E9.3

TABLE A.7 Operations and facilities management – Stage 3 'Test and pilot' (O3)

Element	*Role*	*Example Tasks*	*Item Ref*
Curriculum renewal strategy (E1)	Direct	–	O3.E1.1
	Supporting	–	O3.E1.2
	Assisted	–	O3.E1.3
Identify graduate attributes (E2)	Direct	–	O3.E2.1
	Supporting	–	O3.E2.2
	Assisted	–	O3.E2.3
Map learning pathways (E3)	Direct	–	O3.E3.1
	Supporting	–	O3.E3.2
	Assisted	–	O3.E3.3
Audit learning outcomes (E4)	Direct	–	O3.E4.1
	Supporting	–	O3.E4.2
	Assisted	–	O3.E4.3
Develop and update curriculum (E5)	Direct	Collaborate with faculty to develop and update units that target OFM involvement	O3.E5.1
	Supporting	Provide faculty with clear scope of OFM potential involvement	O3.E5.2
	Assisted	Seek guidance from management regarding level of involvement	O3.E5.3
Implement the program (E6)	Direct	Collaborate to pilot innovations connecting curriculum and campus	O3.E6.1
	Supporting	Pilot supplying campus facilities management data (e.g. energy/water use, procurement) for one or more curriculum renewal initiatives	O3.E6.2
	Assisted	Pilot engaging academics and students in one or more campus design/construction projects, through content/activities/assessment	O3.E6.3
Raise awareness and build capacity (E7)	Direct	Instigate an annual award for initiatives that connect curriculum with campus operations	O3.E7.1
	Supporting	Encourage staff to continue attending the internal seminars	O3.E7.2
	Assisted	Seek briefings on strategy progress	O3.E7.3

TABLE A.7 Continued

Element	*Role*	*Example Tasks*	*Item Ref*
Collaborate internally and externally (E8)	Direct	Make OFM data easily accessible for faculty	O3.E8.1
	Supporting	Provide staff with incentives (resources, task reallocation) to collaborate on connecting curriculum with campus operations	O3.E8.2
	Assisted	Contribute to pilot resources/initiatives for engaging with contractors and suppliers	O3.E8.3
Continually monitor and evaluate (E9)	Direct	Engage suppliers/contractors in induction trial on campus, in collaboration with faculty	O3.E9.1
	Supporting	Continue to keep partnerships and stakeholder engagement informed of contractor discussions	O3.E9.2
	Assisted	Seek feedback from faculty on OFM contributions to curriculum renewal process	O3.E9.3

TABLE A.8 Operations and facilities management – Stage 4 'Integrate' (O4)

Element	*Role*	*Example Tasks*	*Item Ref*
Curriculum renewal strategy (E1)	Direct	Contribute to informing progress on transition	O4.E1.1
	Supporting	Provide update to other activity streams on OFM contributions	O4.E1.2
	Assisted	Seek briefings from other activity streams on progress	O4.E1.3
Identify graduate attributes (E2)	Direct	–	O4.E2.1
	Supporting	–	O4.E2.2
	Assisted	–	O4.E2.3
Map learning pathways (E3)	Direct	–	O4.E3.1
	Supporting	–	O4.E3.2
	Assisted	–	O4.E3.3
Audit learning outcomes (E4)	Direct	–	O4.E4.1
	Supporting	–	O4.E4.2
	Assisted	–	O4.E4.3
Develop and update curriculum (E5)	Direct	Introduce periodic reporting on new and continuing collaborative activities with faculty on campus	O4.E5.1
	Supporting	Contribute to ongoing resources/initiatives for engaging with faculty and students	O4.E5.2
	Assisted	Continue to contribute to resources that engage faculty and students with operations/facilities management	O4.E5.3
Implement the program (E6)	Direct	Allocate staff and resources to guest lectures and other materials development as planned	O4.E6.1
	Supporting	Respond to faculty enquiries about access to campus data as requested	O4.E6.2
	Assisted	Obtain feedback from students and faculty regarding satisfaction with contribution	O4.E6.3

TABLE A.8 Continued

Element	*Role*	*Example Tasks*	*Item Ref*
Raise awareness and build capacity (E7)	Direct	Refine and continue the annual award for connecting curriculum with greening campus operations	O4.E7.1
	Supporting	Encourage staff to continue attending the internal seminars	O4.E7.2
	Assisted	Continue to receive briefings on EfS strategy progress	O4.E7.3
Collaborate internally and externally (E8)	Direct	Create documents to ensure ongoing involvement of contractor/suppliers	O4.E8.1
	Supporting	Continue to keep partnerships and stakeholder engagement informed of contractor discussions	O4.E8.2
	Assisted	Encourage external sponsorship of annual awards on curriculum collaboration	O4.E8.3
Continually monitor and evaluate (E9)	Direct	Monitor external contractor and supplier engagement in integration activities across the program	O4.E9.1
	Supporting	Encourage contractors and suppliers to be involved given graduate recruitment opportunities	O4.E9.2
	Assisted	Introduce faculty directly to contacts, to facilitate collaboration	O4.E9.3

SAMPLE STAGING TABLES FOR 'TEACHING AND LEARNING'

TABLE A.9 Teaching and learning – Stage 1 'Prepare' (T1)

Element	*Role*	*Example Tasks*	*Item Ref*
Curriculum renewal strategy (E1)	Direct	Inform the development of the strategy	T1.E1.1
	Supporting	Review the curriculum renewal strategy	T1.E1.2
	Assisted	Become aware of strategic needs of the institution	T1.E1.3
Identify graduate attributes (E2)	Direct	Determine required human resources and budget (to inform G1)	T1.E2.1
	Supporting	Provide briefing/s on graduate attribute identification	T1.E2.2
	Assisted	Identify possible facilitators and experts to be involved in identifying graduate attributes	T1.E2.3
Map learning pathways (E3)	Direct	Determine required human resources and budget (to inform G1)	T1.E3.1
	Supporting	Provide briefing/s on learning pathway mapping	T1.E3.2
	Assisted	Identify possible facilitators and experts to be involved in pathway mapping	T1.E3.3
Audit learning outcomes (E4)	Direct	Determine required human resources and budget (to inform G1)	T1.E4.1
	Supporting	Provide briefing/s on audit process	T1.E4.2
	Assisted	Identify possible facilitators and experts to be involved in audit	T1.E4.3
Develop and update curriculum (E5)	Direct	Identify existing initiatives and champions in the topic area/s	T1.E5.1
	Supporting	Provide briefing/s on benefits of strategic curriculum renewal	T1.E5.2
	Assisted	Collate information on existing campus initiatives, not located within the core curriculum	T1.E5.3

TABLE A.9 Continued

Element	*Role*	*Example Tasks*	*Item Ref*
Implement the program (E6)	Direct	Identify existing and potential need for resources (to inform G1)	T1.E6.1
	Supporting	Brief senior management (G) on opportunities through curriculum and pedagogy innovation	T1.E6.2
	Assisted	Identify where topics are being implemented in courses elsewhere	T1.E6.3
Raise awareness and build capacity (E7)	Direct	Delegate curriculum renewal actions between faculty, teaching and learning experts, and other parts of the organisation	T1.E7.1
	Supporting	Review draft vision and core actions, considering teaching and learning implications	T1.E7.2
	Assisted	Encourage all staff to participate in seminars and forums	T1.E7.3
Collaborate internally and externally (E8)	Direct	Identify any high achieving stakeholders who are active in the area of the intended curriculum renewal	T1.E8.1
	Supporting	Brief other parts of the organisation on opportunities for engaging suppliers/contractors/schools	T1.E8.2
	Assisted	Identify industry and school stakeholders who are important to the institution, but where no/little capacity exists	T1.E8.3
Continually monitor and evaluate (E9)	Direct	Identify existing monitoring and evaluation tools that could be used to inform this process	T1.E9.1
	Supporting	Brief other parts of the organisation on opportunities for effective and efficient monitoring and evaluation	T1.E9.2
	Assisted	Become aware of monitoring and evaluation tools being used elsewhere in the institution	T1.E9.3

TABLE A.10 Teaching and learning – Stage 2 'Explore' (T2)

Element	*Role*	*Example Tasks*	*Item Ref*
Curriculum renewal strategy (E1)	Direct	Distil findings and recommendations from attributes, learning pathways and audit processes	T2.E1.1
	Supporting	Report findings of attributes, learning pathways and audit processes to other activity streams	T2.E1.2
	Assisted	Seek comments from management regarding findings and recommendations for moving forward	T2.E1.3
Identify graduate attributes (E2)	Direct	Coordinate graduate attribute identification process for each program being considered	T2.E2.1
	Supporting	Engage with activity streams prior to the process regarding expectations	T2.E2.2
	Assisted	Seek feedback from participants regarding process and findings	T2.E2.3
Map learning pathways (E3)	Direct	Coordinate mapping process for each program being considered	T2.E3.1
	Supporting	Engage with activity streams prior to the process regarding expectations	T2.E3.2

TABLE A.10 Continued

Element	*Role*	*Example Tasks*	*Item Ref*
	Assisted	Seek feedback from participants regarding process and findings	T2.E3.3
Audit learning outcomes (E4)	Direct	Undertake a curriculum audit for each existing program, identifying strengths, weaknesses, opportunities and threats	T2.E4.1
	Supporting	Engage with staff prior to the audit, regarding expectations	T2.E4.2
	Assisted	Seek feedback from participants regarding audit findings and recommendations	T2.E4.3
Develop and update curriculum (E5)	Direct	Review existing initiatives for what can be mainstreamed and formalised	T2.E5.1
	Supporting	Prepare a guiding document to assist staff in embedding knowledge and skills, including content and process aspects	T2.E5.2
	Assisted	Seek ideas from other activity streams regarding opportunities for initiatives that involve them	T2.E5.3
Implement the program (E6)	Direct	Prioritise curriculum renewal efforts for testing and piloting	T2.E6.1
	Supporting	Confirm budget and resourcing arrangements with management	T2.E6.2
	Assisted	Seek management approval of proposed testing and piloting	T2.E6.3
Raise awareness and build capacity (E7)	Direct	Brief other activity streams on findings of this stage and renewal opportunities arising	T2.E7.1
	Supporting	Advise other activity streams on curriculum innovations that could involve them	T2.E7.2
	Assisted	Explore opportunities for strategically engaging staff from all activity streams, within this curriculum renewal initiative	T2.E7.3
Collaborate internally and externally (E8)	Direct	Identify internal capacity building needs and opportunities for this to occur through collaborations	T2.E8.1
	Supporting	Discuss with other activity streams the identified campus- and community-based opportunities for curriculum innovations	T2.E8.2
	Assisted	Seek support from other activity streams regarding taking these opportunities forward	T2.E8.3
Continually monitor and evaluate (E9)	Direct	Define key performance measures to evaluate the success of the curriculum renewal process	T2.E9.1
	Supporting	Invite other activity streams to indicate measures that would be useful for them	T2.E9.2
	Assisted	Seek feedback from activity streams G and M regarding effective monitoring and evaluation going forward	T2.E9.3

TABLE A.11 Teaching and learning – Stage 3 'Test and pilot' (T3)

Element	*Role*	*Example Tasks*	*Item Ref*
Curriculum renewal strategy (E1)	Direct	Review strategy for commitments from this activity stream	T3.E1.1
	Supporting	Provide update on needs and progress on request	T3.E1.2
	Assisted	Seek briefing from management on progress of strategy	T3.E1.3
Identify graduate attributes (E2)	Direct	Track the range of initiatives underway to address the prioritised graduate attributes, ensuring a systematic approach	T3.E2.1
	Supporting	Advise other activity streams how they might contribute to initiatives developing the desired graduate attributes	T3.E2.2
	Assisted	Obtain feedback from other activity streams regarding pilot initiatives underway to develop the graduate attributes	T3.E2.3
Map learning pathways (E3)	Direct	Check that the initiatives being tested and piloted address the learning pathway priorities	T3.E3.1
	Supporting	Advise other activity streams how they might contribute to initiatives, targeting gaps	T3.E3.2
	Assisted	Obtain feedback from other activity streams regarding pilot initiatives underway and possible ways to enhance process	T3.E3.3
Audit learning outcomes (E4)	Direct	Review courses as the audit recommendations are completed, with regard to task completion and further recommendations	T3.E4.1
	Supporting	Prepare a briefing to management on the findings of Stage 3 and implications for Stage 4	T3.E4.2
	Assisted	Obtain feedback from other activity streams regarding pilot initiatives underway to address the audit findings	T3.E4.3
Develop and update curriculum (E5)	Direct	Address audit findings, prioritising curriculum actions accordingly	T3.E5.1
	Supporting	Advise other activity streams of their role during this stage	T3.E5.2
	Assisted	Initiate regular reporting from other activity streams, on progress with tasks and the timing of their input	T3.E5.3
Implement the program (E6)	Direct	Run courses with the curriculum innovations	T3.E6.1
	Supporting	Provide ongoing assistance to other activity streams, to deal with obstacles/challenges in participating in the pilot/s	T3.E6.2
	Assisted	Recognise faculty who have additional workload during this stage in particular	T3.E6.3
Raise awareness and build capacity (E7)	Direct	Prepare briefings for internal forums and external seminars, on progress in the test/pilot phase	T3.E7.1
	Supporting	Encourage staff to continue attending the forums	T3.E7.2
	Assisted	Seek regular feedback from other activity streams on their initiatives and learnings	T3.E7.3
Collaborate internally and externally (E8)	Direct	Pilot engaging academics and students in a campus design/ construction project	T3.E8.1
	Supporting	Pilot obtaining campus facilities management data (e.g. energy/ water use) for a particular curriculum renewal initiative	T3.E8.2
	Assisted	Initiate regular reporting from other activity streams, on progress with testing and pilots	T3.E8.3

TABLE A.11 Continued

Element	*Role*	*Example Tasks*	*Item Ref*
Continually monitor and evaluate (E9)	Direct	Periodically report on progress with regard to strategic plan and timelines	T3.E9.1
	Supporting	Brief management on challenges and opportunities that have arisen during the test and pilot stage	T3.E9.2
	Assisted	Consider opportunities for addressing issues that have arisen during the test and pilot stage	T3.E9.3

TABLE A.12 Teaching and learning – Stage 4 'Integrate' (T4)

Element	*Role*	*Example Tasks*	*Item Ref*
Curriculum renewal strategy (E1)	Direct	Contribute to informing progress on transition	T4.E1.1
	Supporting	Provide update to other activity streams on rollout	T4.E1.2
	Assisted	Seek briefings from other activity streams on progress	T4.E1.3
Identify graduate attributes (E2)	Direct	Initiate periodic reporting regarding progress of developing desired graduate attributes	T4.E2.1
	Supporting	Continue to brief new staff on desired graduate attributes, their importance and priority within this initiative	T4.E2.2
	Assisted	Obtain feedback from other activity streams regarding integration initiatives underway	T4.E2.3
Map learning pathways (E3)	Direct	Initiate periodic reporting regarding progress of addressing learning outcomes within courses to develop learning pathways	T4.E3.1
	Supporting	Continue to brief new faculty on learning pathways, their importance and priority within this initiative	T4.E3.2
	Assisted	Obtain feedback from other activity streams regarding integration initiatives underway	T4.E3.3
Audit learning outcomes (E4)	Direct	Audit the renewed subjects with regard to addressing the audit recommendations	T4.E4.1
	Supporting	Continue to brief new faculty on the likelihood of auditing their subject/s as part of the transition	T4.E4.2
	Assisted	Obtain feedback from other activity streams regarding close-out auditing initiatives	T4.E4.3
Develop and update curriculum (E5)	Direct	Undertake curriculum renewal planning activities in order of priority, addressing the audit findings	T4.E5.1
	Supporting	Create supporting 'how to' documents to assist colleagues	T4.E5.2
	Assisted	Receive regular reporting from other activity streams, on progress with integration	T4.E5.3
Implement the program (E6)	Direct	Deliver renewed curriculum according to audit recommendations	T4.E6.1
	Supporting	Host a series of presentations on progress with other activity streams	T4.E6.2
	Assisted	Involve other activity streams as per plans	T4.E6.3

TABLE A.12 Continued

Element	*Role*	*Example Tasks*	*Item Ref*
Raise awareness and build capacity (E7)	Direct	Prepare briefings for forums and seminars on learning and achievements	T4.E7.1
	Supporting	Encourage staff to continue attending the forums	T4.E7.2
	Assisted	Seek feedback from other activity streams on their initiatives and learning	T4.E7.3
Collaborate internally and externally (E8)	Direct	Roll out successful campus-based and external initiatives that use renewed curriculum	T4.E8.1
	Supporting	Liaise with other activity streams with regard to what is being integrated	T4.E8.2
	Assisted	Receive reporting from other activity streams on progress with integration initiatives	T4.E8.3
Continually monitor and evaluate (E9)	Direct	Evaluate student satisfaction with renewed curriculum	T4.E9.1
	Supporting	Provide satisfaction data to other activity streams, with intentions for addressing any challenges arising	T4.E9.2
	Assisted	Run focus groups with students and potential employers regarding changes to curriculum and satisfaction	T4.E9.3

SAMPLE STAGING TABLES FOR 'HUMAN RESOURCES AND CULTURE'

TABLE A.13 Human resources and culture – Stage 1 'Prepare' (H1)

Element	*Role*	*Example Tasks*	*Item Ref*
Curriculum renewal strategy (E1)	Direct	Identify internal documents that incentivise involvement in the proposed initiative	H1.E1.1
	Supporting	Review draft strategy	H1.E1.2
	Assisted	Seek briefing on strategy and potential interactions	H1.E1.3
Identify graduate attributes (E2)	Direct	–	H1.E2.1
	Supporting	–	H1.E2.2
	Assisted	–	H1.E2.3
Map learning pathways (E3)	Direct	–	H1.E3.1
	Supporting	–	H1.E3.2
	Assisted	–	H1.E3.3
Audit learning outcomes (E4)	Direct	–	H1.E4.1
	Supporting	–	H1.E4.2
	Assisted	–	H1.E4.3
Develop and update curriculum (E5)	Direct	–	H1.E5.1
	Supporting	–	H1.E5.2
	Assisted	–	H1.E5.3
Implement the program (E6)	Direct	–	H1.E6.1
	Supporting	–	H1.E6.2
	Assisted	–	H1.E6.3
Raise awareness and build capacity (E7)	Direct	Allocate responsibility for coordinating actions relating to this strategy, with other parts of the organisation	H1.E7.1

TABLE A.13 Continued

Element	*Role*	*Example Tasks*	*Item Ref*
	Supporting	Identify staff with experience in/an interest in engaging with the process	H1.E7.2
	Assisted	Encourage staff to attend seminars hosted within the institution	H1.E7.3
Collaborate internally and externally (E8)	Direct	Identify corporate culture opportunities that arise from this initiative	H1.E8.1
	Supporting	Provide advice on ways to recognise faculty and staff contribution to this initiative	H1.E8.2
	Assisted	Identify internal and external champions who could play a major role in the transition	H1.E8.3
Continually monitor and evaluate (E9)	Direct	Create a database of current faculty with expertise in the topic area of interest	H1.E9.1
	Supporting	Identify corporate reporting initiatives that already include relevant monitoring and evaluation	H1.E9.2
	Assisted	Seek comment from other activity streams on what could be provided by this stream to assist monitoring and evaluation	H1.E9.3

TABLE A.14 Human resources and culture – Stage 2 'Explore' (H2)

Element	*Role*	*Example Tasks*	*Item Ref*
Curriculum renewal strategy (E1)	Direct	Explore how this activity stream can contribute to the resultant curriculum renewal strategy	H2.E1.1
	Supporting	Share the extent of support functions from this activity stream	H2.E1.2
	Assisted	Seek advice from other activity streams regarding key HR actions that they are anticipating over the rollout of the process	H2.E1.3
Identify graduate attributes (E2)	Direct	Create opportunity for interested staff to participate in graduate attributes identification process	H2.E2.1
	Supporting	Have interested staff connect directly with organiser/s	H2.E2.2
	Assisted	Use findings to see how HR can contribute to next stages	H2.E2.3
Map learning pathways (E3)	Direct	Create opportunity for interested staff to participate in graduate attributes identification process	H2.E3.1
	Supporting	Have interested staff connect directly with organiser/s	H2.E3.2
	Assisted	Use findings to see how HR can contribute to next stages	H2.E3.3
Audit learning outcomes (E4)	Direct	Create opportunity for interested staff to participate in graduate attributes identification process	H2.E4.1
	Supporting	Have interested staff connect directly with organiser/s	H2.E4.2
	Assisted	Use findings to see how HR can contribute to next stages	H2.E4.3
Develop and update curriculum (E5)	Direct	Investigate ways to provide academics with access to information regarding institutional innovations (e.g. corporate sustainability data)	H2.E5.1

TABLE A.14 Continued

Element	*Role*	*Example Tasks*	*Item Ref*
	Supporting	Seek input from faculty regarding the need for HR-related information	H2.E5.2
	Assisted	Understand budget and resourcing availability for proposed curriculum renewal activities	H2.E5.3
Implement the program (E6)	Direct	Identify recruitment opportunities that align with the timeframes anticipated for rolling out renewed curriculum	H2.E6.1
	Supporting	Seek input from faculty regarding the need for new appointments	H2.E6.2
	Assisted	Understand budget and resourcing availability for proposed curriculum renewal activities	H2.E6.3
Raise awareness and build capacity (E7)	Direct	Explore opportunities for updating recruitment documents and advertisements to value skills in curriculum renewal	H2.E7.1
	Supporting	Review corporate governance documents for language that promotes the intentions of the curriculum renewal strategy	H2.E7.2
	Assisted	Receive briefing on findings of this stage and consider implications	H2.E7.3
Collaborate internally and externally (E8)	Direct	Investigate grant/funding/scholarship opportunities for staff and students engaging in curriculum renewal initiatives	H2.E8.1
	Supporting	Explore opportunities to create scholarship opportunities in the target knowledge and skill areas to attract future students	H2.E8.2
	Assisted	Seek input from other activity streams regarding opportunities for recruitments and appointments	H2.E8.3
Continually monitor and evaluate (E9)	Direct	Investigate opportunities for using existing monitoring and evaluation data to check on faculty participation in curriculum renewal process	H2.E9.1
	Supporting	Advise other streams of any arising HR issues and opportunities that might relate to curriculum renewal proposal (e.g. appointments)	H2.E9.2
	Assisted	Seek input from other activity streams regarding HR-related monitoring and evaluation needs	H2.E9.3

TABLE A.15 Human resources and culture – Stage 3 'Test and pilot' (H3)

Element	*Role*	*Example Tasks*	*Item Ref*
Curriculum renewal strategy (E1)	Direct	Review strategy for commitments from this activity stream	H3.E1.1
	Supporting	Provide update on needs and progress on request	H3.E1.2
	Assisted	Seek briefing on progress of strategy	H3.E1.3
Identify graduate attributes (E2)	Direct	–	H3.E2.1
	Supporting	–	H3.E2.2
	Assisted	–	H3.E2.3
Map learning pathways (E3)	Direct	–	H3.E3.1
	Supporting	–	H3.E3.2
	Assisted	–	H3.E3.3

TABLE A.15 Continued

Element	*Role*	*Example Tasks*	*Item Ref*
Audit learning outcomes (E4)	Direct	–	H3.E4.1
	Supporting	–	H3.E4.2
	Assisted	–	H3.E4.3
Develop and update curriculum (E5)	Direct	–	H3.E5.1
	Supporting	–	H3.E5.2
	Assisted	–	H3.E5.3
Implement the program (E6)	Direct	–	H3.E6.1
	Supporting	–	H3.E6.2
	Assisted	–	H3.E6.3
Raise awareness and build capacity (E7)	Direct	Trial including reporting on curriculum renewal contributions in annual staff performance review process	H3.E7.1
	Supporting	Create an institutional focus on faculty performance in curriculum renewal for 1–2 years	H3.E7.2
	Assisted	Communicate professional development opportunities to faculty	H3.E7.3
Collaborate internally and externally (E8)	Direct	Promote the early achievements of the institution in recruitment	H3.E8.1
	Supporting	Pilot a fellowship opportunity to attract faculty	H3.E8.2
	Assisted	Seek input from other activity streams regarding the piloted initiative	H3.E8.3
Continually monitor and evaluate (E9)	Direct	Monitor recruitment efforts for evidence of meeting strategy expectations	H3.E9.1
	Supporting	Report regularly on recruitment to management	H3.E9.2
	Assisted	Seek input from other activity streams regarding capacity-related issues that HR could address	H3.E9.3

TABLE A.16 Human resources and culture – Stage 4 'Integrate' (H4)

Element	*Role*	*Example Tasks*	*Item Ref*
Curriculum renewal strategy (E1)	Direct	Contribute to informing progress on transition	H4.E1.1
	Supporting	Provide update to other activity streams on rollout	H4.E1.2
	Assisted	Seek briefings from other activity streams on progress	H4.E1.3
Identify graduate attributes (E2)	Direct	–	H4.E2.1
	Supporting	–	H4.E2.2
	Assisted	–	H4.E2.3
Map learning pathways (E3)	Direct	–	H4.E3.1
	Supporting	–	H4.E3.2
	Assisted	–	H4.E3.3
Audit learning outcomes (E4)	Direct	–	H4.E4.1
	Supporting	–	H4.E4.2
	Assisted	–	H4.E4.3

TABLE A.16 Continued

Element	*Role*	*Example Tasks*	*Item Ref*
Develop and update curriculum (E5)	Direct	Monitor extent of reporting on curriculum innovations during annual performance appraisals	H4.E5.1
	Supporting	Provide annual summary of activities to governance/ management	H4.E5.2
	Assisted	Create role description material clarifying the importance of integrating new knowledge and skill areas in curriculum	H4.E5.3
Implement the program (E6)	Direct	Provide timely assistance with sessional and casual tutor appointments which may change substantially for new units	H4.E6.1
	Supporting	Provide assistance with ensuring workload calculations are current	H4.E6.2
	Assisted	Address faculty resourcing issues as they arise, with governance/management	H4.E6.3
Raise awareness and build capacity (E7)	Direct	Integrate faculty capacity requirements within new curriculum context, into current and future appointments	H4.E7.1
	Supporting	Promote attention to curriculum renewal activities in performance appraisals	H4.E7.2
	Assisted	Seek feedback from other activity streams regarding role descriptions and performance review focus	H4.E7.3
Collaborate internally and externally (E8)	Direct	Seek high-quality faculty and other appointments to address curriculum renewal strategy	H4.E8.1
	Supporting	Encourage participation in forums and awards on curriculum collaboration	H4.E8.2
	Assisted	Create corporate text around desired faculty attributes for job applications	H4.E8.3
Continually monitor and evaluate (E9)	Direct	Monitor job applicant interest in contributing to the newly embedded knowledge and skill areas	H4.E9.1
	Supporting	Provide data to other activity streams	H4.E9.2
	Assisted	Seek feedback on HR contribution to delivering faculty to address curriculum renewal strategy	H4.E9.3

SAMPLE STAGING TABLES FOR 'MARKETING AND COMMUNICATIONS'

TABLE A.17 Marketing and communications – Stage 1 'Prepare' (M1)

Element	*Role*	*Example Tasks*	*Item Ref*
Curriculum renewal strategy (E1)	Direct	Contribute to development of strategy	M1.E1.1
	Supporting	Review strategy for clarity of language	M1.E1.2
	Assisted	Become familiar with technical aspects of the strategy	M1.E1.3
Identify graduate attributes (E2)	Direct	Develop public statement regarding value of graduate attributes in defining program offerings	M1.E2.1
	Supporting	Develop internal communications regarding graduate attributes as an important quality assurance measure	M1.E2.2
	Assisted	Become familiar with accreditation and graduate attribute requirements	M1.E2.3

TABLE A.17 Continued

Element	*Role*	*Example Tasks*	*Item Ref*
Map learning pathways (E3)	Direct	Develop public statement regarding value of mapping learning pathways	M1.E3.1
	Supporting	Develop internal communications regarding learning pathways as an important quality assurance measure	M1.E3.2
	Assisted	Become familiar with learning pathway impacts on curriculum	M1.E3.3
Audit learning outcomes (E4)	Direct	Develop public statement regarding value of audits	M1.E4.1
	Supporting	Coordinate internal communications regarding auditing as an important quality assurance measure for tracking progress	M1.E4.2
	Assisted	Become familiar with the process of curriculum auditing	M1.E4.3
Develop and update curriculum (E5)	Direct	Develop a public statement regarding the value of strategic curriculum renewal for programs	M1.E5.1
	Supporting	Coordinate internal communications regarding the benefits of strategic rather than *ad hoc* content development and renewal	M1.E5.2
	Assisted	Become familiar with opportunities for strategic curriculum renewal in the institution	M1.E5.3
Implement the program (E6)	Direct	Connect with faculty willing to share their journey, considering opportunities for communicating this internally and externally	M1.E6.1
	Supporting	Inform other streams regarding opportunities for assistance with communicating messages internally and externally	M1.E6.2
	Assisted	Become aware of the type of innovation opportunities that could be taken up as part of this process	M1.E6.3
Raise awareness and build capacity (E7)	Direct	Allocate staff resource to interacting with other activity streams	M1.E7.1
	Supporting	Communicate and market introductory seminars and forums	M1.E7.2
	Assisted	Encourage staff to attend seminars/forums within the institution	M1.E7.3
Collaborate internally and externally (E8)	Direct	Coordinate external communications regarding intentions for rapid curriculum renewal (i.e. to stakeholders)	M1.E8.1
	Supporting	Assist other streams consider messaging of bridging and outreach possibilities	M1.E8.2
	Assisted	Consider the type of marketing that will be required for the renewed curriculum	M1.E8.3
Continually monitor and evaluate (E9)	Direct	Identify language that resonates with internal and external stakeholders that could be used for monitoring and evaluation	M1.E9.1
	Supporting	Assist other activity streams develop communications regarding performance progress	M1.E9.2
	Assisted	Seek advice from other activity streams regarding key performance indicators in the transition and determine how these could be used for marketing and communication	M1.E9.3

TABLE A.18 Marketing and communications – Stage 2 'Explore' (M2)

Element	*Role*	*Example Tasks*	*Item Ref*
Curriculum renewal strategy (E1)	Direct	Explore how this activity stream can contribute to the resultant curriculum renewal strategy	M2.E1.1
	Supporting	Share the extent of support functions that this activity stream can offer	M2.E1.2
	Assisted	Seek advice from other activity streams regarding key communications that they are anticipating over the rollout of the process	M2.E1.3
Identify graduate attributes (E2)	Direct	Develop communications (internal and external) regarding desired graduate attributes	M2.E2.1
	Supporting	Assist in communicating the graduate attribute identification process to other activity streams	M2.E2.2
	Assisted	Become familiar with the summary results of the process	M2.E2.3
Map learning pathways (E3)	Direct	Develop communications (internal and external) regarding mapping of learning pathways	M2.E3.1
	Supporting	Assist in communicating the learning pathways process to other activity streams	M2.E3.2
	Assisted	Become familiar with the summary results of the process	M2.E3.3
Audit learning outcomes (E4)	Direct	Develop communications (internal and external) regarding the audit process	M2.E4.1
	Supporting	Assist in communicating the curriculum auditing requirement to faculty	M2.E4.2
	Assisted	Become familiar with the summary results of the process	M2.E4.3
Develop and update curriculum (E5)	Direct	Develop communications (internal and external) regarding existing examples of internal curriculum innovation	M2.E5.1
	Supporting	Assist in communicating the requirement for all curriculum proposals to address desired graduate attributes	M2.E5.2
	Assisted	Seek contributions from academics regarding intentions for testing and piloting innovations	M2.E5.3
Implement the program (E6)	Direct	Explore the potential for developing a database of innovations that could be used to create media articles and stories as the process unfolds	M2.E6.1
	Supporting	Invite other activity streams to request support with communicating their message of renewed curriculum	M2.E6.2
	Assisted	Become familiar with the planned testing and piloting to occur	M2.E6.3
Raise awareness and build capacity (E7)	Direct	Create and promote media stories from seminars and forums	M2.E7.1
	Supporting	Incorporate emerging language into recruitment documents and advertisements	M2.E7.2
	Assisted	Receive briefing on findings from this stage and consider task implications	M2.E7.3
Collaborate internally and externally (E8)	Direct	Investigate marketing strategies for communicating intentions to potential under- and postgraduate recruits	M2.E8.1
	Supporting	Discuss communication options with other activity streams	M2.E8.2
	Assisted	Promote grant/funding/scholarship opportunities for staff and students engaging in initiatives	M2.E8.3

TABLE A.18 Continued

Element	*Role*	*Example Tasks*	*Item Ref*
Continually monitor and evaluate (E9)	Direct	Create a template for communicating progress on this initiative internally	M2.E9.1
	Supporting	Invite other activity streams to request support with communicating their evaluation and monitoring	M2.E9.2
	Assisted	Seek feedback on the template and refine	M2.E9.3

TABLE A.19 Marketing and communications – Stage 3 'Test and pilot' (M3)

Element	*Role*	*Example Tasks*	*Item Ref*
Curriculum renewal strategy (E1)	Direct	Review strategy for commitments from this activity stream	M3.E1.1
	Supporting	Provide update on needs and progress on request	M3.E1.2
	Assisted	Seek briefing on progress of strategy	M3.E1.3
Identify graduate attributes (E2)	Direct	Develop public statement regarding prioritised graduate attributes and early initiatives	M3.E2.1
	Supporting	Assist in internally communicating prioritised graduate attributes to other activity streams	M3.E2.2
	Assisted	Receive briefing on short- and long-term graduate attribute targets from faculty	M3.E2.3
Map learning pathways (E3)	Direct	Develop public statement regarding learning pathways and early initiatives	M3.E3.1
	Supporting	Assist in internally communicating work on learning pathways to other activity streams	M3.E3.2
	Assisted	Receive briefing on short- and long-term curriculum renewal plans from faculty	M3.E3.3
Audit learning outcomes (E4)	Direct	Develop public statement regarding audit recommendations and early initiatives	M3.E4.1
	Supporting	Assist in internally communicating audit recommendations to other activity streams	M3.E4.2
	Assisted	Receive briefing on short- and long-term priorities from academics	M3.E4.3
Develop and update curriculum (E5)	Direct	Develop public statement regarding curriculum innovations	M3.E5.1
	Supporting	Assist developing a checklist-style communication aid for faculty considering how to fulfil curriculum renewal requirements	M3.E5.2
	Assisted	Receive briefing on planned curriculum renewal testing and pilot initiatives	M3.E5.3
Implement the program (E6)	Direct	Develop internal and external media regarding early achievements	M3.E6.1
	Supporting	Offer to assist faculty in communicating their lessons learnt from testing and piloting initiatives	M3.E6.2
	Assisted	Seek stories of experiences from faculty	M3.E6.3

TABLE A.19 Continued

Element	*Role*	*Example Tasks*	*Item Ref*
Raise awareness and build capacity (E7)	Direct	Create a regular internal communiqué regarding testing and piloting projects	M3.E7.1
	Supporting	Create media stories about testing and piloting initiatives	M3.E7.2
	Assisted	Encourage staff to continue attending the internal seminars	M3.E7.3
Collaborate internally and externally (E8)	Direct	Develop marketing strategy for communicating results and plans to potential under- and postgraduate recruits	M3.E8.1
	Supporting	Trial marketing strategies for promoting curriculum for under- and postgraduate programs	M3.E8.2
	Assisted	Pilot marketing options for initiatives with partners and stakeholders	M3.E8.3
Continually monitor and evaluate (E9)	Direct	Pilot a campus and community initiative database, which can be accessed by academics when considering curriculum renewal options	M3.E9.1
	Supporting	Develop internal and external text for reporting on progress	M3.E9.2
	Assisted	Seek update on progress from other activity streams	M3.E9.3

TABLE A.20 Marketing and communications – Stage 4 'Integrate' (M4)

Element	*Role*	*Example Tasks*	*Item Ref*
Curriculum renewal strategy (E1)	Direct	Contribute to informing progress on transition	M4.E1.1
	Supporting	Provide update to other activity streams on rollout	M4.E1.2
	Assisted	Seek briefings from other activity streams on progress	M4.E1.3
Identify graduate attributes (E2)	Direct	–	M4.E2.1
	Supporting	–	M4.E2.2
	Assisted	–	M4.E2.3
Map learning pathways (E3)	Direct	–	M4.E3.1
	Supporting	–	M4.E3.2
	Assisted	–	M4.E3.3
Audit learning outcomes (E4)	Direct	–	M4.E4.1
	Supporting	–	M4.E4.2
	Assisted	–	M4.E4.3
Develop and update curriculum (E5)	Direct	Create media articles regarding innovative curriculum renewal activities, for press release	M4.E5.1
	Supporting	Assist faculty create an internal guide to curriculum renewal	M4.E5.2
	Assisted	Receive regular briefings regarding progress	M4.E5.3
Implement the program (E6)	Direct	Create media articles regarding successful innovations in curriculum renewal that address strategy aspirations	M4.E6.1
	Supporting	Assist faculty write about their successes, for articles and papers	M4.E6.2
	Assisted	Incorporate progress into marketing for renewed curriculum	M4.E6.3

TABLE A.20 Continued

Element	*Role*	*Example Tasks*	*Item Ref*
Raise awareness and build capacity (E7)	Direct	Promote the curriculum renewal process (internally and externally)	M4.E7.1
	Supporting	Promote the forums and seminars	M4.E7.2
	Assisted	Create media stories from the forums and seminars	M4.E7.3
Collaborate internally and externally (E8)	Direct	Generate materials for marketing to potential under- and postgraduate recruits	M4.E8.1
	Supporting	Assist other activity streams with marketing materials to promote curriculum	M4.E8.2
	Assisted	Create marketing materials for external outreach activities	M4.E8.3
Continually monitor and evaluate (E9)	Direct	Create media and marketing materials regarding progress	M4.E9.1
	Supporting	Communicate performance results internally	M4.E9.2
	Assisted	Seek regular briefings on progress	M4.E9.3

SAMPLE STAGING TABLES FOR 'PARTNERSHIPS AND STAKEHOLDER ENGAGEMENT'

TABLE A.21 Partnerships and stakeholder engagement – Stage 1 'Prepare' (P1)

Element	*Role*	*Example Tasks*	*Item Ref*
Curriculum renewal strategy (E1)	Direct	Identify internal opportunities that incentivise involvement in the proposed initiative	P1.E1.1
	Supporting	Review draft strategy	P1.E1.2
	Assisted	Seek briefing on strategy and potential interactions	P1.E1.3
Identify graduate attributes (E2)	Direct	–	P1.E2.1
	Supporting	Consider partners and stakeholders who may be interested in contributing to the process	P1.E2.2
	Assisted	Become aware of what this process is about	P1.E2.3
Map learning pathways (E3)	Direct	–	P1.E3.1
	Supporting	Consider partners and stakeholders who may be interested in contributing to the process	P1.E3.2
	Assisted	Become aware of what this process is about	P1.E3.3
Audit learning outcomes (E4)	Direct	–	P1.E4.1
	Supporting	Consider partners and stakeholders who may be interested in contributing to the process	P1.E4.2
	Assisted	Become aware of what this process is about	P1.E4.3
Develop and update curriculum (E5)	Direct	–	P1.E5.1
	Supporting	Consider partners and stakeholders who may be interested in contributing to the process	P1.E5.2
	Assisted	Become aware of what this process is about	P1.E5.3
Implement the program (E6)	Direct	–	P1.E6.1
	Supporting	Consider partners and stakeholders who may be interested in contributing to the process	P1.E6.2
	Assisted	Become aware of what this process is about	P1.E6.3

TABLE A.21 Continued

Element	*Role*	*Example Tasks*	*Item Ref*
Raise awareness and build capacity (E7)	Direct	Identify staff with experience/an interest in the proposed process	P1.E7.1
	Supporting	Delegate staff to coordinate actions with other parts of the organisation	P1.E7.2
	Assisted	Contribute to coordinating public forums to introduce stakeholders to context for curriculum renewal	P1.E7.3
Collaborate internally and externally (E8)	Direct	Review agreements with partners (government, industry, school) for language around the proposed curriculum renewal	P1.E8.1
	Supporting	Identify external stakeholders who need to be included in this transition, from industry, school and the community	P1.E8.2
	Assisted	Become familiar with curriculum innovation opportunities that may have impacts on what is done in this activity stream	P1.E8.3
Continually monitor and evaluate (E9)	Direct	Consider opportunities to involve community stakeholders in monitoring and/or evaluating the transition	P1.E9.1
	Supporting	Develop a communication protocol for communicating progress to stakeholders	P1.E9.2
	Assisted	Become familiar with proposed monitoring and evaluation for this curriculum renewal transition, and how this could benefit current partnerships and stakeholder engagement activities	P1.E9.3

TABLE A.22 Partnerships and stakeholder engagement – Stage 2 'Explore' (P2)

Element	*Role*	*Example Tasks*	*Item Ref*
Curriculum renewal strategy (E1)	Direct	Explore how this activity stream can contribute to the resultant curriculum renewal strategy	P2.E1.1
	Supporting	Communicate which existing partnerships and relationships could contribute to the proposed process	P2.E1.2
	Assisted	Understand which parts of the strategy can be communicated externally	P2.E1.3
Identify graduate attributes (E2)	Direct	Consider what contacts may be most useful to the institution in participating in this process	P2.E2.1
	Supporting	Provide possible invitee lists to organiser/s for consideration	P2.E2.2
	Assisted	Seek feedback regarding the usefulness of connecting stakeholders and partners	P2.E2.3
Map learning pathways (E3)	Direct	Consider what contacts may be most useful to the institution in participating in this process	P2.E3.1
	Supporting	Provide possible invitee lists to organiser/s for consideration	P2.E3.2
	Assisted	Seek feedback regarding the usefulness of connecting stakeholders and partners	P2.E3.3
Audit learning outcomes (E4)	Direct	Consider what contacts may be most useful to the institution in participating in this process	P2.E4.1
	Supporting	Provide possible invitee lists to organiser/s for consideration	P2.E4.2

TABLE A.22 Continued

Element	*Role*	*Example Tasks*	*Item Ref*
	Assisted	Seek feedback regarding the usefulness of connecting stakeholders and partners	P2.E4.3
Develop and update curriculum (E5)	Direct	Consider what contacts may be most useful to the institution in participating in this process	P2.E5.1
	Supporting	Provide possible invitee lists to organiser/s for consideration	P2.E5.2
	Assisted	Seek feedback regarding the usefulness of connecting stakeholders and partners	P2.E5.3
Implement the program (E6)	Direct	Review existing partnerships and relationships for opportunities to involve them in future curriculum innovations	P2.E6.1
	Supporting	Inform other activity streams of potential opportunities to consider	P2.E6.2
	Assisted	Seek briefing about findings of this stage, regarding intentions for curriculum renewal	P2.E6.3
Raise awareness and build capacity (E7)	Direct	Continue coordinating public forums involving inspiring speakers and dignitaries of interest to stakeholders	P2.E7.1
	Supporting	Canvas stakeholders regarding topics of confusion, to inform the planning of public seminars and forums	P2.E7.2
	Assisted	Receive briefing on progress of the curriculum transition and consider implications	P2.E7.3
Collaborate internally and externally (E8)	Direct	Organise a forum for academics and key community stakeholders to discuss curriculum collaboration opportunities	P2.E8.1
	Supporting	Inform other activity streams of external perceptions on progress	P2.E8.2
	Assisted	Seek input from stakeholders regarding possible collaboration initiatives	P2.E8.3
Continually monitor and evaluate (E9)	Direct	Monitor stakeholder and partner interest in the curriculum renewal process	P2.E9.1
	Supporting	Communicate feedback from stakeholders and partners to other activity streams as relevant	P2.E9.2
	Assisted	Seek feedback from other activity streams regarding how well this activity stream is connecting the community with faculty	P2.E9.3

TABLE A.23 Partnerships and stakeholder engagement – Stage 3 'Test and Pilot' (P3)

Element	*Role*	*Example Tasks*	*Item Ref*
Curriculum renewal strategy (E1)	Direct	Review strategy for commitments from this activity stream	P3.E1.1
	Supporting	Provide update on needs and progress on request	P3.E1.2
	Assisted	Seek briefing on progress of strategy	P3.E1.3
Identify graduate attributes (E2)	Direct	–	P3.E2.1
	Supporting	–	P3.E2.2
	Assisted	–	P3.E2.3

TABLE A.23 Continued

Element	*Role*	*Example Tasks*	*Item Ref*
Map learning pathways (E3)	Direct	–	P3.E3.1
	Supporting	–	P3.E3.2
	Assisted	–	P3.E3.3
Audit learning outcomes (E4)	Direct	–	P3.E4.1
	Supporting	–	P3.E4.2
	Assisted	–	P3.E4.3
Develop and update curriculum (E5)	Direct	Seek contribution from stakeholders on tests/pilots as they are designed	P3.E5.1
	Supporting	Report feedback to other activity streams	P3.E5.2
	Assisted	Develop feedback communication to stakeholders	P3.E5.3
Implement the program (E6)	Direct	Seek feedback from stakeholders regarding tests/pilots as they proceed	P3.E6.1
	Supporting	Respond to requests for information as they arise	P3.E6.2
	Assisted	Develop feedback communication to stakeholders	P3.E6.3
Raise awareness and build capacity (E7)	Direct	Brief other activity streams regarding stakeholder perceptions, to inform testing and piloting options	P3.E7.1
	Supporting	Continue running a number of public forums discussing and generating interest in pilot collaborations	P3.E7.2
	Assisted	Receive briefing on testing and piloting	P3.E7.3
Collaborate internally and externally (E8)	Direct	Liaise with stakeholders (visits/mail-outs etc.) to promote renewed curriculum, seeking comments for improvement	P3.E8.1
	Supporting	Identify opportunities for involving stakeholders in giving guest lectures and so on within curriculum renewal process	P3.E8.2
	Assisted	Seek requests from faculty looking for external contributions to curriculum renewal process	P3.E8.3
Continually monitor and evaluate (E9)	Direct	Monitor stakeholder engagement in curriculum renewal process	P3.E9.1
	Supporting	Provide monitoring data to management for consideration and direction	P3.E9.2
	Assisted	Seek reflections from stakeholders on opportunities for further engagement	P3.E9.3

TABLE A.24 Partnerships and stakeholder engagement – Stage 4 'Integrate' (P4)

Element	*Role*	*Example Tasks*	*Item Ref*
Curriculum renewal strategy (E1)	Direct	Contribute to informing progress on transition	P4.E1.1
	Supporting	Provide update to other activity streams on rollout	P4.E1.2
	Assisted	Seek briefings from other activity streams on progress	P4.E1.3
Identify graduate attributes (E2)	Direct	–	P4.E2.1
	Supporting	–	P4.E2.2
	Assisted	–	P4.E2.3

TABLE A.24 Continued

Element	*Role*	*Example Tasks*	*Item Ref*
Map learning pathways (E3)	Direct	–	P4.E3.1
	Supporting	–	P4.E3.2
	Assisted	–	P4.E3.3
Audit learning outcomes (E4)	Direct	–	P4.E4.1
	Supporting	–	P4.E4.2
	Assisted	–	P4.E4.3
Develop and update curriculum (E5)	Direct	Initiate periodic guest lectures by knowledgeable stakeholders	P4.E5.1
	Supporting	Report findings from evaluation to other activity streams	P4.E5.2
	Assisted	Develop feedback communication to stakeholders	P4.E5.3
Implement the program (E6)	Direct	Seek feedback from stakeholders regarding curriculum renewal initiatives as they proceed, involving formal evaluation (i.e. surveys) where practical	P4.E6.1
	Supporting	Set up a communication tool to alert academics regarding possible keynote speakers	P4.E6.2
	Assisted	Create opportunity for potential students to experience renewed curriculum	P4.E6.3
Raise awareness and build capacity (E7)	Direct	Brief other activity streams regarding stakeholder perspectives, to inform integration initiatives	P4.E7.1
	Supporting	Host a number of public forums discussing curriculum renewal	P4.E7.2
	Assisted	Receive briefing on integration initiatives	P4.E7.3
Collaborate internally and externally (E8)	Direct	Continue to liaise with stakeholders (visits/mail-outs etc.) to promote renewed curriculum, seeking feedback for improvement	P4.E8.1
	Supporting	Set up an interface for stakeholders to express interest and provide scholarships/funding opportunities	P4.E8.2
	Assisted	Promote seminars and awards event to partners and stakeholders	P4.E8.3
Continually monitor and evaluate (E9)	Direct	Monitor stakeholder and partner interest in curriculum renewal initiatives	P4.E9.1
	Supporting	Share findings with other activity streams to inform progress	P4.E9.2
	Assisted	Inform stakeholders and partners of progress	P4.E9.3

GLOSSARY

Accreditation A review of one or more 'programs' by a professional authority (e.g. engineering), focusing on the delivery of 'graduate outcomes' using criteria that cover the teaching and learning environment, the structure and content of the program, and the quality assurance framework. Typically program accreditation is cyclical, over a period of 3–5 years.

Activity stream Key organisational parts of an 'institution', made up of one or more 'departments'. This text refers to six activity streams: governance (G), operations (O), teaching and learning (T), human resources and culture (H), marketing and communications (M), and partnerships and stakeholder engagement (P).

Armada course A term used in this text to refer to a 'course' that connects with other 'courses' to develop 'graduate attributes' in a 'program'. This may comprise consideration of common and intersecting 'knowledge and skill' areas to reinforce the student's development of a given 'graduate attribute'. The armada approach needs to be carefully coordinated to ensure that 'courses' are aligned with regard to language, terms, definitions and overarching message, to provide a clear learning pathway for students, and avoiding duplication.

Attribute See 'Graduate attribute'

Capability A specific skill set that forms part of a 'graduate attribute'. These may be described in a number of ways, for example technical, specialist, process, non-technical, common or generic.

College (institution) See 'Institution'

College (professional body) Eight colleges are distinguished by Engineers Australia to broadly cover all areas of practice in engineering, and are supported by a number of 'technical societies'. The colleges are Biomedical, Chemical, Civil, Electrical, Environmental, Information Telecommunications and Electronics (ITEE), Mechanical and Structural.

Competency In the field this term is used interchangeably at the level of graduate capability (see 'Graduate attribute'), as well as component knowledge and skill capabilities (see 'knowledge and skills'). In this text, for every 'graduate attribute' there will be one or more component competencies ('knowledge and skills') that need to be embedded into student learning, across one or more 'courses'.

Competency standard A standard against which a 'program' can be evaluated during 'accreditation'. For example, the Engineers Australia 'Stage 1' competency standards consist of three broad competencies: knowledge and skills base, engineering application ability, and professional and personal attributes. Each competency is broken down into several 'elements', each elaborated with a number of 'indicators of attainment'. Together, each 'competency' and its 'elements of competency' represent the profession's expression of the knowledge and skill base, engineering application abilities, and professional skills, values and attitudes that must be demonstrated at the point of entry to practice.

Course (Also referred to as a 'unit', 'subject', or 'module'.) A piece of work undertaken, which is part of the overall 'program' of study (i.e. 1/8 of a nominal full study year). It may have anything from 3 to 12 'credit points' of value.

Credit points A metric used to indicate the amount of work required to complete a 'course' of study within a 'program'. Depending on the university metrics, a 'program' will have an allocated number of credit points to distribute among the 'courses'.

Cross-disciplinary See 'Trans-disciplinary'

Curriculum (singular) or curricula (plural) All of the learning that is developed and implemented for a given 'program'. This includes a 'syllabus' (i.e. what is taught) and 'pedagogy' considerations (i.e. how it is taught). In this text, 'curriculum' is at times interchanged with 'curricula' in accordance with conversational use in the academic community.

Curriculum-ready A 'resource' that can be immediately and easily incorporated into a 'course' with minimal modification.

Curriculum renewal Curriculum renewal is defined to mean the redevelopment of curriculum, which may involve one or more existing or new 'courses' in a 'program', the review of past 'syllabi', and 'pedagogy'.

Dean (Also referred to as 'head of department' and 'head of school'.) The manager of a 'department', responsible for overseeing the delivery of one or more 'programs'.

Degree See 'Program'

Department (Also referred to as 'school' and 'faculty'.) A part of an institution responsible for one or more degree program offerings in a particular discipline area or cluster, for example engineering or business, and to which 'faculty' (also referred to as 'lecturers', 'staff' or educators) belong.

Education for sustainability Curriculum that is directed towards the development of 'graduate attributes' that build capacity for delivering sustainable solutions for humanity in the 21st century.

Education provider See 'Institution'

Element (curriculum renewal) In this text we refer to a number of 'elements' that are significant considerations within deliberative and dynamic curriculum renewal. Namely, develop a curriculum renewal strategy, identify graduate attributes, map learning pathways, audit learning outcomes, develop and update curriculum, implement the program, raise awareness and build capacity, collaborate internally and externally, and continually monitor and evaluate.

Element (graduate attribute) An aspect of a 'graduate attribute' that needs to be developed for the attribute to be achieved, comprising one or more 'indicators' of consideration. One 'graduate attribute' may have several 'elements'. An element may be developed through one or more 'learning outcomes', in one or more 'courses'. For example, there are sixteen elements in the Engineers Australia Stage 1 Competency Standard that must be demonstrated for the accreditation of an engineering program to be achieved, or demonstrated by an individual seeking admission to membership of Engineers Australia without an accredited qualification.

Faculty (organisation) See 'Department'

Faculty (personnel) (Also referred to as 'lecturer', 'staff' and educator.) Those employees in a 'department' who teach or manage (e.g. convene) some aspect of teaching.

Flagship course A term used in this text to refer to a 'course' that focuses on priority 'graduate attributes' in a 'program'. Such a course may distinguish the 'program' from other programs, highlighting important messages that the 'institution' is committed to, for example advanced computer modelling, or project management. The course could be placed anywhere in a program, for example as a first-year common 'course', an elective 'course', or in the form of a PhD or a Master's 'course'.

Formative engineering qualification See 'Stage 1 Competency Standard'. The accredited qualification for entry to practice in each engineering occupation. Generic award titles for formative qualifications for each occupation are: Professional Engineer: Bachelor of Engineering or Master of Engineering; Engineering Technologist: Bachelor of Engineering Technology; and Engineering Associate: Advanced Diploma or Associate Degree.

Graduate attribute (Also referred to as a 'graduate competency', 'program outcome', 'graduate capability', and 'competency'.) A desirable quality that a graduate engineer will possess by the time they complete their 'program' of study. This may be a common attribute that is shared with one or more other disciplines.

Graduate capability See 'Graduate attribute'

Graduate competency See 'Graduate attribute'

Graduate outcomes See 'Graduate attribute'

Greenhouse gas Any atmospheric gas that contributes to the greenhouse effect by absorbing infrared radiation.

Head of department See 'Dean'

Head of school See 'Dean'

Indicator of attainment Each of the 'elements' of Engineers Australia's 'Competency Standards' is elaborated with 'indicators' of attainment. These provide insight into the breadth and depth of ability expected for each 'element', guiding the demonstration and assessment process, as well as curriculum design. The indicators are not discrete sub-elements of competency mandated for individual audit. Rather, each element is tested holistically, and there may well be additional indicator statements that could complement those listed.

Institution The organisation responsible for employing 'faculty' and 'staff' to deliver 'programs' of study, comprising a number of 'departments' and 'activity streams'. (Also referred to as 'college', 'university', 'education provider'.)

Inter-disciplinary See 'Trans-disciplinary'

Inverse learning pathway An alternative 'learning pathway' or method of instruction where students learn as a direct result of experience. Learning occurs in the order of 'demonstrate', 'practise', 'learn', rather than 'learn', then 'practise' and 'demonstrate'.

Knowledge and skills Components of 'graduate attribute' goals that need to be developed in one or more 'courses' over a 'program' of study.

Laboratory A scheduled period of teaching and learning, usually held in a laboratory room, involving activities such as construction, testing and analysis of equipment, machinery or materials.

Learning outcome A statement of what 'knowledge and skills' a student should have developed and to what extent, by the time they complete a 'course' within a 'program' of study. The statement usually begins with a phrase such as, 'By the end of this course, you will be able to ...'

Learning pathway The way/s in which students are led through a 'program' to develop a 'graduate attribute'. This comprises one or more sequences of 'courses' with 'learning outcomes' that target the development of 'knowledge and skills'.

Lecture A scheduled period of teaching and learning, held face to face in a flat or tiered room or online, involving the largely one-way sharing of knowledge by faculty (also referred to as 'lecturer' and educator).

Lecturer See 'Faculty'

Major The primary specialisation within a program of study, which is usually comprised of a sequence of 'courses' that the student selects as part of their 'program'.

Map of learning pathways A qualitative and stylised representation of what is planned for the 'curriculum', comprising a map showing how each 'graduate attribute' will be developed. It involves creating a calendar/year-based chart, or map, for each 'course', which tracks the integration of 'knowledge and skills' taught in each 'course' in each year of study for a 'program'.

Materials Information that can be drawn upon to create a 'resource' for a 'course', for example policy documents, industry standards and regulations.

Meta-narrative (Also referred to as meta-theme.) An overarching statement that summarises the intent or key point of a 'course' or 'program'.

Minor A secondary specialisation within a program of study, which is usually comprised of a sequence of 'courses' that the student selects as part of their 'program'.

Module (Also referred to as a 'sub-topic'.) A piece of study that is part of a 'course' (i.e. associated with learning outcomes and assessment items for that course) and which may be taught over a period of one or more weeks within the course.

Niche program A term used in this text to refer to specialisations existing either within a larger discipline context (for example a Bachelor of Sustainable Energy Systems as a major specialisation within engineering) or as a 'trans-disciplinary' degree that cuts across traditional boundaries (for example a Bachelor of Environmental Management involving Science, Engineering, Business and the Arts).

Package See 'Resource'

Pedagogy The way in which a 'course' is taught, otherwise referred to as the strategy or style of instruction.

Program (Also referred to as a 'course' and 'degree'.) The award that a student works towards, and which is made up of a certain number of approved 'courses' (also known as 'subjects' or 'units').

Program outcome See 'Graduate attribute'

Resource (Also referred to as a 'module', 'unit' or 'package'.) A discreet set of teaching and learning materials, targeted at specific 'knowledge and skill' area/s, which can be integrated within existing 'courses'. This could include, for example, a lecture, an assignment, an assessment item, lecture notes, slides, discussion points etc.

School See 'Department'

Situation report A term used in this text to describe a summary of current external drivers, opportunities and trends, and to provide context for the institutions' intended activities.

Staff Employees of an 'institution' responsible for administrative and managerial functions that support 'faculty' to deliver 'programs' of study. This includes for example media and marketing personnel, facilities managers, secretaries and some teaching and learning support personnel. In some institutions 'staff' is used to cover all types of employees.

Stage 1 Competency Standard See 'Competency Standard'. In Australia, the Stage 1 Competency Standard sets the outcome expectations for a formative educational qualification accredited or recognised by Engineers Australia. Each Stage 1 Competency Standard comprises three 'competencies', 16 'elements' and a number of elaborating 'indicators of attainment'.

Stage 2 Competency Standard See 'Competency Standard'. In Australia, the Stage 2 Competency Standard defines requirements for independent practice. They are used as the basis of assessment for chartered membership of Engineers Australia and, for Professional Engineers, for registration on the

National Professional Engineers Register. The Stage 2 Competency Standard embodies both the enabling (Stage 1) and practice competencies relevant to a field of engineering.

Status report A term used in this text to refer to a response-style document to the 'situation report', which considers how well the 'institution' is positioned to engage in the intended area of 'curriculum renewal'.

Sub-topic See 'Module'

Subject See 'Course'

Syllabus/syllabi A document that includes statements of the aims and objectives of a 'course' or 'program' of study, and its content.

Technical society Engineers Australia uses 'technical societies' to connect engineering and other professions, providing a forum for technical development, networking, and knowledge sharing and expansion.

Threshold learning concepts Concepts that create a transformational shift in students' approach to the rest of their curriculum. Once learned, such concepts cannot be 'unlearned'.

Time (t) The time at which there is a sudden market and regulatory shift in requirements for emerging 'knowledge and skills'.

Time lag dilemma A situation where the usual or standard timeframe to update 'curriculum' for professional disciplines is too long to meet changing market and regulatory requirements for emerging 'knowledge and skills'.

Trans-disciplinary (Also referred to as 'inter-disciplinary'.) Study that bridges more than one traditional discipline of study, for example engineering and business.

Tutorial A scheduled period of teaching and learning, held face to face in a flat or tiered room, or online, involving student interaction with each other and one or more tutors, regarding a topic of focus.

Unit See 'Course' and 'Resource'

University See 'Institution'

Whole-of-system approach An approach that takes into account the entirety of the system, including individual components and their relationships. In this text we refer to a whole-of-system approach to curriculum renewal within institutions, involving consideration of a number of 'elements' and direct, supported and assisted roles by each 'activity stream' in addressing the 'elements' at each 'stage'.

Workshop A scheduled period of teaching and learning, held face to face in a flat room or online, involving students working on one or more problems with some lecturer or tutor assistance.

NOTES

Chapter 1

1 Perkin, H. (2007) 'History of universities', *International Handbook of Higher Education*, Springer, London, pp159–205.

2 Concoran, P. and Wals, A. (2008) *Higher Education and the Challenge of Sustainability – Problematics, promise, and practice*, Kluwer Academic Publishers, Boston; Wals, A. (ed.) (2008) *From Cosmetic Reform to Meaningful Integration: Implementing education for sustainable development in higher education institutes – the state of affairs in six European countries*, DHO, Amsterdam; Jones, P., Selby, D. and Sterling, S. (2010) *Sustainability Education: Perspectives and practice across higher education*, Renouf Publishing, London.

3 Erdelen, W. (2009) 'Plenary Session 1 – Trends in Global Higher Education', Co-Chair Introduction, 2009 World Conference on Higher Education: The New Dynamics of Higher Education and Research for Societal Change and Development, Paris, 6–9 July 2009.

4 Orr, D. (1994) *Earth in Mind*, Island Press, Washington DC, p27.

5 O'Connor, I. (2006) 'Earth Dialogues Dinner', Keynote Address, Brisbane, Australia, 21–24 July 2006.

6 Higher Education Funding Council for England (2008) 'HEFC strategic review of sustainable development in higher education in England', *Report to the HEFC*, PA Consulting Group and the Centre for Research in Education and the Environment, Bath.

7 See Table 2.2 for a summary of publications on these experiences.

8 See Hargroves, K. and Smith, M. (2005) *The Natural Advantage of Nations: Business opportunities, innovation and governance in the 21st century*, The Natural Edge Project, Earthscan, London. (Section 1: The Need for a New Paradigm).

9 Smith, M., Hargroves, K. and Desha, C. (2010) *Cents and Sustainability – Securing our common future by decoupling economic growth from environmental pressures*, The Natural Edge Project, Earthscan, London (Chapter 1: Securing our Common Future).

10 Stasinopoulos, P., Smith, M., Hargroves, K. and Desha, C. (2008) *Whole System Design: An integrated approach to sustainable engineering*, The Natural Edge Project, Earthscan, London.

11 Smith, M., Hargroves, K., Stasinopoulos, P., Stephens, R., Desha, C. and Hargroves, S. (2007) *Energy Transformed: Sustainable Energy solutions for climate change mitigation*, The Natural Edge Project, CSIRO, and Griffith University.

12 Smith, M., Hargroves, K., Desha, C. and Stasinopoulos, P. (2009) *Water Transformed: Sustainable water solutions for climate change adaptation*, The Natural Edge Project (TNEP), Griffith University, and Australian National University, Australia.

13 von Weizsäcker, E., Hargroves, K., Smith, M., Desha, C. and Stasinopoulos, P. (2009) *Factor 5: Transforming the global economy through 80% increase in resource productivity*, Earthscan, London and Droemer, Germany.

14 Smith, M., Hargroves, K. and Desha, C. (2010) *Cents and Sustainability – Securing our common future by decoupling economic growth from environmental pressures*, The Natural Edge Project, Earthscan, London (Chapter 1: Securing our Common Future; Foreword).

15 Stern, N. (2006) *The Stern Review: The economics of climate change*, Cambridge University Press, Cambridge, Chapters 3–6, cited in Hargroves, K. and Smith, M. (2005) *The Natural Advantage of Nations: Business opportunities, innovation and governance in the 21st century*, The Natural Edge Project, Earthscan, London, p218.
16 Stern, N. (2006) *The Stern Review: The economics of climate change*, Cambridge University Press, Cambridge.
17 Davidson and Janssens (2006), Gedney *et al.* (2004) and Archer (2005) cited in Stern, N. (2007) *The Stern Review: The economics of climate change*, Cambridge Press, Cambridge, p14, cited in Smith, M. *et al.* (2010) *Cents and Sustainability – Securing our common future by decoupling economic growth from environmental pressures*, The Natural Edge Project, Earthscan, London.
18 IPCC (2007) *Climate Change 2007: Synthesis report*, Contribution of Working Groups I, II and III to the Fourth Assessment Report of the Intergovernmental Panel on Climate Change, Cambridge University Press, Cambridge.
19 Hargroves, K. and Smith, M. (2005) *The Natural Advantage of Nations: Business opportunities, innovation and governance in the 21st century*, The Natural Edge Project, Earthscan, London, pp323–324.
20 IPCC (2007) *Climate Change 2007: Synthesis report*, Contribution of Working Groups I, II and III to the Fourth Assessment Report of the Intergovernmental Panel on Climate Change, Cambridge University Press, Cambridge.
21 Thomas, C. *et al.* (2004) 'Feeling the heat: Climate change and biodiversity loss', *Nature*, 8 January, Vol 427, No 6970, pp87–180, cited in Smith, M. *et al.* (2010) *Cents and Sustainability – Securing our common future by decoupling economic growth from environmental pressures*, Earthscan, London, p262.
22 FAO (2001) *The State of World Fisheries and Aquaculture 2006*, Rome, p29, cited in Brown, L. R. (2008) *Plan B 3.0, Mobilizing to Save Civilization*, Earth Policy Institute, New York, p97.
23 Myers, R. A., and Worm, B. (2003) 'Rapid worldwide depletion of predatory fish communities', *Nature*, Vol 432, pp280–83; Crosby, C. (2003) '"Blue Frontier" is Decimated', *Dalhousie News*, 11 June 2003, cited in Brown, L. R. (2008) *Plan B 3.0, Mobilizing to Save Civilization*, Earth Policy Institute, New York, p98.
24 Science News (2007) 'Declining bee numbers raise concerns over plant pollination', *Science Daily*, 11 May, cited in Smith, M. *et al.* (2010) *Cents and Sustainability – Securing our common future by decoupling economic growth from environmental pressures*, Earthscan, London, p265.
25 White, A. *et al.* (2006) *China and the Global Market for Forest Products*, Washington, DC; Forest Trends, cited in Brown, L. R. (2008) *Plan B 3.0, Mobilizing to Save Civilization*, Earth Policy Institute, New York, p88.
26 Fearnside, P. M. quoted in Fraser, B. J. (2002) 'Putting a price on the forest', *LatinamericaPress.org*, 10 November 2002; Fearnside, P. M. (1997) 'The main resources of Amazonia', paper for presentation at the Latin American Studies Association XX International Congress, Guadalajara, Mexico, 17–19 April 1997; Lean, G., 'Dying forest: One year to save the Amazon', *The Independent*, 23 July 2006; Lean, G. (2006) 'A disaster to take everyone's breath away', *The Independent*, 24 July 2006, cited in Brown, L. R. (2008) *Plan B 3.0, Mobilizing to Save Civilization*, Earth Policy Institute, New York, p90.
27 Secretariat of the Convention on Biological Diversity (2010) *Global Biodiversity Outlook 3*, Montréal, Canada.
28 Brown, L. R. (2008) *Plan B 3.0, Mobilizing to Save Civilization*, Earth Policy Institute, New York, p90.
29 Government of Nigeria (2002) *Combating Desertification and Mitigating the Effects of Drought in Nigeria*, Revised National Report on the Implementation of the United Nations Convention to Combat Desertification, Nigeria, cited in Brown, L. R. (2008) *Plan B 3.0, Mobilizing to Save Civilization*, Earth Policy Institute, New York, p95.
30 Secretariat of the Convention on Biological Diversity (2010) *Global Biodiversity Outlook 3*, Montréal, Canada.
31 Brown, L. R. (2008) *Plan B 3.0, Mobilizing to Save Civilization*, Earth Policy Institute, New York, pp68–69.
32 Velut, C. (2011) 'Yemen's water crisis', *Yemen Today Magazine*, 7 January, yemen-today.com/go/investigations/8197.html, accessed 11 January 2011.
33 'Pakistan: Focus on water crisis', *U.N. Integrated Regional Information Networks News*, 17 May 2002; Garstang quoted in 'Water crisis threatens Pakistan: Experts', *Agence France-Presse*, 26 January 2001, cited in Brown, L. R. (2008) *Plan B 3.0, Mobilizing to Save Civilization*, Earth Policy Institute, New York, p74.
34 Khan, S. R. A. (2004) 'Declining land resource base', *Dawn*, Pakistan, 27 September 2004, cited in Brown, L. R. (2008) *Plan B 3.0, Mobilizing to Save Civilization*, Earth Policy Institute, New York, p73.
35 Carlton, J. (2003) 'Shrinking lake in Mexico threatens future of region', *Wall Street Journal*, 3 September 2003; UN Population Division (2005) *World Urbanization Prospects: 2005 Revision*, electronic database, at esa.un.org/unup, updated October 2006, cited in Brown, L. R. (2008) *Plan B 3.0, Mobilizing to Save Civilization*, Earth Policy Institute, New York, p78.
36 Heng, L. (2002) '20 natural lakes disappear each year in China', *People's Daily*, 21 October 2002; (2004) 'Glaciers receding, wetlands shrinking in river fountainhead area', *China Daily*, 7 January 2004, cited in Brown, L. R. (2008) *Plan B 3.0, Mobilizing to Save Civilization*, Earth Policy Institute, New York, p78.

37 Bates, B. C., Jundewicz, Z. W., Wu, S. and Palutikof, J. P. (eds) (2008) *Climate Change and Water*, Technical Paper of the Intergovernmental Panel on Climate Change, IPCC Secretariat, Geneva, cited in Smith, M. *et al.* (2010) *Cents and Sustainability – Securing our common future by decoupling economic growth from environmental pressures*, The Natural Edge Project, Earthscan, p286.
38 American Society of Civil Engineers (2005) 'Solid waste C+', Infrastructure Report Card, ASCE, cited in Smith, M. *et al.* (2010) *Cents and Sustainability – Securing our common future by decoupling economic growth from environmental pressures*, Earthscan, London, pp310–311.
39 Lipton, E. (2001) 'The long and winding road now followed by New York City's trash', *New York Times*, 24 March 2001, cited in Brown, L. R. (2008) *Plan B 3.0, Mobilizing to Save Civilization*, Earth Policy Institute, New York, p115.
40 Hoornweg, D., Lam, P. and Chaudhry, M. (2005) 'Waste management in China: issues and recommendations', Urban Development Working Papers, No. 9, World Bank, Washington DC, p6.
41 Calculated from U.S. Geological Survey (2007) *Mineral Commodity Summaries 2007*, Washington, DC, U.S. Government Printing Office, cited in Brown, L. R. (2008) *Plan B 3.0, Mobilizing to Save Civilization*, Earth Policy Institute, New York, p115.
42 WHO (2005) 'Air quality guidelines for particulate matter, ozone, nitrogen dioxide and sulfur dioxide: global update 2005', Summary of Risk Assessment, WHO, cited in Smith, M. *et al.* (2010) *Cents and Sustainability – Securing our common future by decoupling economic growth from environmental pressures*, The Natural Edge Project, Earthscan, p334.
43 Sigman, R. *et al.* (2012) 'Health and environment', in OECD, *OECD Environmental Outlook to 2050: The Consequences of Inaction*, OECD Publishing.
44 IPCC (2007) *Climate Change 2007: Synthesis report*, Contribution of Working Groups I, II and III to the Fourth Assessment Report of the Intergovernmental Panel on Climate Change, Cambridge University Press, Cambridge.
45 Li, Z. (2006) 'Acid rain affects one-third of China; main pollutants are sulfur dioxide and particulate matter', *Worldwatch Institute*, August 31, worldwatch.org/node/4496, accessed 12 January 2010.
46 Indian Express (2011) 'Acid rain makes life hard in 258 Chinese cities', *Indian Express*, 12 January, indianexpress.com/news/acid-rain-makes-life-hard-in-258-chinese-cities/736487/0, accessed 13 January 2011.
47 Indian Express (2011) 'Acid rain makes life hard in 258 Chinese cities', *Indian Express,* 12 January, indianexpress.com/news/acid-rain-makes-life-hard-in-258-chinese-cities/736487/0, accessed 13 January 2011.
48 Bradsher, K. and Barboza, D. (2006) 'Pollution from Chinese coal casts a global shadow', *The New York Times*, June 11, nytimes.com/2006/06/11/business/worldbusiness/11chinacoal.html?_r=1&ref=acidrain, accessed 13 January 2011.
49 Summarised from: Smith, M., Hargroves, K. and Desha, C. (2010) *Cents and Sustainability – Securing our common future by decoupling economic growth from environmental pressures*, The Natural Edge Project, Earthscan, London (Chapter 1: Securing our Common Future).
50 Porrit, J. (2007) 'Short interview with Mr Jonathon Porrit', Global Sustainability Forum: The Future for Engineering Education, Imperial College, London, 18 September 2007, www3.imperial.ac.uk/global sustainability, accessed 20 January 2011.
51 Talberth, J., Cobb, C. and Slattery, N. (2006) *The Genuine Progress Indicator 2006: A tool for sustainable development*, Redefining Progress, Oakland, California.
52 Smith, M., Hargroves, K. and Desha, C. (2010) 'Figure 2.7 Gross domestic product compared with estimated environmental costs (billions) for the US, 1950–2004', *Cents and Sustainability – Securing our common future by decoupling economic growth from environmental pressures*, p49.
53 Based on research undertaken by Dr Michael Smith as part of his thesis development at the Australian National University.
54 Stern, N. (2006) *The Stern Review: The economics of climate change*, Cambridge University Press, Cambridge, p10.
55 CBD (2006) *Global Biodiversity Outlook 2*, Secretariat of the Convention on Biological Diversity (CBD), Montreal.
56 Hutton, G. and Haller, L. (2004) *Evaluation of the Costs and Benefits of Water and Sanitation Improvements at the Global Level, Water, Sanitation and Health, Protection of the Human Environment*, World Health Organization, Geneva.
57 Lovins, A. (n.d.) 'Amory Lovin's Natural Capitalism Lecture', *ABC: The Slab*, Australia.
58 CSE (1996) Press release, Centre for Science and Environment, 29 September 1996.

59 Holland, M., Kinghorn, S., Emberson, L., Cinderby, S., Ashmore, M., Mills, G. and Harmens, H. (2006) 'Development of a framework for probabilistic assessment of the economic losses caused by ozone damage to crops in Europe', CEH project No. C02309NEW, Centre for Ecology and Hydrology, Natural Environment Research Council, Bangor, Wales.
60 Diamond, J. (2005) *Collapse: How societies choose to fail or succeed*, Penguin Books, New York.
61 Shriberg, M. and Tallent, H. (2003) 'Beyond principles: Implementing the Talloires Declaration', Report and Declaration of the President's Conference 1990, Association of University Leaders for a Sustainable Future, Ball State University, Muncie, IN, pp1, www.ulsf.org/pdf/ShribergTallen, accessed 21 January 2011.
62 Copernicus Alliance (2012) People's Sustainability Treaty on Higher Education, presented for endorsement at Rio+20, June 2012, www.uncsd2012.org/index.php?page=view&type=1006&menu=153&nr=135, accessed 30 July 2012.
63 ULSF (1990) 'The Talloires Declaration: 10 point action plan', Association of University Leaders for a Sustainable Future, http://www.ulsf.org/pdf/TD.pdf, accessed 27 January 2011.
64 United Nations (2002) 'Report of the World Summit on Sustainable Development', World Summit on Sustainable Development, United Nations, Johannesburg, South Africa, 26 August–4 September 2002.
65 Parliamentary Commissioner for the Environment (2004) *See Change: Learning and education for sustainability*, Wellington, New Zealand, 2004.
66 UNESCO (1997) 'Thessaloniki Declaration', International Conference on Environment and Society Education and Public Awareness for Sustainability, UNESCO and the Government of Greece, 8–12 December 1997.
67 UNESCO (1998) 'World Declaration on Higher Education for the Twenty-First Century: Vision and action', World Conference on Higher Education, UNESCO, Paris, 9 October 1998.
68 United Nations (2000) 'The Earth Charter', United Nations.
69 GHESP (2002) 'The Lüneburg Declaration on Higher Education for Sustainable Development', International COPERNICUS Conference, Global Higher Education for Sustainability Partnership, University of Lüneburg, 8–10 October 2001.
70 United Nations Department of Economic and Social Affairs (2002) 'Ubuntu Declaration on Education and Science and Technology for Sustainable Development', World Summit on Sustainable Development, Johannesburg, September 2002.
71 UN General Assembly (2002) 'Proclamation of the Decade of Education of Sustainable Development (2005–2014)'.
72 The G8 University Summit (2009) 'Torino Declaration on Education and Research for Sustainable and Responsible Development', G8 University Summit, Torino, Italy, 17–19 May 2009.
73 UNESCO (2009) 'Communique – The new dynamics of higher education and research for societal change and development', 2009 World Conference on Higher Education, UNESCO, Paris, 8 July 2009.
74 Association for the Advancement of Sustainability in Higher Education (2010) *Sustainability Curriculum in Higher Education: A call to action*, Denver, Colorado.
75 Copernicus Alliance (2012) People's Sustainability Treaty on Higher Education, presented for endorsement at Rio+20, June 2012, www.uncsd2012.org/index.php?page=view&type=1006&menu=153&nr=135, accessed 30 July 2012.
76 Based on research undertaken by Michael Smith as part of his thesis development at ANU.
77 Department for Communities and Local Government (2006) 'Code for Sustainable Homes: A step-change in sustainable home building practice', Crown Copyright; Rajgor, G. (2007) 'Countdown to zero emissions', *Refocus*, Vol 8, No 1, pp60–61.
78 Chong, S. K. (2005) 'Anmyeon-do recreation forest: A millennium of management', in Durst, P. *et al.* (eds), *In Search of Excellence: Exemplary forest management in Asia and the Pacific*, Asia-Pacific Forestry Commission, FAO Regional Office for Asia and the Pacific, Bangkok, pp251–259.
79 Rosegrant, M. W. (2001) 'Dealing with water scarcity in the 21st century', in Pinstrup-Andersen, P. and Pandya-Lorch, R. (eds), *The Unfinished Agenda: Perspectives on overcoming hunger, poverty, and environmental degradation*, International Food Policy Research Institute, Washington, DC, p149.
80 European Union (2003) 'Directive 2002/95/EC of the European Parliament and of the Council of 27 January 2003 on the restriction of the use of certain hazardous substances in electrical and electronic equipment', *Official Journal of the European Union*.
81 Global Sources (2005) *RoHS Compliance Readiness Survey: Mainland China, Taiwan, Hong Kong and South Korea*, Global Sources, p3.
82 Ministry of Commerce (2006) *Administrative Measure on the Control of Pollution Caused by Electronic Information Products*, People's Republic of China, Beijing.

83 Yun, J. And Park, I. (2007) *Act for Resource Recycling of Electrical and Electronic Equipment and Vehicles: English translation*, Eco-Frontier, Seoul.
84 UNECE (1979–2005) *The Convention on Long-range Transboundary Air Pollution*, United Nations Economic Commission for Europe, Geneva.
85 UNECE (1994) *The 1994 Oslo Protocol on Further Reduction of Sulphur Emissions*, United Nations Economic Commission for Europe, Geneva.
86 OECD (2002) *Indicators to Measure Decoupling of Environmental Pressure and Economic Growth*, OECD, Paris.
87 See Table 6.1 in *Cents and Sustainability* for a detailed listing of energy consumption reduction enforcement measures in the 10 largest economies in the world.
88 Developed by The Natural Edge Project in considering opportunities for companies and the HEI sector; see also Smith, M., Hargroves, C. and Desha, C. (2010) 'Figure 3.1 A stylized representation of the level of commitment to reducing environmental pressure over time, for three performance scenarios', *Cents and Sustainability – Securing our common future by decoupling economic growth from environmental pressures*, Earthscan, London, p69.
89 Burnett, N. (2008) Mr Nick Burnett, Assistant Director-General for Education, UNESCO, Opening speech, 13th General Conference, International Universities Association, The Netherlands, 15 July 2008.
90 Australian Federal Department of Resources, Energy and Tourism (DRET) 'Energy efficiency opportunities', www.ret.gov.au/energy/efficiency/eeo/pages/default.aspx, accessed 17 July 2009.
91 Victorian Environmental Protection Agency (n.d.) 'Environment and resource efficiency plans – EREP'.
92 DRET (2010) 'Energy efficiency opportunities program: Mid-cycle review final report', Department of Resources Energy and Tourism, Australian Federal Government.
93 DRET (2010) 'Energy efficiency opportunities program: Mid-cycle review final report', Department of Resources Energy and Tourism, Australian Federal Government.
94 GHD (2010) *Report for Long Term Training Strategy for the Development of Energy Efficiency Assessment Skills – Functional skills analysis report*, GHD.
95 Smith, M., Hargroves, K., Stasinopoulos, P., Stephens, R., Desha, C. and Hargroves, S. (2007) *Energy Transformed: Sustainable energy solutions for climate change mitigation*, The Natural Edge Project (TNEP), Australia.
96 Australian Deans of Built Environment and Design (2008) *Review of Australian Higher Education*, ADBED, Canberra, p4.
97 HEFCE (2008) *HEFCE Strategic Review of Sustainable Development in Higher Education in England*, Higher Education Funding Council for England, Bristol, www.hefce.ac.uk/pubs/rdreports/2008/rd03_08/rd03_08.doc, accessed 18 August 2011.
98 Australian Deans of Built Environment and Design (2008) *Review of Australian Higher Education*, ADBED, Canberra, p4.
99 RAE (2007) *Educating Engineers for the 21st Century*, The Royal Academy of Engineering, London, www.raeng.org.uk/news/publications/list/reports/Educating_Engineers_21st_Century.pdf, accessed 18 August 2011.
100 RAE (2005) *Engineering for Sustainable Development: Guiding principles*, The Royal Academy of Engineering, London, www.raeng.org.uk/events/pdf/Engineering_for_Sustainable_Development.pdf, accessed 18 August 2011.
101 Byrne, E., Desha, C., Fitzpatrick, J. and Hargroves, K. (2010) 'Engineering education for sustainable development: A review of international progress', workshop paper for the 3rd International Symposium for Engineering Education, Cork, 30 June–2 July 2010, available at: http://www.ucc.ie/academic/processeng/isee2010/pdfs/primer.pdf, accessed 18 August 2011.
102 Byrne, E., Desha, C., Fitzpatrick, J. and Hargroves, C. (in press) 'Exploring sustainability themes in engineering accreditation and curricula, *International Journal of Sustainability in Higher Education*, Vol 15, No 1.
103 For a detailed explanation of this concept refer to Chapter 2 of Smith, M., Hargroves, K. and Desha, C. (2010) *Cents and Sustainability – Securing our common future by decoupling economic growth from environmental pressures*, The Natural Edge Project, Earthscan, London.
104 Smith, M., Hargroves, C. and Desha, C. (2010) 'Figure 2.2 Conceptual and stylized representation of a decoupling graph', *Cents and Sustainability – Securing our common future by decoupling economic growth from environmental pressures*, Earthscan, London. p32.
105 Based on research undertaken by Michael Smith as part of his thesis development at ANU.
106 Stern, N. (2006) *The Stern Review: The economics of climate change*, Cambridge University Press, Cambridge.
107 Stern, N. (2006) *The Stern Review: The economics of climate change*, Cambridge University Press, Cambridge.
108 See Smith, M., Hargroves, K., Stasinopoulos, P., Stephens, R., Desha, C. and Hargroves, S. (2007) *Energy Transformed: Sustainable energy solutions for climate change mitigation*, The Natural Edge Project, CSIRO, Griffith University, Australia.

109 von Weizsäcker, E., Hargroves, K., Smith, M., Desha, C. and Stasinopoulos, P. (2009) *Factor 5: Transforming the global economy through 80% increase in resource productivity*, Earthscan, London.
110 CBD (2006) *Global Biodiversity Outlook 2*, Secretariat of the Convention on Biological Diversity (CBD), Montreal.
111 Briscoe, J. (2008) *India's Water Economy: Bracing for a turbulent future*, World Bank, Washington DC.
112 Chaves, M., Santos, T., Souza, R., Ortuno, M., Rodrigues, M., Lopes, C., Maroco, J. and Pereira, J. (2007) 'Deficit irrigation in grapevine improves water use efficiency while controlling vigour and production quality', *Annals of Applied Biology*, Vol 150, pp237–252.
113 Savory, A. and Butterfield, J. (1988) *Holistic Management: A new framework for decision making* (2nd edition), Island Press, Washington DC.
114 Smith, M., Hargroves, K., Stasinopoulos, P. and Desha, C. (2010) *Water Transformed: Sustainable water solutions for climate change adaptation*, The Natural Edge Project (TNEP), Australia.
115 OECD (2008) *OECD Environmental Outlook to 2030*, OECD, Paris, pp246, 249.
116 CSE (1999) 'Sick of air pollution', press release, Centre for Science and Environment, 5 June 1999.
117 United Nations Environment Programme (2008) *Green Jobs: Towards decent work in a sustainable, low-carbon world*, United Nations.
118 Hargroves, K. and Smith, M. (2005) *The Natural Advantage of Nations: Business opportunities, innovation and governance in the 21st century*, The Natural Edge Project, Earthscan, London, p17.
119 Hargroves, K. and Smith, M. (2005) *The Natural Advantage of Nations: Business opportunities, innovation and governance in the 21st century*, The Natural Edge Project, Earthscan, London, Figure 1.1, p17.
120 Brown, L. (2007) *Plan B 3.0: Mobilizing to save civilization*, W.W. Norton & Company, New York.
121 UNSD (2007) *International Civil Aviation Yearbook: Civil aviation statistics of the world*, UNSD.
122 OECD (2001) *OECD Environmental Outlook for the Chemicals Industry*, Organisation for Economic Co-operation and Development, Paris.
123 UNCTAD (2006) 'Review of maritime transport 2006', United Nations Conference on Trade and Development, New York and Geneva, p48.
124 Kirby, R. (1990) *Engineering in History*, Dover Publications, New York, pp207–213.
125 Bagwell, P. and Lyth, P. (2002) *Transport in Britain, 1750–2000*. Hambledon and London, pp10–11.
126 Flinn, M. 'Abraham Darby and the coke smelting process', new series, Vol. 26, No. 101 (Feb., 1959), pp54–59, published by Blackwell Publishing on behalf of The London School of Economics and Political Science Article Stable.
127 Ryall, M. *et al.* (2000) *The Manual of Bridge Engineering*, pp18–19, Thomas Telford Publishing, London.
128 Ryall, M. *et al.* (2000) *The Manual of Bridge Engineering*, pp18–20, Thomas Telford Publishing, London.
129 Ripon Historical Society (2011) *American Oil Lamp Exhibit*, http://www.riponhistory.com/museum-and-archives/american-oil-lamp-exhibit/
130 Weissenbacher, M. (2009) *Sources of Power: How Energy forges human history*, Greenwood Publishing Group, California, pp207–209.
131 Weissenbacher, M. (2009) *Sources of Power: How Energy forges human history*, Greenwood Publishing Group, California, p212.
132 Solomon, B. (2000) *The American Diesel Locomotive*, MBI Publishing Company, p46.
133 Schramm, J. (2010) *Out of Steam: Deiselization and American railroads, 1920–1960,* Rosemont Publishing and Printing Corp., New Jersey, p297.
134 Churella, A. (1998) *From Steam to Diesel: Managerial customs and organizational capabilities in the twentieth century American locomotive industry*, Princeton University Press, pp3–4.
135 No author, 'A technical history of photography', http://www.photo101.org/docs/history%20of%20 photography.pdf, accessed 19 January 2011.
136 Lester, P. (2006) *Visual Communication: Images with messages*, Thomas Wadsworth, USA.
137 Shreve, R. and Austin, T. (1956) *Shreve's Chemical Process Industries*. McGraw Hill, Singapore, p225.
138 Hirth, L. (2007) *State, Cartels and Growth: The German chemical industry*, Grin Verlag, Germany, p14.
139 'Johannes Gutenberg.' Encyclopædia Britannica Online. Encyclopædia Britannica, 2011.
140 Moran, J. (1973) *Printing Presses: History and development from the fifteenth century to modern times*, University of California Press, p123.
141 Cost, F. (2005) *The New Medium of Print: Material communication in the internet age*, RIT Cary Graphic Arts Press, pp54–55.
142 Clever, D. (2009) 'Linotype remembered fondly, but print technology continues to leap ahead', http://www.goskagit.com/home/article/the_linotype_is_remembered_fondly_but_print_technology_continues_to_leap_ah/, accessed 20 January 2011.

143 Cost, F. (2005) *The New Medium of Print: Material communication in the internet age*, RIT Cary Graphic Arts Press, p55.
144 Winston, B. (1998) *Media Technology and Society: A history: From the telegraph to the internet*, Routledge, p23.
145 'History of communication', http://inventors.about.com/library/inventors/bl_history_of_communication.htm.
146 Britiannica Academic Edition, http://www.britannica.com/EBchecked/topic/1350805/history-of-technology/10459/Chemicals.
147 Winston, B. (1998) *Media Technology and Society: A history: From the telegraph to the internet*, Routledge, p166.
148 Gindis, E. (2010) *Up and Running with AutoCAD*, p463.
149 Crawlet, E., Malmqvist, J. and Ostlund, S. (2007) *Rethinking Engineering Education: The CDIO approach*, pp231–232.
150 US Department of Energy (2008) *Energy Efficiency Trends in Residential and Commercial Buildings.*
151 OECD (2003) *Environmentally Sustainable Buildings: Challenges and policies*, Organisation for Economic Co-operation and Development, Paris.
152 Factor Four.
153 IPCC (2007) *Climate Change 2007: Mitigation of climate change*, contribution of Working Group III to the 4th Assessment Report of the Intergovernmental Panel on Climate Change, Cambridge University Press, Cambridge.
154 US Department of Energy (2008) *Energy Efficiency Trends in Residential and Commercial Buildings.*
155 Chui, Y. (2010) *An Introduction to the History of Project Management*, Eburon Uitgeverij B.V., pp172–173.
156 Britiannica Online, 'Industry and Innovation' http://www.britannica.com/EBchecked/topic/1350805/history-of-technology/10459/Chemicals.
157 Wals, A. (2012) *Shaping the Education of Tomorrow, 2012*, full-length report on the UN Decade of Education for Sustainable Development, DESD Monitoring and Evaluation 2012, United Nations Educational, Scientific and Cultural Organization, Education Sector.
158 See UNESCO (2010) *Engineering: Issues, challenges and opportunities for development*, UNESCO Publishing, Paris.
159 Desha, C. and Hargroves, K. (2011) 'Informing engineering education for sustainable development using a deliberative dynamic model for curriculum renewal', Proceedings of the Research in Engineering Education Symposium 2011, Madrid.
160 Rensburg, E., Lotz Sisitka, H. and Mosidi, S. (2001) *Environmental Learning Across NQF*, Department of Environmental Affairs and Tourism, South Africa.
161 Parliamentary Commissioner for the Environment (2004) *See Change: Learning and education for sustainability*, Wellington, New Zealand.
162 Sustainable Development Education Panel (2002) *Learning to Last: Business and sustainable development – A learning guide for sector skills councils*, www.dcsf.gov.uk/aboutus/sd/docs/LearningtoLast03.doc, accessed 21 January 2011.
163 Desha, C. (2010) 'An investigation into the strategic application and acceleration of curriculum renewal in engineering education for sustainable development' (doctoral dissertation), Chapter 3, Griffith University, Brisbane.
164 Desha, C. and Hargroves, K. (2009) 'Re-engineering higher education for energy efficiency solutions', *ECOS, CSIRO*, vol 151, October–November.
165 Higher Education Funding Council (2013) 'Sustainable development', summary of project funding for Leadership, Governance and Management, UK, http://www.hefce.ac.uk/whatwedo/lgm/sd/.
166 Luna, H., Martin, S., Scott, W., Kemp, S. and Robertson, A. (2012) *Universities and the Green Economy: Graduates for the future – Higher Education Academy policy think tank report 2012*, report to the Higher Education Academy.
167 Desha, C. and Hargroves, K. (2012) 'Engaging in a multiple-track approach to building capacity for 21st century engineering, opportunities and challenges for rapid curriculum renewal', in proceedings of the 8th International CDIO Conference, Queensland University of Technology, Brisbane, July 1–4, 2012, Figure 5.
168 Forum for the Future (2004) 'Tip for top in higher education: Forum's Learning in Action conference pushes the SD solutions', GreenFutures Magazine, http://www.forumforthefuture.org/greenfutures/articles/tip-top-higher-ed.
169 The US Partnership for Education for Sustainable Development (n.d.) 'The US Partnership for Education for Sustainable Development', www.uspartnership.org.
170 DANS (n.d.) 'Mission and Action Plan for the Disciplinary Associations Networks for Sustainability', http://dans.aashe.org/.
171 Desha, C. and Hargroves, K. (2011) 'Informing engineering education for sustainable development using a deliberative dynamic model for curriculum renewal', Proceedings of the Research in Engineering Education Symposium 2011, Madrid.

Chapter 2

1 Jones, D. (2004) 'Sustainability – The greatest challenge', opening address, 8th International Conference on Bulk Materials Storage, Handling and Transportation, Wollongong, Australia.
2 UNESCO (2010) 'Introduction' in *Engineering: Issues, challenges and opportunities for development*, produced in conjunction with the World Federation of Engineering Organisations (WFEO), the International Council of Academies of Engineering and Technological Sciences (CAETS) and the International Federation of Consulting Engineers (FIDIC), UNESCO Publishing, Paris.
3 Rajai, M. and Johnson, K. (2001) 'Creating new engineers for the new millennium', *Industry and Higher Education*, Vol 15, No 5, pp349–352; The Royal Academy of Engineering (2007) *Educating Engineers for the 21st Century*, The Royal Academy of Engineering, London; Cortese, A. (2007) 'Higher education leadership in reversing global warming and creating a healthy, just and sustainable society', a presentation to the Annual Meeting of the Annapolis Group, 19 June 2007.
4 Grear, B. (2008) Personal communications with Barry Grear, Former President of the Institution of Engineers Australia, and incoming President for the World Federation of Engineering Organisations, 29 August 2008.
5 WFEO (2002) *Engineers and Sustainable Development*, World Federation of Engineering Organisations' Committee on Technology.
6 Yusoff, S. (2005) 'The need to review engineering education for achieving sustainable development', *Journal Pendidikan*, Universiti Malaya, pp41–49, 45.
7 Hargroves, K. and Smith, M. (eds) (2005) *The Natural Advantage of Nations: Business opportunities, innovation and governance in the 21st century*, The Natural Edge Project, Earthscan, London (with co-authors including Dr Cheryl Desha).
8 Stasinopoulos, P., Smith, M., Hargroves, K. and Desha, C. (2008) *Whole System Design: An integrated approach to sustainable engineering*, The Natural Edge Project, Earthscan, London.
9 von Weizsäcker, E., Hargroves, K., Smith, M., Desha, C. and Stasinopoulos, P. (2009) *Factor 5: Transforming the global economy through 80% increase in resource productivity*, Earthscan, London and Droemer, Germany (ranked 12th in the University of Cambridge Programme for Sustainability Leadership 'Top 40 books of 2010').
10 Smith, M., Hargroves, K. and Desha, C. (2010) *Cents and Sustainability – Securing our common future by decoupling economic growth from environmental pressures*, The Natural Edge Project, Earthscan, London (ranked 5th in the University of Cambridge Programme for Sustainability Leadership 'Top 40 books of 2010').
11 Desha, C. and Hargroves, K. (2012) 'Applying threshold learning theory to teach sustainable business practice in post-graduate engineering education', in proceedings of the 2012 American Society for Engineering Education Conference, Texas, 10–13 July.
12 Boyd, J. (2010) 'Acknowledgements' in UNESCO (2010) *Engineering: Issues, challenges and opportunities for development*, produced in conjunction with the World Federation of Engineering Organisations (WFEO), the International Council of Academies of Engineering and Technological Sciences (CAETS) and the International Federation of Consulting Engineers (FIDIC), UNESCO Publishing, Paris.
13 Desha, C. and Hargroves, K. (2009) 'Surveying the state of higher education in energy efficiency in Australian engineering curriculum', *Journal of Cleaner Production*, Vol 18, No 7, pp652–658.
14 Lattuca, L., Terenzine, P. and Fredricks Volkwein, J. (2006) *Engineering Change: A study of the impact of EC2000, Centre for the Study of Higher Education*, The Pennsylvania State University, ABET, Baltimore, MD.
15 Desha, C., Hargroves, K. and Reeves, A. (2009) *Engineering Curriculum Renewal for EE: Barriers and benefits analysis*, report to the National Framework for Energy Efficiency, The Natural Edge Project (TNEP), Australia.
16 GHD (2010) *Long Term Training Strategy for the Development of EE Assessment Skills*, final report to the National Framework for EE, Melbourne.
17 DRET (2011) Research Project 2 'Energy efficiency resources for undergraduate engineering education', University of Adelaide, report to the Energy Efficiency Advisory Group, Canberra.
18 Desha, C., Hargroves, K. and El Baghdadi, O. (2012) 'Review of postgraduate energy efficiency course content and recommendations for use of existing course: Vocational graduate certificate in building energy analysis (non-residential)', report to the National Framework for Energy Efficiency, The Natural Edge Project (TNEP), Australia.
19 UNEP, WFEO, WBCSD, ENPC (1997) 'Engineering education and training for sustainable development', report of the joint UNEP, WFEO, WBCSD, ENPC Conference, Paris, France, 24–26 September 1997, p42.
20 Azapagic, A., Perdan, S. and Shallcross, D. (2005) 'How much do engineering students know about sustainable development? The findings of an international survey and possible implications for the engineering curriculum', *European Journal of Engineering Education*, Vol 30, No 1, pp1–19.

21 Azapagic, A., Perdan, S. and Shallcross, D. (2005) 'How much do engineering students know about sustainable development? The findings of an international survey and possible implications for the engineering curriculum', *European Journal of Engineering Education*, Vol 30, No 1, p1.
22 Thomas, I. and Nicita, J. (2002) 'Sustainability education in Australian universities', *Environmental Education Research*, Vol 8, No 4, pp475–492.
23 The Alliance for Global Sustainability (2006) *The Observatory: Status of engineering education for sustainable development in European higher education*, EfS-Observatory, Technical University of Catalonia, Spain, p3, www.upc.edu./eesd-observatory/, accessed 17 July 2008.
24 Meddings, L. and Thorne, T. (2008) 'Engineers of the 21st Century: Engineering Education Project "What I wish I'd learnt at university"', report to the Forum for the Future's Engineers of the 21st Century programme, ARUP, Australia.
25 Desha, C. and Hargroves, K. (2009) 'Surveying the state of higher education in energy efficiency, in Australian engineering curriculum', *Journal of Cleaner Production*, Vol 18, No 7, pp652–658.
26 Allen, D., Allenby, B., Bridges, M., Crittenden, J., Davidson, C., Hendrickson, C., Matthews, C., Murphy, C. and Pijawka, D. (2008) *Benchmarking Sustainability Engineering Education: Final report*, EPA Grant X3-83235101-0, Centre for Sustainable Engineering, Pittsburgh, America, www.csengin.org/benchmark.htm, accessed 12 March 2009.
27 Allen, D., Allenby, B., Bridges, M., Crittenden, J., Davidson, C., Hendrickson, C., Matthews, C., Murphy, C. and Pijawka, D. (2008) *Benchmarking Sustainability Engineering Education: Final report*, EPA Grant X3-83235101-0, Centre for Sustainable Engineering, Pittsburgh, America, p3, www.csengin.org/benchmark.htm, accessed 12 March 2009.
28 The Alliance for Global Sustainability (2008) *The Observatory: Status of engineering education for sustainable development in European higher education*, EfS-Observatory, Technical University of Catalonia, Spain.
29 Department of Resources, Energy and Tourism (2011) *Energy Efficiency Resources for Undergraduate Engineering Education*, University of Adelaide, Australia.
30 Jorgensen, U. (2007) 'Historical accounts of engineering education', in Crawley, E., Malmqvist, J., Ostlund, S. and Brodeur, D. (eds) *Rethinking Engineering Education: The CDIO approach*, Springer Press, New York.
31 NAE (2005) 'Educating the engineer of 2020: Adapting engineering education to the new century', Committee on the Engineer of 2020, Phase II, Committee on Engineering Education, National Academy of Engineering of the National Academies, Washington DC.
32 Marjoram, T. (2006) 'Report and recommendations of workshop', International Workshop: Engineering Education for Sustainable Development, Tsinghua University, Beijing, 31 October–2 November, 2006, www.wfeo.org/documents/download/EfSTsinghuaRecommsActionNov06V3.doc, accessed 6 August 2008a.
33 The Royal Academy of Engineering (2007) *Educating Engineers for the 21st Century*, The Royal Academy of Engineering, London.
34 HEFC (2007) 'HEFC strategic review of sustainable development in higher education in England', report to the Higher Education Funding Council for England by the Policy Studies Institute, PA Consulting Group and the Centre for Research in Education and the Environment.
35 The Chinese Academy of Engineering (2004) The International Forum on Engineering Technology and Sustainable Development, held in conjunction with the 8th East Asia Round Table Meeting of Engineering Academics Suzhou, China, 31 October–1 November, 2004.
36 King, R. (2008) 'Addressing the supply and quality of engineering graduates for the new century', report to the Carrick Institute for Learning and Teaching in Higher Education Ltd, Sydney.
37 Institution of Engineers Australia (1996) *Changing the Culture: Engineering education into the future*, Institution of Engineers Australia, Canberra.
38 Sauvé, L. (1996) 'Environmental education and sustainable development: Further appraisal', *Canadian Journal of Environmental Education*, Vol 1, pp1–34.
39 Fien, J. (2002) 'Advancing sustainability in higher education – Issues and opportunities for research', *International Journal of Sustainability in Higher Education*, Vol 3, No 3, pp243–253.
40 Leal Filho, W. (ed.) (2002) *Teaching Sustainability at Universities: Towards curriculum greening*, Peter Lang.
41 Sterling, S. (2003) *Sustainable Education: Re-visioning learning and change*, Schumacher Briefings 6, Green Books, Darlington.
42 Corcoran, P. and Wals, A. (eds) (2004) *Higher Education and the Challenges of Sustainability: Problematics, promise, and practice*, Kluwer Academic Publishers, The Netherlands.
43 Parkin, S., Johnston, A., Buckland, H., Brookes, F. and White, E. (2004) *Learning and Skills for Sustainable Development: Developing a sustainability literate society – guidance for higher education institutions*, Higher Education Partnership for Sustainability (HEPS)/Forum for the Future, London.

44 Cortese, A. (2003) 'The critical role of higher education in creating a sustainable future', *Planning for Higher Education*, March–May 2003.
45 Blewitt, J. and Cullingford, C. (2004) *The Sustainability Curriculum: The challenge for higher education*, Earthscan, London.
46 Dawe, G., Jucker, R. and Martin, S. (2005) *Sustainable Development in Higher Education: Current practice and future developments*, a report for The Higher Education Academy, November 2005.
47 Jansen, L. (2002) 'The challenge of sustainable development', *Journal of Cleaner Production*, Vol 11, No 3, pp231–245.
48 Mulder, K. (2005) 'Engineering education in sustainable development', guest editorial, *International Journal of Sustainability in Higher Education*, Vol 5, No 3.
49 Mulder, K. (2004) 'Engineering education in sustainable development: Sustainability as a tool to open up the windows of engineering education', *International Journal of Business Strategy and the Environment,* Vol 13, No 4, pp275–285.
50 Ferrer-Balas, D. and Mulder, K. (2005) 'Engineering education in sustainable development', guest editorial, *International Journal of Sustainability in Higher Education*, Vol 6, No 3.
51 Holmberg, J., Svanström, M., Peet, D., Mulder, K., Ferrer-Balas, D. and Segalàs, J. (2008) 'Embedding sustainability in higher education through interaction with lecturers: Case studies from three European technical universities', *European Journal of Engineering Education*, Vol 33, No 3, pp271–282.
52 Allenby, B., Folsom Murphy, C., Allen, D. and Davidson, C. (2009) 'Sustainable engineering education in the United States', *Journal of Sustainability Science*, Vol 4, No 1, pp7–15.
53 Carroll, W. (1993) 'World Engineering Partnership for Sustainable Development', *Journal of Professional Issues in Engineering Education and Practice*, Vol 119, No 3, pp238–240.
54 Cortese, A. (1997) 'Engineering education for a sustainable future', Engineering Education and Training for Sustainable Development Conference, Paris, France.
55 Crofton, F. (2000) 'Educating for sustainability: opportunities in undergraduate engineering', *Journal of Cleaner Production*, Vol 8, pp397–405.
56 Ashford, N. (2004) 'Major challenges to engineering education for sustainable development: What has to change to make it creative, effective, and acceptable to the established disciplines?', *International Journal of Sustainability in Higher Education*, Vol 4, No 3, pp239–250.
57 Azapagic, A., Perdan, S. and Clift, R. (2000) 'Teaching sustainable development to engineering students', *International Journal of Sustainability in Higher Education*, Vol 1, No 3, pp267–279.
58 Azapagic, A., Perdan, S. and Clift, R. (eds) (2004) *Sustainable Development in Practice: Case studies for engineers and scientists*, Wiley, New York.
59 McKeown, R., Hopkins, C. and Rizzi, R. (2002) *Education for Sustainable Development Toolkit*, Version 2, July 2002, www.esdtoolkit.org, accessed 20 December 2008.
60 Pritchard, J. and Baillie, C. (2006) 'How can engineering education contribute to a sustainable future?', *European Journal of Engineering Education*, Vol 31, No 5, pp555–565.
61 Allenby, B., Allen, D. and Davidson, C. (2007) 'Teaching sustainable engineering', *Journal of Industrial Ecology*, Vol 11, No 1, pp8–10.
62 Timpson, W., Dunbar, B., Kimmel, G., Bruyere, B., Newman, P. and Mizia, H. (2009) *147 Practical Tips for Teaching Sustainability: Connecting the environment, the economy, and society*, Atwood Publishing.
63 Newman, J. and Fernandez, L. (2007) 'Strategies for institutionalizing sustainability in higher education', report on the Northeast Campus Sustainability Consortium 3rd Annual Conference and International Symposium, Yale School of Forestry and Environmental Studies, April 2007.
64 Steinemann, A. (2003) 'Implementing sustainable development through problem-based learning: Pedagogy and practice', *Journal of Professional Issues in Engineering Education and Practice*, Vol 129, No 4, pp216–224.
65 Lehmann, M., Christensen, P., Du, X. and Thrane, M. (2008) 'Problem-oriented and project-based learning (POPBL) as an innovative learning strategy for sustainable development in engineering education', *European Journal of Engineering Education*, Vol 33, No 3, pp283–295.
66 Crawley, E., Malmqvist, J., Östlund, S. and Brodeur, D. (2007) *Rethinking Engineering Education: The CDIO approach*, Springer, New York.
67 Rowe, D. (2007) 'Policy forum sustainability: Education for a sustainable future', *Science*, Vol 317, No 5836, pp323–324.
68 Stephens, J. and Graham, A. (2009) 'Toward an empirical research agenda for sustainability in higher education: exploring the transition management framework', *Journal of Cleaner Production*, Vol 18, No 7, pp611–618.
69 Steinfeld, J. and Takashi, M. (2009) 'Special Feature Editorial: Education for sustainable development – the challenge of trans-disciplinarity', *Journal of Sustainability Science*, Vol 4, No 1, pp1–2.

70 Holdsworth, S., Wyborn, C., Bekessy, S. and Thomas, I. (2008) 'Professional development for education for sustainability: How advanced are Australian universities?', *International Journal of Sustainability in Higher Education*, Vol 9, No 2, pp131–146.

71 Holmberg, J., Svanström, M., Peet, D., Mulder, K., Ferrer-Balas, D. and Segalàs, J. (2008) 'Embedding Sustainability in higher education through interaction with lecturers: Case studies from three European technical universities', *European Journal of Engineering Education*, Vol 33, No 3, pp271–282; The Alliance for Global Sustainability (2006) *The Observatory: Status of engineering education for sustainable development in European higher education*, EfS-Observatory, Technical University of Catalonia, Spain, p8, www.upc.edu./eesd-observatory/, accessed 17 July 2008.

72 World Federation of Engineering Organisations (1997) 'Achievements – educational programs', www.wfeo.org/index.php?page=archives, accessed 2 March 2010.

73 National Academy of Engineering (1998) 'The urgency of engineering education reform', *The Bridge – Frontiers of Engineering*, Vol 28, No 1, www.nae.edu/nae/bridgecom.nsf/weblinks/NAEW-4NHMKV?Open Document, accessed 8 February 2009.

74 NAE (2004) *The Engineer of 2020: Visions of engineering in the new century*, National Academy of Engineering, Washington, DC, USA, The National Academies Press.

75 See The Natural Edge Project website.

76 Desha, C., Hargroves, K., Smith, M., Stasinopoulos, P., Stephens, R. and Hargroves, S. (2007) *State of Education for Energy Efficiency in Australian Engineering Education – Summary of questionnaire results*, The Natural Edge Project (TNEP), Australia.

77 Smith, M., Hargroves, K., Stasinopoulos, P., Stephens, R., Desha, C. and Hargroves, S. (2007) *Energy Transformed: Sustainable energy solutions for climate change mitigation*, The Natural Edge Project, CSIRO, and Griffith University, Australia.

78 Desha, C., Hargroves, K., Smith, M., Stasinopoulos, P., Stephens, R., and Hargroves, S. (2007) *State of Education for Energy Efficiency in Australian Engineering Education – Summary of questionnaire results*, The Natural Edge Project (TNEP), Australia.

79 Desha, C., Hargroves, K., and Reeves, A. (2009) 'Engineering curriculum renewal for EE: Barriers and benefits analysis', report to the National Framework for Energy Efficiency, The Natural Edge Project (TNEP), Australia.

80 GHD (2010) 'Long term training strategy for the development of energy efficiency assessment skills', final report to the National Framework for EE, Melbourne.

81 DRET (2011) Research Project 1 'Energy efficiency graduate attributes project', Queensland University of Technology, report to the Energy Efficiency Advisory Group, Canberra.

82 DRET (2011) Research Project 2 'Energy efficiency resources for undergraduate engineering education', University of Adelaide, report to the Energy Efficiency Advisory Group, Canberra.

83 Desha, C., Hargroves, K. and El Baghdadi, O. (2012) 'Review of postgraduate energy efficiency course content and recommendations for use of existing course: Vocational graduate certificate in building energy analysis (non-residential)', report to the National Framework for Energy Efficiency, The Natural Edge Project (TNEP), Australia.

84 Von Weisacker, E., Hargroves, K., Smith, M. and Desha, C. (2010) *Factor 5: Achieving 80 percent improvements in resource productivity*, Earthscan, London.

85 Stasinopoulos, P., Hargroves, K., Smith, M. and Desha, C. (2010) *Whole System Design: A systems approach to engineering*, Earthscan, London.

86 Birkeland, J. (2010) *Net Positive Development*, Earthscan, London.

87 RET (2011) Research Project 1 'Energy efficiency graduate attributes project', Queensland University of Technology, report to the Energy Efficiency Advisory Group, Canberra.

88 RET (2011) Research Project 2 'Energy efficiency resources for undergraduate engineering education', University of Adelaide, report to the Energy Efficiency Advisory Group, Canberra.

89 Stern, N. (2006) *The Stern Review: The economics of climate change*, Cambridge University Press, Cambridge.

90 Barnett, R. and Coate, K. (2004) *Engaging the Curriculum in Higher Education*, Society for Research into Higher Education (SRHE) and Open University Press, Buckingham, UK.

91 Carew, A. and Cooper, P. (2008) 'Engineering curriculum review: Processes, frameworks and tools', Proceedings of the Annual Meeting of SEFI July, Aalborg, Denmark.

92 King, R. (2008) 'Addressing the supply and quality of engineering graduates for the new century', report on Engineering Education in Australia for the Carrick Institute for Teaching and Learning in Higher Education.

93 Hargroves, K. and Smith, M. (2005) *The Natural Advantage of Nations: Business opportunities, innovation and governance in the 21st century*, Earthscan, London, p61.

94 Marjoram, T. (ed.) (2012) *UNESCO Report: Engineering – Issues and challenges for development*, produced in conjunction with World Federation of Engineering Organisations (WFEO), International Council Academies of Engineering and Technological Sciences (CAETS), International Federation of Consulting Engineers (FIDIC).
95 The Business Council Sustainable Growth Task Force (2004) *Beyond the Horizon: Short-termism in Australia*, Business Council of Australia, www.bca.com.au/Content/101402.aspx, accessed 4 September 2008.
96 The Business Council Sustainable Growth Task Force (2004) *Beyond the Horizon: Short-termism in Australia*, Business Council of Australia, www.bca.com.au/Content/101402.aspx, accessed 4 September 2008.
97 Sudan Virtual Engineering Library (SudVEL) (n.d.) www.sudvel-uofk.net/, accessed 25 October 2008. This pilot project was supported by UNESCO in partnership with the Foundation Ecole d'Ingenieurs (EPF) through the International Institute of Women in Engineering (IIWE), Australian Virtual Engineering Library (AVEL-SKN), World Federation of Engineering Organizations (WFEO), Sustainable Alternatives Network (SANet) and the Natural Edge Project (TNEP).
98 King, R. (2008) 'Addressing the supply and quality of engineering graduates for the new century', report to the Carrick Institute for Learning and Teaching in Higher Education Ltd, Sydney, pp63–64.
99 Stern, N. (2006) *The Stern Review: The economics of climate change*, Cambridge University Press, Cambridge.
100 Heywood, J. (2005) *Engineering Education Research and Development in Curriculum and Instruction*, Institute of Electrical and Electronics Engineers (IEEE), p459.
101 Wulf, W. (1998) 'The urgency of engineering education reform', *The Bridge – Frontiers of Engineering*, Vol 28, No 1, www.nae.edu/nae/bridgecom.nsf/weblinks/NAEW-4NHMKV?OpenDocument, accessed 9 April 2009.
102 Heywood, J. (2005) *Engineering Education Research and Development in Curriculum and Instruction*, Institute of Electrical and Electronics Engineers (IEEE), pp5–6.
103 Beavis, C. (1997) 'Lovely literature: Teacher subjectivity and curriculum change', 1997 AARE Conference, The Australian Association for Research in Education, Brisbane; King, R. (2008) 'Addressing the supply and quality of engineering graduates for the new century', report to the Carrick Institute for Learning and Teaching in Higher Education Ltd, Sydney.
104 Mulder, K. (2008) personal communication with Dr Karel Mulder, Delft University of Technology, 28 August 2008.
105 Association for the Advancement of Sustainability in Higher Education (AASHE), 'Professional development opportunities', http://www.aashe.org/profdev/profdev.php, accessed 3 February 2010.
106 Greenwood, P. (2007) 'Helping small business survive the skill shortage – an Australian perspective', a presentation by Peter Greenwood from the Australian Institution of Engineers, at the International Conference Supporting Small- and Medium-Sized Enterprises in Engineering and Technological Innovation Activity, Cracow, Poland. Hosted by WFEO-CET in cooperation with the Polish Federation of Engineering Associations.
107 Desha, C., Hargroves, K. and El Baghdadi, O. (2012) 'Review of postgraduate energy efficiency course content and recommendations for use of existing course: Vocational graduate certificate in building energy analysis (non-residential)', report to the National Framework for Energy Efficiency, The Natural Edge Project (TNEP), Australia.
108 Smith, M., Hargroves, K. and Desha, C. (2010) Table 6.1 'Energy consumption reduction and renewable energy targets set by the ten largest economies' in *Cents and Sustainability – Securing our common future by decoupling economic growth from environmental pressures*, Earthscan, London.
109 NDRC (2007) *National Climate Change Program*, National Development and Reform Commission, China.
110 Liu, Y. (2005) 'Shanghai embarks on 100,000 solar roofs initiative', China watch, 10 November, Worldwatch Institute, Washington, DC.
111 European Union (2002) 'Directive on the restriction of the use of certain hazardous substances in electrical and electronic equipment', European Union.
112 Environment Agency (2006) Statutory Instrument 2006 No. 3315: The Waste Electrical and Electronic Equipment (Waste Management Licensing) (England and Wales) Regulations 2006, Environment Agency, UK, www.opsi.gov.uk/si/si2006/20063315.htm, accessed 6 August 2010.
113 UN Global Compact, UNEP and WBCSD (2007) 'Caring for climate: The business leadership platform', Business leaders statement issued at the UN Global Compact Leaders Summit, Geneva July, 2007, www.unglobalcompact.org/Issues/Environment/Climate_Change/index.html, accessed 3 July 2008.
114 United Nations Global Compact (n.d.) 'What is the Global Compact?' http://www.unglobalcompact.org/, accessed 19 January 2011.
115 World Business Council for Sustainable Development (n.d.), www.wbcsd.org/, accessed 31 July 2008.

116 WBCSD (2010) *Vision 2050: The new agenda for business*, World Business Council for Sustainable Development.
117 World Business Council for Sustainable Development (2010) 'Vision 2050 Lays a Pathway to Sustainable Living Within Planet', document details, http://www.wbcsd.org/Plugins/DocSearch/details.asp?DocTypeId=33&ObjectId=MzcOMDE, accessed 19 January 2011.
118 Porrit, J. (2007) 'The future for engineering education', keynote speech, Global Sustainability Forum, http://www3.imperial.ac.uk/globalsustainability, accessed 20 August 2008.
119 van der Veer, J. (2006) 'The importance and need for EfS: Re-engineering the engineers', in The Alliance for Global Sustainability, *The Observatory: Status of Engineering Education for Sustainable Development in European Higher Education, 2006*, EfS-Observatory, Technical University of Catalonia, Spain, p8, www.upc.edu./eesd-observatory/, accessed 17 July 2008.
120 Environmental Leader (2009) 'Sustainability knowledge in demand', *Environmental Leader*, 25 March 2009; James, R. (2002) 'Students' changing expectations of higher education and the "consequences of mismatches with reality"', in *Responding to Student Expectations*, Paris, OECD.
121 Meddings, L. and Thorne, T. (2008) 'Engineers of the 21st century: Engineering education project "What I wish I'd learnt at university"', report to the Forum for the Future's Engineers of the 21st Century programme, Final Report, Ove Arup and Partners, London.
122 Meddings, L. and Thorne, T. (2008) 'Engineers of the 21st century: Engineering education project "What I wish I'd learnt at university"', report to the Forum for the Future's Engineers of the 21st Century programme, Final Report, Ove Arup and Partners, London.
123 Leckstrom, J. (2008) 'Kaplan releases new College Guide 2009 featuring 25 green colleges and 10 hot green ...' *PRNewswire,* Tuesday 5 August 2008, New York, www.reuters.com/article/pressRelease/idUS135812+05-Aug-2008+PRN20080805, accessed 30 October 2008.
124 Alpay, E., Ahearn, A., Graham, R. and Bull, A. (2008) 'Student enthusiasm for engineering: Charting changes in student aspirations and motivation', *European Journal of Engineering Education*, Vol 33, No 5, pp573–585.
125 Alpay, E. (2008) Personal communication with Dr Esat Alpay, Senior Lecturer in Engineering Education, Imperial College, London, 13 September 2008.
126 Grant, M. (2008) Personal communication with Ms Michelle Grant, Project Manager, ETH Sustainability, Zurich, 14 August 2008.
127 Marjoram, T. (2010) 'Foreword' in Desha, C. and Hargroves, K. (2010) *Engineering Education and Sustainable Development – A guide to rapid curriculum renewal*, Earthscan, London.
128 IFEES (2008) 'Aligning engineering education initiatives for a knowledge economy', 2nd IFEES Global Engineering Education Summit, Cape Town, South Africa, 19–10 October 2008, www.aeea.co.za/i/i//IFEES2008SummitProgram.pdf, accessed 1 September 2008.
129 Higher Education Funding Council for England (2010) 'Leading sustainable development in HE,' http://www.hefce.ac.uk/lgm/build/lgmfund/lead.htm, accessed 19 January 2011.
130 Scottish Funding Council, Sustainable Development, http://www.sfc.ac.uk/effective_institutions/sustainable development/sustainable_development.aspx, accessed 20 January 2011.
131 Higher Education Funding Council for Wales (HEFCW), Education for Sustainable Development and Global Citizenship, http://www.hefcw.ac.uk/about_he_in_wales/wag_priorities_and_policies/edu_sustainable_dev_glob_cit.aspx, accessed 20 January 2010.
132 Department of Education, Employment and Workplace Relations (n.d.) 'Education Investment Fund', http://www.deewr.gov.au/HigherEducation/Programs/EIF/Pages/FutureRounds.aspx, accessed 19 January 2011.
133 National Climate Change Adaptation Research Facility. Research Grants, http://www.nccarf.edu.au/, accessed 20 January 2011.
134 Sustainable Agriculture Research and Education (SARE), For Educators, http://www.sare.org/coreinfo/pdp.htm, accessed 20 January 2011.
135 Department of Education, United States of America, Budget FY2010, http://www2.ed.gov/about/overview/budget/budget11/justifications/u-highered.pdf, accessed 21 January 2011.
136 US Department of Energy, RE-ENERGYSE, http://energyprograms.energy.gov/initiativesRenergyse.html, accessed 21 January 2011.
137 National Science Foundation (NSF), Climate Change Education (CCE): Climate Change Education Partnership (CCEP) Program, http://www.nsf.gov/funding/pgm_summ.jsp?pims_id=503465&org=ERE&sel_org=ERE&from=fund, accessed 21 January 2011.
138 National Science Foundation (NSF), National STEM Education Distributed Learning (NSDL), http://www.nsf.gov/funding/pgm_summ.jsp?pims_id=5487&org=ERE&sel_org=ERE&from=fund, accessed 21 January 2011.

139 King, R. (2008) 'Addressing the supply and quality of engineering graduates for the new century', report to the Carrick Institute for Learning and Teaching in Higher Education Ltd, Sydney, p43.
140 Kamp, L. (2006) 'Engineering education in sustainable development at Delft University of Technology', *Journal of Cleaner Production*, Vol 14, No 9–11, pp928–931.
141 Mulder, K. (2004) 'Engineering education in sustainable development: Sustainability as a tool to open up the windows of engineering education', *International Journal of Business Strategy and the Environment,* Vol 13, No 4, pp275–285.
142 Lundqvist, U. and Svanström, M. (2008) 'Inventory of content in basic courses in environment and sustainable development at Chalmers University of Technology in Sweden', *European Journal of Engineering Education*, Vol 33, No 3, pp355–364.
143 Fenner, R., Ainger, C., Cruickshank, H. and Guthrie, P. (2005) 'Embedding sustainable development at Cambridge University Engineering Department', *International Journal of Sustainability in Higher Education*, Vol 6, No 3, pp229–241.
144 Humphries-Smith, T. (2008) 'Sustainable design and the design curriculum', *Journal of Design Research*, Vol 7, No 3, pp259–274.
145 Lozano, R. (2009) 'Diffusion of sustainable development in universities' curricula: An empirical example from Cardiff University', *Journal of Cleaner Production*, Vol 18, No 7, pp637–644.
146 Fletcher, J., Drahun, G., Davies, P. and Knowles, P. (2008) 'The teaching of sustainable development at Aston University', proceedings of the 2008 International Symposium for Engineering Education, Dublin City University, Ireland.
147 Ferrer-Balas, D., Bruno, J., de Mingo, M. and Sans, R. (2004) 'Advances in education transformation towards sustainable development at the Technical University of Catalonia, Barcelona', *International Journal of Sustainability in Higher Education*, Vol 5, No 3, pp251–266.
148 Epstein, A., Bras, R. and Bowring, S. (2009) 'Building a freshman-year foundation for sustainability studies: Terrascope, a case study', *Journal of Sustainability Science*, Vol 4, No 1, pp37–43.
149 Mihelcic, J., Phillips, L. and Watkins, D. (2006) 'Integrating a global perspective into education and research: engineering international sustainable development', *Environmental Engineering Science*, Vol 23, No 3, pp426–438.
150 Lozano-Garcià, F., Gàndara, G., Perrni, O., Manzano, M., Hernàndez, D. and Huisingh, D. (2008) 'Capacity building: a course on sustainable development to educate the educators', *International Journal of Sustainability in Higher Education*, Vol 9, No 3, pp257–281.
151 Wright, S., Habit, E., Adlerstein, S., Parra, O. and Semrau, J. (2009) 'Graham Scholars Program: Sustainability education through an interdisciplinary international case study', *Journal of Sustainability Science*, Vol 4, No 1, pp29–36.
152 Onuki, M. and Mino, T. (2009) 'Sustainability education and a new master's degree, the master of sustainability science: the Graduate Program in Sustainability Science (GPSS) at the University of Tokyo', *Journal of Sustainability Science*, Vol 4, No 1, pp55–59.
153 Uwasu, M., Yabar, H., Hara, K., Simoda, Y. and Saijo, T. (2009) 'Educational initiative of Osaka University in sustainability science: Mobilizing science and technology towards sustainability', *Journal of Sustainability Science*, Vol 4, No 1, pp45–53.
154 Xu, K. (2008) 'Engineering education and technology in a fast-developing China', *Journal of Technology in Society,* Vol 30, pp265–274.
155 Olorunfemi, A. and Dahunsi, B. (2004) 'Towards a sustainable engineering education and practice in Nigeria', proceedings of the SEFI 2004 Annual Congress, The Golden Opportunity for Engineering Education.
156 Ramjeawon, T. (2008) 'Sustainable development: The enabling role of the engineer' in Institution of Engineers Mauritius 60th Anniversary Commemorative Issue, *The Journal of the Institution of Engineers Mauritius*, October 2008, pp12–17.
157 Davis, R. and Savage, S. (2008) 'Built environment and design in Australia: Challenges and opportunities for professional education', proceedings of the 20th Australasian Association of Engineering Education conference, 6–9 December, Adelaide.
158 Goh, S. (2009) 'A new paradigm for professional development framework and curriculum renewal in engineering management education: A proposal for reform', proceedings of the 20th Australasian Association of Engineering Education conference, 6–9 December, Adelaide.
159 Bryce, P., Johnston, S. and Yasukawa, K. (2004) 'Implementing a program in sustainability for engineers at University of Technology, Sydney: A story of intersecting agendas', *International Journal of Sustainability in Higher Education*, Vol 5, No 3, pp267–277.
160 Mitchell, C. (2000) 'Integrating sustainability in chemical engineering practice and education: Concentricity and its consequences', *Trans IChemE*, Vol 78, Part B, July 2000.

161 Carew, A. and Therese, S. (2007) 'EMAP: Outcomes from regional forums on graduate attributes in engineering', proceedings of the 2007 AaeE Conference, Melbourne.
162 Koth, B. and Woodward, M. (2009) 'Civil engineering education for sustainability: Faculty perceptions and result of an Australian course audit', proceedings of the 20th Australasian Association of Engineering Education conference, 6–9 December, Adelaide, pp776–782.
163 Daniell, T. and Maier, H. (2005) 'Embedding sustainability in civil and environmental engineering courses', proceedings of the 2005 ASEE/AaeE 4th Global Colloquium on Engineering Education, Sydney.
164 Carew, A. and Lindsay, E. (2009) 'Curriculum lifeboat: A process for rationalising engineering course content', proceedings of the 20th Australasian Association of Engineering Education conference, 6–9 December, Adelaide.
165 Short case study summaries of the University of Technology Sydney (UTS), and the Royal Melbourne Institute of Technology (RMIT) are provided through the UTS Centre for Learning and Teaching website, http://www.clt.uts.edu.au/Theme.ident.grad.att.htm, accessed 16 July 2008.
166 Monash Sustainability Institute (n.d.) 'Embedding sustainability through unit renewal: A professional development program for academic staff', www.monash.edu/research/sustainability-institute/programs/efs-unit-renewal.html.
167 Huisingh, D. (2008) Personal communications with Professor Don Huisingh, Senior Scientist in Sustainable Development, the University of Tennessee, 6 November 2008.
168 EfS (2004) 'Engineering Education for Sustainable Development: Declaration of Barcelona', International Conference Barcelona, 27–29 October, 2004, www.upc.edu/eesd-observatory/BCN%20Declaration%20 EfS_english.pdf, accessed 12 July 2008.
169 American Society of Engineering Education (2008) 7th Global Colloquium on Engineering Education, American Society of Engineering Education, Cape Town, South Africa. Program, www.asee.org/ conferences/international/2008/Program.cfm, accessed 1 September 2008.
170 International Network for Engineering Education and Research (2008) 'ICEE 2008: New challenges in engineering education and research in the 21st century', International Network for Engineering Education and Research, Pecs-Budapest, Hungary, 27–31 July 2008, http://icee2008hungary.net/main.php?menu=1, accessed 1 September 2008.
171 IIES (1992) 'Arusha Declaration: Statement by the World Federation of Engineering Organisations (WFEO) Environment and Development', The UNCED Conference, www.iies.es/FMOI-WFEO/ desarrollosostenible/main/assets/ArushaDeclaration.doc, accessed 31 July 2008.
172 World Commission on Environment and Development (1987) *Our Common Future*, Oxford University Press, Oxford.
173 Elms, D. and Wilkinson, D. (1995) 'The environmentally educated engineer: Focus on fundamentals,' Centre for Advanced Engineering, University of Canterbury, New Zealand.
174 UNEP, WFEO, WBCSD, ENPC (1997) 'Engineering education and training for sustainable development', report of the joint UNEP, WFEO, WBCSD, ENPC Conference, Paris, France, 24–26 September 1997, p4.
175 Sustainability Taskforce (1997) *The Engineer's Response to Sustainable Development*, World Federation of Engineering Organizations.
176 National Academy of Sciences (2002) 'Dialogue on the engineer's role in sustainable development – Johannesburg and beyond', National Academy of Sciences in affiliation with the American Society of Civil Engineers, Engineers International Round Table, and WFEO-ComTech.
177 EfS (2004) 'Engineering Education for Sustainable Development: Declaration of Barcelona', International Conference Barcelona, 27–29 October, 2004, www.upc.edu/eesd-observatory/BCN%20Declaration% 20EfS_english.pdf, accessed 12 July 2008.
178 UNESCO (2004) 'The Shanghai Declaration on Engineering and the Sustainable Future', World Engineers Convention, Shanghai, 5 November 2004, http://portal.unesco.org/science/en/ev.php-URL_ID= 4146&URL_DO=DO_PRINTPAGE&URL_SECTION=201.html, accessed 6 August 2008.
179 Marjoram, T. (2006) 'Report and recommendations of the workshop', International Workshop: Engineering Education for Sustainable Development, Tsinghua University, Beijing, 31 October–2 November, 2006, www.wfeo.org/documents/download/EfSTsinghuaRecommsActionNov06V3.doc, accessed 6 August 2008.
180 World Commission on Environment and Development (1987) *Our Common Future*, Oxford University Press, Oxford.
181 Carroll, W. (1992) 'World Engineering Partnership, for Sustainable Development', *Journal of Professional Issues in Engineering, Education and Practice*, Vol 119, No 3, pp238–240.
182 Ridley, T. and Ir. Lee Yee-Chong, D. (2002) 'Engineering and technology for sustainable development', 2002 World Summit for Sustainable Development, Johannesburg and World Federation of Engineering Organizations Committee on Technology, www.wfeo-comtech.org, accessed 18 July 2008.

183 JCEETSD (1997) Joint conference report, engineering education and training for sustainable development, Joint UNEP, WFEO, WBCSD, ENPC Conference, Paris, France, 24–26 September, 1997.
184 Sustainability Alliance (2003) 'Learning the sustainability lesson: Environmental audit committee inquiry into education for sustainable development (ESD)', response from members of The Sustainability Alliance, http://www.sustainabilityalliance.org.uk/index.asp.
185 EfS (2004) 'Engineering Education for Sustainable Development: Declaration of Barcelona', Engineering Education for Sustainable Development International Conference, Barcelona, Spain, 27–29 October, 2004, www.upc.edu/eesd-observatory/BCN%20Declaration%20EfS_english.pdf.
186 UNESCO (2004) The Shanghai Declaration on Engineering and the Sustainable Future, World Engineers Convention, Shanghai, China, 5 November 2004, www.eccenet.org/Activities/Environ/ENV-Shanghai.pdf.
187 WFEO (1997) *The Engineer's Response to Sustainable Development*, The World Federation of Engineering Organizations.
188 King, J. (2007) 'Introduction', in RAE Working Party, *Educating Engineers for the 21st Century*, RAE, London.
189 Taylor, P. (2008) Personal communications with Mr Peter Taylor, CEO, Institution of Engineers Australia.
190 Xu, K. (2008) 'Engineering education and technology in a fast-developing China', *Technology in Society*, Vol 30, p268.
191 Grear, B. (2008) Personal communications with Mr Barry Grear, Former President of the Institution of Engineers Australia, and incoming President for the World Federation of Engineering Organizations, 29 August 2008.
192 Van Oortmerssen, G. (2008) 'Statement – International Council of Academies of Engineering and Technological Sciences', in *Engineering: Issues, challenges and opportunities for development*, UNESCO Report, p7.
193 WFEO (2001) The WFEO Model Code of Ethics, World Federation of Engineering Organizations, Tunis, Tunisia, www.wfeo.org/index.php?page=ethics.
194 Engineers Australia (2000) *Engineering Code of Ethics*, Institution of Engineers Australia, Canberra.
195 IPENZ (2005) *Code of Ethics*, Institution of Professional Engineers, New Zealand.
196 Engineers Ireland (2010) *Code of Ethics*, The Institution of Engineers Ireland.
197 FIDIC (2004) Policy Statement: Consulting Engineers and the Environment, International Federation of Consulting Engineers, http://www1.fidic.org/about/statement04.asp.
198 UNEP, undated. Environment Management and Performance, www.unepie.org/scp/business/emp/index.htm.
199 Carew, A. L. and Mitchell, C. A. (2006) 'Metaphors used by some engineering academics in Australia for understanding and explaining sustainability', *Environmental Education Research*, Vol 12, No 2, pp217–231.
200 Batterham, R. J. (2003) 'Ten years of sustainability: Where do we go from here?', *Chemical Engineering Science*, Vol 58, pp2167–2179.
201 JCEETSD (1997) Joint conference report, engineering education and training for sustainable development, Joint UNEP, WFEO, WBCSD, ENPC Conference, Paris, France, 24–26 September, 1997.
202 WFEO (2007) 'Engineers response to sustainable development', World Federation of Engineering Organizations.
203 RAE (2005) *Engineering for Sustainable Development: Guiding principles*, London, England.
204 WFEO (2001) The WFEO Model Code of Ethics, World Federation of Engineering Organizations, Tunis, Tunisia, www.wfeo.org/index.php?page=ethics.
205 NAE (2004) *The Engineer of 2020: Visions of engineering in the new century*, National Academy of Engineering, Washington, DC, The National Academies Press.
206 RAE (2005) *Engineering for Sustainable Development: Guiding principles*, London, England.
207 CCPE (2006) *Canadian Engineering Qualifications Board National Guideline on Environment and Sustainability*, Ottawa, ON, Canada, http://www.engineerscanada.ca/e/files/guideline_enviro_with.pdf.
208 IChemE (2007) *A Roadmap for 21st Century Chemical Engineering*, Rugby, England, http://www.icheme.org/roadmap2007.pdf; IChemE (2008) *Driving in the Right Direction: Technical strategy roadmap*, progress report, Rugby, England, Institution of Chemical Engineers.
209 Engineers Australia (2007) 'Engineers Australia Sustainability Charter', Canberra, Australia.
210 National Academy of Engineering (2008) 'Grand challenges for engineering', http://www.engineeringchallenges.org/, accessed 15 January 2011.
211 ECUK (2009) *Guidance on sustainability for the engineering profession*, London, England, www.engc.org.uk/documents/EC0018_SustainabilityGuide.pdf.
212 Engineers Australia (2000) *Engineering Code of Ethics*, Institution of Engineers Australia, Canberra.
213 ECUK (2009) *Guidance on Sustainability for the Engineering Profession*, London, England, www.engc.org.uk/documents/EC0018_SustainabilityGuide.pdf.

214 ECUK (2009) *Guidance on Sustainability for the Engineering Profession*, London, England, www.engc.org.uk/documents/EC0018_SustainabilityGuide.pdf.
215 Byrne, E., Desha, C., Fitzpatrick, J. and Hargroves, K. (2010) 'Engineering education for sustainable development: A review of international progress', in proceedings of the 2010 International Symposium on Engineering Education, Cork, Ireland.
216 King, R. (2008) 'Addressing the supply and quality of engineering graduates for the new century', report to the Carrick Institute for Learning and Teaching in Higher Education Ltd, Sydney, p29; Leonard, M., Beasley, D., Scales, K. and Elzinga, D. (1998) 'Planning for curriculum renewal and accreditation under ABET engineering criteria 2000', ASEE Annual Conference and Exposition, Seattle, Washington.
217 Phillips, W., Peterson, G. and Aberle, K. (2000) 'Quality assurance for engineering education in a changing world', *International Journal of Engineering Education*, Vol 15, No 2, pp97–103; Speicher, A. (1994) 'ASEE project report: Engineering education for a changing world using partnerships to respond to new needs in engineering education', *ASEE Prism*, pp20–27; Leonard, M. and Beasley, D. (1998) 'Planning for curriculum renewal and accreditation under ABET engineering criteria 2000', 1998 ASEE Annual Conference and Exposition, Seattle.
218 Brown, K. (1998) 'SARTOR 97: The background and the looming shake-out for university engineering departments', *Journal of Engineering Science Education,* Vol 7, No 1, pp41–48.
219 European Network for Accreditation of Engineering Education (ENAEE) (n.d.) 'Presentation – Accreditation/EUR-ACE', http://www.feani.org/webfeani/, accessed 21 December 2009.
220 Sustainability Alliance (2003) 'Learning the sustainability lesson: Environmental Audit Committee inquiry into education for sustainable development (ESD)', response from members of the Sustainability Alliance, http://www.sustainabilityalliance.org.uk/index.asp.
221 Lype, G. (2006) 'Engineering education: Can India overtake China?', *Rediff News*, 9 June 2007, www.rediff.com/money/2006/jun/09bspec.htm, accessed 3 July 2008.
222 The Chinese Academy of Engineering (2004) The International Forum on Engineering Technology and Sustainable Development, held in conjunction with the 8th East Asia Round Table Meeting of Engineering Academics Suzhou, China, 31 October–1 November, 2004.
223 McKinsey Global Institute (2005) *The Emerging Global Labor Market*, McKinsey Global Institute, www.mckinsey.com/mgi/publications/emerginggloballabormarket/index.asp, accessed 3 July 2008.
224 ABET (2007) *Criteria for Accrediting Engineering Programs: Effective for evaluations during the 2008–2009 accreditation cycle*, Engineering Accreditation Commission. Incorporates all changes approved by the ABET Board of Directors as of November 3, 2007.
225 UK Engineering Council (2004) *UK Standard for Professional Engineering Competence*, Chartered Engineer and Incorporated Engineer Standard, UK Engineering Council (reprinted in 2005).
226 The authors' research group, The Natural Edge Project, was incubated by Engineers Australia for three years before transferring to Griffith University and Australian National University in 2006.
227 ACED (2012) *Students, Graduates and Academic Staffing in Engineering: Trends to 2011*, Australian Council of Engineering Deans Inc., November 2012.
228 Engineers Australia (n.d.) 'Accreditation Management System', www.engineersaustralia.org.au/about-us/accreditation-management-system-professional-engineers, accessed 23 March 2013.
229 Engineers Australia (2011) *Stage 1 Competency Standard for Professional Engineers*, The Institution of Engineers Australia, Canberra.
230 International Engineering Alliance (n.d.) 'Introduction', www.washingtonaccord.org/, accessed 15 June 2008.
231 International Engineering Alliance (2007) *Rules and Procedures: International Educational Accords*, International Engineering Alliance, pp40–41.
232 International Engineering Alliance (2007) *Rules and Procedures: International Educational Accords*, International Engineering Alliance, pp40–41.
233 Basri, H., Che Man, A., Wan Badruzzaman, W. and Nor, M. (2004) 'Malaysia and Washington Accord: What it takes for full membership', *International Journal of Engineering and Technology*, Vol 1, No 1, pp64–73.

Chapter 3

1 Heywood, J. (2005) *Engineering Education Research and Development in Curriculum and Instruction*, Institute of Electrical and Electronics Engineers (IEEE), p16.
2 Carson, R. (1962) *Silent Spring*, Houghton Mifflin, Boston.
3 UNEP (1992) *Rio Declaration on Environment and Development*, United Nations Environment Program, Conference on Environment and Development, Rio de Janeiro, 3–14 June.

4 Porter, M. (1990) *The Competitive Advantage of Nations*, The Free Press, New York.
5 Von Weizsäcker, E., Lovins, A. and Lovins, L. (1998) *Factor Four: Doubling wealth, halving resource use – The new report to the Club of Rome*, Earthscan, London.
6 Hawken, P., Lovins, A. and Lovins, L. (1997) *Natural Capitalism, The Next Industrial Revolution*, Earthscan, London.
7 Hargroves, K. and Smith, M. (2005) *The Natural Advantage of Nations: Business opportunities, innovation and governance in the 21st century*, The Natural Edge Project, Earthscan, London.
8 von Weizsäcker, E., Hargroves, K., Smith, M., Desha, C. and Stasinopoulos, P. (2009) *Factor 5: Transforming the global economy through 80% increase in resource productivity*, The Natural Edge Project, Earthscan, London.
9 Hargroves, K. and Smith, M. (2005) *The Natural Advantage of Nations: Business opportunities, innovation and governance in the 21st century*, The Natural Edge Project, Earthscan, London.
10 Stasinopoulos, P., Smith, M., Hargroves, K. and Desha, C. (2008) *Whole System Design: An integrated approach to sustainable engineering*, The Natural Edge Project, Earthscan, London.
11 Smith, M., Hargroves, K., and Desha, C. (2010) *Cents and Sustainability – Securing our common future by decoupling economic growth from environmental pressures*, The Natural Edge Project, Earthscan, London.
12 The Natural Edge Project (n.d.) 'Our books and companions', www.naturaledgeproject.net.
13 The Natural Edge Project (n.d.) 'Curriculum and course notes', www.naturaledgeproject.net.
14 Desha, C. (2010) 'Chapter 4: Personal Narrative', in 'An investigation into the strategic application and acceleration of curriculum renewal in engineering education for sustainable development' (doctoral dissertation), Griffith University, Brisbane.
15 See Desha, C. (2010) 'Chapter 3: Literature Review' in 'An investigation into the strategic application and acceleration of curriculum renewal in engineering education for sustainable development' (doctoral dissertation), Griffith University, Australia.
16 Engineers Australia (2011) Stage 1 Competency Standards: Elements of Competency Set 1 'Knowledge and Skills Base', Competency 1.3.
17 Tyler, R. (1949) *Basic Principles of Curriculum and Instruction*, University of Chicago Press, Chicago, IL.
18 Taba, H. (1962) *Curriculum Development: Theory and practice*, Harcourt Brace and World, New York, NY.
19 Wheeler, D. K. (1967). *Curriculum Process*. London, University of London Press.
20 Kerr, J. (1968) *Changing the Curriculum*, Unibooks, University of London Press.
21 Walker, D. F. (1978). 'A naturalistic model for curriculum development', in J. R. Gress and D. E. Purpel (eds), *Curriculum: An introduction to the field*, Berkeley, California, McCutchan Publishing, pp268–280.
22 Stenhouse, L. (1975) *An Introduction to Curriculum Research and Development*, Heinemann, London, pp4–5.
23 Egan, K. (1978) 'What is Curriculum?' *Curriculum Inquiry*, Vol 8, No 1, pp66–72.
24 Desha, C. and Hargroves, K. (2011) 'Informing engineering education for sustainable development using a deliberative dynamic model for curriculum renewal', proceedings of the Research in Engineering Education Symposium 2011, Madrid.
25 Reynolds, J. and Skilbeck, M. (1976) *Culture in the Classroom*, Open Books, London.
26 Cornbleth, C. (1990) *Curriculum in Context*, Falmer Press, Basingstoke, p5.
27 See Desha, C. (2010) 'Chapter 3: Literature Review' in 'An investigation into the strategic application and acceleration of curriculum renewal in engineering education for sustainable development' (doctoral dissertation), Griffith University, Australia.
28 Taba, H. (1962) *Curriculum Development: Theory and practice*, Harcourt Brace and World, New York, NY, p448.
29 Marsh, C. (2004) *Key Concepts for Understanding Curriculum*, 3rd Edition, Routledge, New York.
30 Kift, S. (2008) '21st century climate for change: Curriculum design for quality learning engagement in law', *Legal Education Review*, Vol 18, No 1 & 2, Australia, p1.
31 Billet, S. and Stevens, J. (2007) 'Study Guide', 4141/8144VTA Curriculum Development in Adult and Vocational Education, Griffith University.
32 Franklin, B. (n.d.) 'Education Encyclopedia: Franklin Bobbitt', www.answers.com/topic/franklin-bobbitt, accessed 12 June 2009.
33 Bobbitt, F. (1918) *Curriculum*, Houghton Mifflin, Boston.
34 Bobbitt, F. (1924) *How to Make a Curriculum*, Houghton Mifflin, Boston.
35 Howard, J. (2007) *Curriculum Development*, Department of Education, Elon University.
36 Tyler, R. (1949) *Basic Principles of Curriculum and Instruction*, University of Chicago Press, Chicago, IL.
37 Stenhouse, L. (1975) *An Introduction to Curriculum Research and Development*, Heinemann, London.
38 Cornbleth, C. (1990) *Curriculum in Context*, Falmer Press, Basingstoke.
39 Kerr, J. (1968) *The Changing Curriculum*, Heinemann, London.

40 Kerr, J. (1968) *The Changing Curriculum*, Heinemann, London.
41 Tyler, R. (1949) *Basic Principles of Curriculum and Instruction*, University of Chicago Press, Chicago, IL; Taba, H. (1962) *Curriculum Development: Theory and practice*, Harcourt Brace and World, New York, NY; Bobbitt, F. (1971) *The Curriculum*, Houghton Mifflin, Boston, MA; Stenhouse, L. (1975) *An Introduction to Curriculum Research and Development*, Heinemann, London; MacDonald, B. and Walker, R. (1976) *Changing the Curriculum*, Open Books, London; Grundy, S. (1987) *Curriculum: Product or praxis?*, Falmer Press, Lewes; Cornbleth, C. (1990) *Curriculum in Context*, Falmer Press, Basingstoke; Marsh, C. (2004) *Key Concepts for Understanding Curriculum*, 3rd Edition, Routledge, New York, NY.
42 Taba, H. (1962) *Curriculum Development: Theory and practice*, Harcourt Brace and World, New York, NY.
43 Brady, L. (1995) *Curriculum Development*, 5th Edition, Prentice Hall, New York, NY, pp74–76.
44 Smith, M. K. (2000) 'Curriculum theory and practice', www.infed.org/biblio/b-curric.htm, accessed 21 January 2011.
45 Wheeler, D. K. (1967) *Curriculum Process*. London, University of London Press.
46 Wheeler, D. K. (1967) *Curriculum Process*. London, University of London Press.
47 Billet, S. and Stevens, J. (2007) 'Study Guide', 4141/8144VTA Curriculum Development in Adult and Vocational Education, Griffith University, Brisbane, p73.
48 Stenhouse, L. (1975) *An Introduction to Curriculum Research and Development*, Heinemann, London, p11.
49 Howard, J. (2007) *Curriculum Development*, Department of Education, Elon University.
50 Stenhouse, L. (1975) *An Introduction to Curriculum Research and Development*, Heinemann, London, pp4–5.
51 Stenhouse, L. (1975) *An Introduction to Curriculum Research and Development*, Heinemann, London, pp4–5.
52 Grundy, S. (1987) *Curriculum: Product or praxis?*, Falmer Press, Lewes.
53 Reynolds, J. and Skilbeck, M. (1976) *Culture in the Classroom*, Open Books, London.
54 Reynolds, J. and Skilbeck, M. (1976) *Culture in the Classroom*, Open Books, London.
55 Coaldrake, P. and Stedman, L. (1999) *Academic Work in the Twenty-first Century: Changing roles and policies*, DEST Occasional Paper Series 99-H, p13, www.dest.gov.au/highered/occpaper.htm, accessed 5 February 2010.
56 James, R. (2002) 'Students' changing expectations of higher education and the consequences of mismatches with the reality' in OECD, *Responding to Student Expectations*, OECD, p81.

Chapter 4

1 Hargroves, K. and Smith, M. (2005) Chapter 10 'Operationalizing natural advantage through the Sustainability Helix', in *The Natural Advantage of Nations: Business opportunities, innovation and governance in the 21st century*, Earthscan, London, pp161–164.
2 Desha, C. and Hargroves, K. (2011) 'Fostering rapid transitions to Education for Sustainable Development through a whole system approach to curriculum and organizational change', in proceedings of the World Symposium on Sustainable Development at Universities, Rio de Janeiro, 5–7 June 2012.
3 Hargroves, K. and Smith, M. (2005) 'Operationalizing natural advantage through the Sustainability Helix', in *The Natural Advantage of Nations: Business opportunities, innovation and governance in the 21st century*, Earthscan, London, pp161–164.
4 Wals, A. (2012) 'Shaping the education of tomorrow', full-length report on the UN Decade of Education for Sustainable Development, DESD Monitoring and Evaluation 2012, United Nations Educational, Scientific and Cultural Organization, Education Sector.
5 Bobbitt, F. (1971) *The Curriculum*, Houghton Mifflin, Boston, Preface.
6 Desha, C., Hargroves, K. and Smith, M. (2009) 'Addressing the time lag dilemma in curriculum renewal towards engineering education for sustainable development', *International Journal of Sustainability in Higher Education*, Vol 10, No 2, pp184–199.
7 Shriberg, M. and Tallent, H. (2003) 'Beyond principles: Implementing the Talloires Declaration', report and Declaration of the Presidents Conference 1990, Association of University Leaders for a Sustainable Future, Ball State University, Muncie, IN.
8 AASHE (2010) *Sustainability Curriculum in Higher Education: A call to action*, Association for the Advancement of Sustainability in Higher Education, Denver, Colorado; Corcoran, P. and Wals, A. (2008) *Higher Education and the Challenge of Sustainability – Problematics, promise, and practice*, Kluwer Academic Publishers, Boston; Jones, P., Selby, D. and Sterling, S. (2010) *Sustainability Education: Perspectives and practice across higher education*, Renouf Publishing, London; Wals, A. (ed.) (2008) *From Cosmetic Reform to Meaningful Integration: Implementing education for sustainable development in higher education institutes – the state of affairs in six European countries*, DHO, Amsterdam.

9 Tilbury, D. and Ryan, A. (2012) *The Online Guide to Quality and Education for Sustainability in HE*, University of Gloucestershire, http://efsandquality.glos.ac.uk/.
10 Kemp, S. (2013) Personal communications with Mr Simon Kemp, Academic Lead Education for Sustainable Development, Higher Education Academy, UK, 19 February 2013.
11 McCoshan, A. and Martin, S. (2012) *Evaluation of the Impact of the Green Academy Programme and Case Studies*, Higher Education Academy, UK.
12 Desha, C. and Hargroves, K. (2012) 'Applying threshold learning theory to teach sustainable business practice in post-graduate engineering education', Proceedings of the American Society of Engineering Education, 10–12 June, Texas.
13 Hargroves, K. and Smith, M. (2005) 'Operationalizing natural advantage through the Sustainability Helix', in Hargroves, K. and Smith, M. (eds) (2005) *The Natural Advantage of Nations: Business opportunities, innovation and governance in the 21st century*, Earthscan, London. The Sustainability Helix was co-developed with Hunter Lovins and the team at Natural Capitalism Solutions, and was informed by Global Academy and the TABATI Group, and MBA candidates at the Presidio School of Management under the Supervision of Hunter Lovins.
14 Lovins, H. (2006) 'Green Technology Interview: Hunter Lovins', *Green Technology* magazine, www.green-technology.org/green_technology_magazine/hunter_lovins.htm, accessed 21 August 2012.
15 Desha, C. and Hargroves, K. (2006) 'Griffith University Senior IT Planning Workshop Sustainability Session: Sustainability Session Report', 17 November 2006, facilitated by The Natural Edge Project.
16 Hargroves, K. and Smith, M. (2005) 'Operationalizing natural advantage through the Sustainability Helix', in *The Natural Advantage of Nations: Business opportunities, innovation and governance in the 21st century*, Earthscan, London, pp161–164.
17 Sharp, L. (2009) Personal communications with Dr Leith Sharp, Founding Director of the Harvard Green Campus Initiative (2000–2008, Office for Sustainability), visiting scholar at the Harvard School of Public Health, and Faculty at the Harvard Extension School.
18 Hargroves, K. and Smith, M. (2005) 'Operationalizing natural advantage through the Sustainability Helix', in *The Natural Advantage of Nations: Business opportunities, innovation and governance in the 21st century*, Earthscan, London, pp161–164.
19 The Campus Sustainability Assessment Project (n.d.) 'Introduction: The sustainability imperative and higher education: The challenge of sustainability', http://csap.envs.wmich.edu/pages/intro_imperative.html, accessed 21 January 2011.
20 King, R. (2008) *Addressing the Supply and Quality of Engineering Graduates for the New Century*, The Carrick Institute for Learning and Teaching in Higher Education Ltd, Sydney, p30.

Chapter 5

1 Whitfield, R. (1978) 'The changing school curriculum in Europe', *Paedagogica Europaea*, Vol 6, Blackwell Publishing, p116.
2 Holmberg, J., Svanström, M., Peet, D., Mulder, K., Ferrer-Balas, D. and Segalàs, J. (2008) 'Embedding sustainability in higher education through interaction with lecturers: Case studies from three European technical universities', *European Journal of Engineering Education*, Vol 33, No 3, pp271–282.
3 Ferrer-Balas, D., Adachi, J., Banas, A., Davidson, C., Hoshikoshi, A., Mishra, A., Motodoa, Y., Onga, M. and Ostwald, M. (2008) 'An international comparative analysis of sustainability transformation across seven universities', *International Journal of Sustainability in Higher Education*, Vol 9, No 3.
4 Hargroves, K. and Smith, M. (2005) *The Natural Advantage of Nations: Business opportunities, governance and innovation in the 21st century*, Earthscan, London, Figure 4.3, http://www.naturaledgeproject.net/NAON1 Chapter4.3.aspx.
5 Kronholz, J. (2005) 'US universities face turbulent times ahead', College Journal, from The Wall Street Journal Online, 1 March.
6 Atkinson, C. (2008) 'Niche programs branch out into the unexpected', *Canadian Globe and Mail*, 14 February 2008.
7 Koth, B., Woodward, M. and Iversen Y. (2009) 'Civil engineering education for sustainability: Faculty perceptions and result of an Australian course audit', 20th Australasian Association of Engineering Education Conference, Adelaide, 6–9 December, p781.
8 Desha, C. and Hargroves, K. (2012) 'Applying threshold learning theory to teach sustainable business practice in post-graduate engineering education', Proceedings of the American Society of Engineering Education, 10–12 June, Texas.

9 HEFC (2007) 'HEEPI Green Gown Awards Winners', www.heepi.org.uk/green_gown_past_winners.htm, accessed 24 January 2011.
10 Personal communications with Professor Geoff Rose 25 October 2012, discussing Rose, G., Codner, G. and Griggs, D. (2012) 'Integrating sustainability into higher education: Insight from a case study of the first year engineering curriculum at Monash University', Proceedings of the LaTrobe Symposium on Advancing Sustainability in Education and Research, 20 February, LaTrobe, Australia.
11 Higher Education Funding Council for England (2008) 'HEFC strategic review of sustainable development in higher education in England', report to the HEFC, PA Consulting Group and the Centre for Research in Education and the Environment, Bath.
12 Royal Academy of Engineering (2005) *Engineering for Sustainable Development: Guiding principles*, The Royal Academy of Engineering, London.
13 Spinks, N., Silburn, N. and Birchall, D. (2006) 'Educating engineers for the 21st century: The industry view', report by the Henley Management College, Royal Academy of Engineering, Oxfordshire, UK.
14 Royal Academy of Engineering (2010) 'Visiting Professor Scheme', www.raeng.org.uk/education/vps/sustdev/default.htm, accessed 27 January 2011.
15 Leckstrom, J. (2008) 'Kaplan releases new College Guide 2009 featuring 25 green colleges and 10 hot green . . . ', *PRNewswire*, 5 August.
16 NewsWire (2008) 'Mater to direct OSU engineering sustainability programs', http://media-newswire.com/release_1077459.html, accessed 27 January 2011.
17 Steiner, S. and Penlington, R. (2010) *An Introduction to Sustainable Development in the Engineering Curriculum – An engineering subject centre guide*, Higher Education Academy Engineering Subject Centre, United Kingdom.
18 Holmberg, J. and Samuelsson, B. (eds) (2006) 'Drivers and barriers for implementing sustainable development in higher education', *Education for Sustainable Development in Action*, Technical Paper No. 3, UNESCO Education Sector.

Chapter 6

1 de la Harpe, B., Radloff, A., Scoufis, M., Dalton, H., Thomas, J., Lawson, A., David, C. and Girardi, A. (2009) *The B Factor Project: Understanding academic staff beliefs about graduate attributes*, RMIT University, Melbourne.
2 King, R. (2008) 'Addressing the supply and quality of engineering graduates for the new century', Report to the Carrick Institute for Learning and Teaching in Higher Education Ltd, Sydney.
3 ABET (2008) 'Download Accreditation Criteria and Forms', www.abet.org/forms.shtml, accessed 28 January 2011.
4 Svanstrom, M., Lozano-García, F. and Rowe, D. (2008) 'Learning outcomes for sustainable development in higher education', *International Journal of Sustainability in Higher Education*, Vol 9, No 3, pp339–351.
5 Institution of Engineers Australia (2011) 'Management system for professional engineers', www.engineersaustralia.org.au/ieaust/index.cfm?6E4B2E32-F580-52C6-EA25-F543C9B4767E, accessed 28 January 2011.
6 Bowden, J., Hart, G., King, B., Trigwell, K. and Watts, O. (2000) 'Generic capabilities of ATN university graduates', www.clt.uts.edu.au/ATN.grad.cap.project.index.html, accessed 28 January 2011.
7 Zou, P. (2008) 'Working together to achieve graduate attributes of our students', *Centre for Education in the Built Environment (CEBE) Transactions*, Vol 5, No 1, pp25–42.
8 Drayson, R., Bone, E. and Agombar, J. (2012) 'Student attitudes towards and skills for sustainable development', a report for the Higher Education Academy, UK.
9 The Natural Edge Project (2010) 'Workshop facilitation methods', TNEP.
10 Hargroves, K. and Smith, M. (2005) *The Natural Advantage of Nations (Vol. I): Business opportunities, innovation and governance in the 21st century*, www.naturaledgeproject.net/NAON_ch23.aspx, accessed 28 January 2011.
11 Sheehan, M., Schneider, P. and Desha, C. (2012) 'Implementing a systematic process for rapidly embedding sustainability within chemical engineering education: A case study of James Cook University, Australia', *Chemistry Education Research and Practice*, Vol 13, pp112–119.
12 Sheehan, M., Desha, C., Schneider, P. and Turner, P. (2012) 'Embedding sustainability into chemical engineering education: Content development and competency mapping', in proceedings of CHEMICA 2012, 23–26 September, Wellington, New Zealand.
13 Rowe, D. (2010) Personal communication with Dr Debra Rowe, President, US Partnership for Education for Sustainable Development, 29 September 2010.

14 Triggers summarised from a review of projects funded by the Australian Learning and Teaching Council on graduate atttibutes, www.altcexchange.edu.au/content/2010-snapshot-altc-resources-graduate-attributes, accessed 28 January 2011.

Chapter 7

1 English, F. (2000) *Deciding What to Teach and Test*, Corwin Press, California.
2 Frase, L., English, F. and Poston, W. (1995) *The Curriculum Management Audit: Improving school quality*, Technomic, Virginia.
3 Jacobs, H. (1997) *Mapping the Big Picture: Integrating curriculum and assessment K–12*, Association for Supervision and Curriculum Development, Alexandria, VA.
4 Jacobs, H. (ed.) (2004) *Getting Results with Curriculum Mapping*, Association for Supervision and Curriculum Development, Alexandria, VA.
5 Jacobs, H. (2008) 'Foreword', in Hale, J. (ed.) *A Guide to Curriculum Mapping: Planning, implementing, and sustaining the process*, Corwin Press, Thousand Oaks, CA.
6 Harden, R. M. (2001) 'AMEE Guide No. 21: Curriculum mapping: a tool for transparent and authentic teaching and learning', *Medical Teacher*, Vol 23, No 2, pp123–137.
7 Sumsion, J. and Goodfellow, J. (2004) 'Identifying generic skills through curriculum mapping: A critical evaluation', *Higher Education Research & Development*, Vol 23, No 3, pp329–346.
8 Svanstrom, M., Lozano-García, F. and Rowe, D. (2008) 'Learning outcomes for sustainable development in higher education', *International Journal of Sustainability in Higher Education*, Vol 9, No 3, pp339–351.
9 Zou, P. (2008) 'Working together to achieve graduate attributes of our students', *Centre for Education in the Built Environment Transactions*, Vol 5, No 1, pp25–42.
10 Carew, A. and Therese, S. (2007) 'EMAP outcomes from regional forums on graduate attributes in engineering', 2007 Australasian Association for Engineering Education Conference, Melbourne, Australia, 9–12 December.
11 Campbell, D., Beck, H., Buisson, D. and Hargreaves, D. (2009) 'Graduate attribute mapping with the extended CDIO framework', 20th Australasian Association of Engineering Education Conference, 6–9 December, pp598–604.
12 Popp, A. and Levy, D. (2009) 'A comparison and evaluation of the CDIO Reference Syllabus against the Engineers Australia competency standards and the development of a new compact framework' 20th Australasian Association of Engineering Education conference, Adelaide, Australia, 6–9 December, pp581–586.
13 Davis, R. and Savage, S. (2009) 'Built environment and design in Australia: challenges and opportunities for professional education', 20th Australasian Association of Engineering Education conference, Adelaide, Australia, 6–9 December, p800.
14 Curtin University (n.d.) 'Curriculum 2010', http://c2010.curtin.edu.au/, accessed 28 January 2011.
15 Oliver, B. (2010) 'Curriculum maps' http://web.me.com/beverleyoliver1/benchmarking/Curriculum_maps.html, accessed 27 January 2011.
16 Lowe, K. and Marshall, L. (2004), 'Plotting renewal: Pushing curriculum boundaries using a web based graduate attribute mapping tool', cited in Atkinson, R., McBeath, C., Jonas-Dwyer, D. and Phillips, R. (eds), *Beyond the Comfort Zone: Proceedings of the 21st ASCILITE Conference*, Perth, Australia, 5–8 December, pp548–557.
17 Australian Universities Quality Agency (n.d.) 'Embedding graduate attributes in course curricula – Murdoch University', AUQA Good Practice Database, www.auqa.edu.au/gp/search/detail.php?gp_id=2795, accessed 28 January 2011.
18 Academy of Technical Societies and Engineering (1997) *Government Submission on Education and Sustainable Development*, ATSE.
19 Example extracted from a set of learning outcomes in a first year subject 'ENB100 Engineering and Sustainability', Queensland University of Technology, facilitated by the authors.
20 RET (2012) 'Figure 2', in *Consultation: Energy efficiency and engineering education*, a report to the Energy Efficiency Advisory Group, Department of Resources, Energy and Tourism (RET), Canberra.
21 RET (2012) *Consultation: Energy efficiency and engineering education*, a report to the Energy Efficiency Advisory Group, Department of Resources, Energy and Tourism (RET), Canberra.
22 RET (2012) 'Figure 3', in *Consultation: Energy efficiency and engineering education*, a report to the Energy Efficiency Advisory Group, Department of Resources, Energy and Tourism (RET), Canberra.

23 Sheehan, M., Desha, C., Schneider, P. and Turner, P. (2012) 'Embedding sustainability into chemical engineering education: Content development and competency mapping', in proceedings of CHEMICA 2012, 23–26 September, Wellington, New Zealand.

Chapter 8

1 Boks, C. and Diehl, J. (2006) 'Integration of sustainability in regular courses: experiences in industrial design engineering', *Journal of Cleaner Production*, Vol 14, No 9–11, pp932–939.
2 Lourdel, N., Gondran, N., Laforest, V. and Brodhag, C. (2005) 'Introduction of sustainable development in engineers' curricula – Problematic and evaluation methods', *International Journal of Sustainability in Higher Education*, Vol 6, No 3, pp254–264.
3 Peet, D., Mulder, K. and Bijma A. (2004) 'Integrating sustainable development into engineering courses at the Delft University of Technology: The individual interaction method', *International Journal of Sustainability in Higher Education*, Vol 5, No 3, pp278–288.
4 Lozano, R. (2006) 'A tool for a graphical assessment of sustainability in universities (GASU)', *Journal of Cleaner Production*, Vol 14, No 9–11, pp963–972.
5 Roorda, N. (2001) AISHE: Auditing Instrument for Sustainable Higher Education, Dutch Committee for Sustainable Higher Education.
6 Desha, C. and Hargroves, K. (2008) 'Education for Sustainable Development Curriculum Audit (E4SD Audit): A curriculum diagnostic tool for quantifying requirements to embed SD into higher education – Demonstrated through a focus on engineering education', UNESCO International Centre for Engineering Education (UICEE), *World Transactions on Engineering and Technology Education*, Vol 6, No 2, pp365–372.
7 Barnett, R. and Coate, K. (2004) *Engaging the Curriculum in Higher Education*, Society for Research into Higher Education and Open University Press, Buckingham, UK.
8 Cardiff University (2008) 'Helping schools assess their sustainability', *Cardiff News*, February, Vol 15, No 5, p4.
9 Lozano, R. (2008) 'Audit of contributions of Cardiff University curricula to sustainable development', International Advanced Research Workshop on Higher Education for Sustainable Development, Slovenia.
10 Holdsworth, S., Bekessy, S. and Thomas, I. (2009) 'Evaluation of curriculum change at RMIT: Experiences of the BELP project', *Reflecting Education*, Vol 5, No 1, pp51–72.
11 Koth, B., Woodward, M. and Iversen, Y. (2009) 'Civil engineering education for sustainability: Faculty perceptions and result of an Australian course audit', proceedings of the 20th Australasian Association of Engineering Education Conference, 6–9 December, Adelaide, pp776–782.
12 For an example of chemical engineering staff publishing on curriculum renewal, see Sheehan, M., Desha, C., Schneider, P. and Turner, P. (2012) 'Embedding sustainability into chemical engineering education: Content development and competency mapping', in proceedings of CHEMICA 2012, 23–26 September, Wellington, New Zealand.

Chapter 9

1 Carew, A., Therese, S., Barrie, S., Bradley, A., Cooper, P., Currie, J., Hadgraft, R., McCarthey, T., Nightingale, S. and Radcliffe, D. (2009) 'CG623 teaching and assessment of meta-attributes in engineering: Identifying, developing and disseminating good practice (EMAP)', Australian Learning and Teaching Council, Sydney.
2 Carew, A. L. and Cooper, P. (2008) 'Engineering curriculum review: Processes, frameworks and tools', European Society for Engineering Education (SEFI) annual conference, Aalborg, Denmark, 1–4 July.
3 Desha, C., Hargroves, K., Smith, M., Stasinopoulos, P., Stephens, R. and Hargroves, S. (2007) *Energy Transformed: Australian University Survey summary of questionnaire results*, The Natural Edge Project (TNEP), Australia.
4 Desha, C., Hargroves, K. and El Baghdadi, O. (2012) 'Review of postgraduate energy efficiency course content and recommendations for use of existing course: Vocational graduate certificate in building energy analysis (non-residential)', report to the National Framework for Energy Efficiency, The Natural Edge Project (TNEP), Australia.
5 Desha, C. and Hargroves, K. (2009) 'Surveying the state of higher education in energy efficiency in Australian engineering curriculum', *Journal of Cleaner Production*, Vol 18, No 7, pp652–658.

6 Desha, C., Hargroves, K., Smith, M., Stasinopoulos, P., Stephens, R. and Hargroves, S. (2007) *Energy Transformed: Australian University Survey summary of questionnaire results*, The Natural Edge Project (TNEP), Australia.
7 Desha, C., Hargroves, K., Smith, M., Stasinopoulos, P., Stephens, R. and Hargroves, S. (2007) *Energy Transformed: Australian University Survey summary of questionnaire results*, cited in Desha, C. and Hargroves, K. (2009) 'Surveying the state of higher education in energy efficiency in Australian engineering curriculum', *Journal of Cleaner Production*, Vol 8, No 7, pp652–658.
8 U-Now (n.d.) 'Learning Resources', University of Nottingham Open Couseware.
9 King, R. (2008) *Addressing the Supply and Quality of Engineering Graduates for the New Century*, University of Technology Sydney, Sydney, Australia, p87.
10 University of Cambridge Programme for Sustainability Leadership and The World Business Council for Sustainable Development (2010) 'Welcome to Chronos', www.sdchronos.org/ImmChronos/chronos_in_english.html, accessed 25 January 2011.
11 The Natural Edge Project (n.d.) 'Curriculum and Unit Notes', www.naturaledgeproject.net, accessed 25 January 2011.
12 Stibbe, A. (ed.) (2010) *The Handbook of Sustainability Literacy: Skills for a changing world*, Green Books, www.sustainability-literacy.org/, accessed 24 January 2011.
13 Second Nature (n.d.) 'Catalysing sustainable strategies for Higher Education', www.secondnature.org, accessed 25 January 2011.
14 United Nations Environment Program (2007) 'Sustainability communications: A toolkit for marketing and advertising units', http://opentraining.unesco-ci.org/cgi-bin/page.cgi?g=Detailed%2F1235.html;d=1, accessed 25 January 2011.
15 United Nations Environment Program (2008) 'Resource kit on sustainable consumption and production', http://www.unep.fr/scp/publications/details.asp?id=WEB/0008/PA, accessed 25 January 2011.
16 Brown, L. (2007) *Plan B 3.0: Mobilizing to save civilization*, W.W Norton & Company, New York, NY.
17 Hawken, P., Lovins, A. and Lovins, L. (1999) *Natural Capitalism: Creating the next industrial revolution*, Earthscan, London.
18 Stasinopoulos, P., Smith, M., Hargroves, K. and Desha, C. (2008) *Whole System Design: An integrated approach to sustainable engineering*, Earthscan, London.
19 Sterling, S. (2012) 'The Future Fit Framework: An introductory guide to teaching and learning for sustainability in HE', www.heacademy.ac.uk/assets/documents/esd/The_Future_Fit_Framework.pdf.
20 Fien, J., Heck, D. and Ferreira, J. (eds) (1997) *Learning for a Sustainable Environment: Professional development for teacher educators*, UNESCO-ACEID, Bangkok.
21 Fien, J. and Tilbury, D. (1996) *Learning for a Sustainable Environment: An agenda for teacher education in Asia and the Pacific*, UNESCO-ACEID, Bangkok.
22 McKeown, R. (2002) 'Education for sustainable development toolkit', Version 2.0, http://www.esdtoolkit.org/, accessed 25 January 2011.
23 This website and functionality are created and maintained by the Society for College and University Planning (SCUP), in collaboration with the US Partnership for Education for Sustainable Development and the Association for the Advancement of Sustainability in Higher Education (AASHE).
24 Massachusetts Institute of Technology (2008) 'Open Unitware', http://ocw.mit.edu/index.htm, accessed 24 January 2011.
25 Loughborough University and the Higher Education Academy Engineering Subject Centre (2004) 'Toolbox for sustainable design education' www.lboro.ac.uk/research/susdesign/LTSN/introduction/Introduction.htm, accessed 25 January 2011.
26 Royal Academy of Engineers (n.d.) 'Visiting professor scheme for sustainable development – Case studies', www.raeng.org.uk/education/vps/sustdev/background.htm, accessed 24 January 2011.
27 The Natural Edge Project (n.d.) 'Curriculum and Unit Notes', www.naturaledgeproject.net, accessed 3 March 2010.
28 Takashi, M. (2008) Personal communications with Professor Mino Takashi, Director IR3S Program, University of Tokyo, 6 September 2008.
29 ETHZurich (2008) 'YES Program', www.sustainability.ethz.ch/en/activities/braunwald.cfm, accessed 17 September 2008.
30 International Urban Training Center (2008) 'Welcome to IUTC', www.iutc.org/, accessed 27 January 2011.
31 Forum for the Future (2008) 'Masters in Leadership for Sustainable Development: Prospectus 2009–2010', www.forumforthefuture.org/files/Masters_Prospectus_2010-2011.pdf, accessed 27 January 2011.

32 International Water Center Postgraduate Programs in Integrated Water Management, www.watercentre.org/education/masters, accessed 27 January 2011.
33 Monash Sustainability Institute (n.d.) 'Green Steps Program', Monash University, www.monash.edu.au/research/sustainability-institute/green-steps/, accessed 27 January 2011.
34 Systems Design (n.d.) 'Systems Design – Motto', www.systemsdesign.polito.it, accessed 27 January 2011.
35 LEAD (2011) 'Training and Capacity Development' www.lead.org/training/, accessed 27 January 2011.
36 UNESCO (n.d.) 'Modialogo' http://portal.unesco.org/es/ev.php-URL_ID=37095&URL_DO=DO_TOPIC&URL_SECTION=201.html, accessed 27 January 2011.
37 US Department of Energy (n.d.) 'Solar Decathlon' http://www.solardecathlon.gov/, accessed 27 January 2011.
38 Bullen, F., Webb, E. and Brodie, L. (2007) 'Developing a national design competition through collaborative partnerships', proceedings of the Connected 2007 International Conference on Design Education, University of New South Wales, 9–12 July.
39 Australian Government Department of Environment, Water, Heritage and the Arts (2009) *Living Sustainably: The Australian Government's National Action Plan for Education for Sustainability*, DEWHA, Canberra.
40 Thomas, I. (2006) 'Sustainability and universities', cited in Farrell, R. (ed.) *Education for Sustainability, Encyclopedia of Life Support Systems (EOLSS)*, UNESCO, EOLSS Publishers, Oxford, UK, www.eolss.net, accessed 27 January 2011.
41 Thomas, I. (2006) '1. Introduction/The Relationship of Universities and Sustainability', in *Encyclopedia of Life Support Systems*, UNESCO, EOLSS Publishers, Oxford, UK, www.eolss.net, accessed 27 January 2011.
42 Thomas, I. (2006) 'Sustainability and universities', in Farrell, R. (ed.) *Education for Sustainability, Encyclopedia of Life Support Systems (EOLSS)*, UNESCO, EOLSS Publishers, Oxford, UK, www.eolss.net, accessed 27 January 2011.
43 Koester, R., Eflin, J. and Vann, J. (2006) 'Greening of the campus: A whole-systems approach', *International Journal of Cleaner Production*, Vol 14, No 9–11, pp769–779.
44 Roberts, C. and Roberts, J. (eds) (2007) 'Chapter C5: Learning, living and leading green? Curriculum policy for sustainability', in *Greener by Degrees: Exploring sustainability through higher education curriculum*, Geography Discipline Network, Gloucestershire, p321.
45 HEEPI (2004) The Green Gown Awards 2004 http://www.heepi.org.uk/documents/gg_brochure%20march%2025.pdf.
46 Sharp, L. (2009) 'Greening your organisation', Griffith University special seminar presentation, Toohey Forest EcoCentre, hosted by the Australian Campuses Towards Sustainability, 19 October.
47 Association for the Advancement of Sustainability in Higher Education (2008) 'AASHE 2008 Round Up!', www2.aashe.org/conf2008/, accessed January 27, 2011.
48 Campus Community Partnership Foundation (2011) 'The Foundation', www.c2pf.org/about.phtml, accessed 27 January 2011.
49 McNall, S., Pushnik, J., Riley, M., Stallman, J. and Stemen, M. (2008) 'Chico and Butte: Working together for sustainability on campus and beyond', AASHE 2008, Working Together for Sustainability on Campus and Beyond, 9–11 November, Raleigh, NC.
50 Allen-Gil, S., Walker, L., Thomas, G., Shevory, T. and Elan, S. (2005) 'Forming a community partnership to enhance education in sustainability', *International Journal of Sustainability in Higher Education*, Vol 6, No 4, pp392–402.
51 Brown, M. (2011) Personal correspondence with Ms Marian Brown, Special Assistant to the Provost for Sustainability, Ithaca College, New York, 5 January.
52 Stasinopoulos, P., Smith, M., Hargroves, K. and Desha, C. (2008) *Whole System Design – An integrated approach to sustainable engineering*, Earthscan, London.
53 Compston, P. (2012) Personal correspondence with Associate Professor Paul Compston, Associate Director, Coursework, School of Engineering, Australian National University, Canberra, 11 October.
54 Koth, B., Woodward, M. and Iversen Y. (2009) 'Civil engineering education for sustainability: Faculty perceptions and result of an Australian unit audit', 20th Australasian Association of Engineering Education Conference, 6–9 December, Adelaide, Australia, p776.

Chapter 10

1 Frankic, A. (2012) 'Connecting the dots . . .', Association for the Advancement of Sustainability in Higher Education, AASHE Bulletin, 23 October.

2 Allan, M. (2009) 'The formation of the engineer for the 21st century – A global perspective', 20th Australasian Association of Engineering Education conference, 6–9 December, pp447–452.
3 King, R. (2008) *Addressing the Supply and Quality of Engineering Graduates for the New Century*, University of Technology Sydney, Sydney, Australia, p106.
4 Wallace, K. (ed.) (2005) 'Educating engineers in design – Lessons learnt from the visiting professors scheme', The Royal Academy of Engineering, p4.
5 Smith, M. (1999) 'Andragogy', The Encyclopaedia of Informal Education, www.infed.org/lifelonglearning/b-andra.htm, accessed 27 January 2011.
6 Powell, M. (2008) Personal communications with Professor Michael Powell, Pro Vice Chancellor (Business) Executive, Office of the PVC (Business), Griffith University, 27 August 2008.
7 Bournemouth University (2012) 'Green Economy MSc', Course Overview.
8 Shiel, C. (2013) Personal communications with Professor Chris Shiel, 19 February 2013.
9 See Desha, C., Hargroves, K., Smith, M. and Stasinopoulos, P. (2008) *Sustainability Education for High Schools: Year 10–12 – Subject Supplements*, The Natural Edge Project, Australia; Desha, C., Hargroves, K. and Farr, A. (2011) *Sustainability Education for High Schools: Year 10–12 – Teacher Supplements*, The Natural Edge Project, Australia.
10 Custer, R. and Daugherty, J. (2009) 'Professional development for teachers of engineering: Research and related activities', *The Bridge*, Vol 39, No 3.
11 Grant, M. (2008) Personal communications with Michelle Grant, Director of the YES Program at the Centre for Sustainability, ETHSustainability, ETH Zurich, 20 August 2008.
12 Wallace, K. (ed.) (2005) 'Educating engineers in design – Lessons learnt from the visiting professors scheme', The Royal Academy of Engineering, p4.
13 Tongji University (2008) *Training Program Brochure*, Tongji University.
14 Tongji University (n.d.) 'College of Architecture and Urban Planning', www.tongji-caup.org/en/index.asp, accessed 27 January 2011.
15 Zhiqiang, W. (2008) 'Preface', *Training Program Brochure*, Tongji University.
16 Lozano, R. (2006) 'Incorporation and institutionalization of sustainable development into universities: Breaking through barriers to change', *International Journal of Cleaner Production*, Vol 14, No 9–11, pp787–796.
17 Lozano, F., Huisingh, D. and Delgado, M. (2006) 'An integrated, interconnected, multi-disciplinary approach for fostering sustainable development at the Monterrey Institute of Technology, Monterrey Campus', cited in Holmberg, J. and Samuelsson, B. (eds) *Drivers and Barriers for Implementing Sustainable Development in Higher Education,* Education for Sustainable Development in Action Technical Paper No. 3, UNESCO, pp37–48.
18 Koth, B., Woodward, M. and Iversen Y. (2009) 'Civil engineering education for sustainability: Faculty perceptions and result of an Australian course audit', 20th Australasian Association of Engineering Education Conference, 6–9 December, Adelaide, p777.
19 Goricanec, J. and Hadgraft, R. (2006) 'Sustainable Practice "in action"', 17th Annual Conference of the Australasian Association for Engineering Education, AUT Engineering, Auckland, NZ, 10–13 December.
20 Chu, C. (2008) Personal communications with Professor Cordia Chu, Director, International Centre for Development, Environment and Population Health, Griffith School of Environment, Griffith University, 15 August 2008.
21 Queensland Education and Training International (2008) '08 QETI Awards Booklet'.
22 Svanström, M. (2010) Personal communications with Professor Magdalena Svanström, Director of the Chalmers University Learning Centre, Chalmers University, Sweden.
23 Mulder, K. (2010) Personal communications with Dr Karel Mulder, Faculty of Technology, Process & Management, Delft University of Technology, the Netherlands.
24 Ferrer-Balas, F. (2010) Personal communications with Dr Didac Ferrer-Balas, Technical Director, Centre for Sustainability, Polytechnic University of Catalonia, Spain.
25 Holmberg, J. and Samuelsson, B. (eds) (2006) *Drivers and Barriers for Implementing Sustainable Development in Higher Education*, Education for Sustainable Development in Action, Technical Paper No. 3, UNESCO Education Sector.
26 Ferrer-Balas, D., Cruz, Y. and Segalas, J. (2009) 'Lessons learned from our particular "Decade" of Education for Sustainable Development (1996–2005) at UPC', in Holmberg, J. and Samuelsson, B. (eds), *Drivers and Barriers for Implementing Sustainable Development in Higher Education*, UNESCO, Paris, http://unesdoc. Unesco.org/images/0014/001484/148466E.pdf.
27 Lourdel, N., Gondran, N., Laforest, V. and Brodhag, C. (2005) 'Introduction of sustainable development in engineers' curricula – problematic and evaluation methods', *International Journal of Sustainability in Higher Education*, Vol 6, No 3, pp254–264.

28 Lourdel, N., Gondran, N., Laforest, V., Debray, B. and Brodhag, C. (2007) 'Sustainable development cognitive map: A new method of evaluating student understanding', *International Journal of Sustainability in Higher Education*, Vol 8, No 2, pp170–182.

29 QAA (2012) 'Quality Code – Chapter B3: Learning and Teaching', UK Quality Assurance Agency for Higher Education.

INDEX